D1408814

Mutagens in food.

Mutagens in Food:
Detection
and
Prevention

Editor

Hikoya Hayatsu

Professor Bioorganic Chemistry
Faculty of Pharmaceutical Sciences
Okayama University
Okayama, Japan

CRC Press
Boca Raton   Ann Arbor   Boston

Library of Congress Cataloging-in-Publication Data

Mutagens in food : detection and prevention / editor, Hikoya Hayatsu.
    p. cm.
  Includes bibliographical references and index.
  ISBN 0-8493-5877-9
  1. Food--Toxicology. 2. Mutagenicity testing. 3. Carcinogenicity
testing. I. Hayatsu, Hikoya, 1934- .
  RA1258.M88 1990
  615.9'54--dc20

90-20621
CIP

Direct all inquiries to CRC Press, Inc., 2000 Corporate Blvd., N.W., Boca Raton, Florida 33431.

© 1991 by CRC Press, Inc.

International Standard Book Number 0-8493-5877-9

Library of Congress Card Number 90-20621
Printed in the United States

PREFACE

Food is essential for human life. On the other hand, it has become increasingly clear that food can also be a source of health hazard. Epidemiological studies show that the dietary habit is one of the major factors for causing cancers.

It is believed that cancer is initiated by mutagenesis of certain genes. Consequently, mutagens, the substances that cause mutation, are the major focus of concern as something that should be gotten rid of from human contact. During the past 15 years, a large amount of knowledge has been rapidly accumulated about mutagens in food, particularly about those produced by cooking. In the process of heating proteinaceous food, many kinds of mutagens are formed. A major portion of these mutagenic substances have now been identified and their carcinogenic potencies revealed. I believe, therefore, that reviewing the present status of mutagens in food is a timely undertaking.

Because food is a complex mixture of components, it may be expected that a variety of interactions among these components can take place either in the food itself or after being taken into the human body. Some compounds in food may react to each other, chemically or biologically, to form mutagens; or otherwise, a component may antagonize food-borne mutagens, abolishing their toxic actions.

This book deals with various mutagens that arise in and from foods; they may be either naturally occurring or man-made. In the subsequent sections of the book, detection methods of the mutagens, preventative measures against mutagen-formations and mutagen-actions, and the risk assessment of the mutagens are presented.

I hope that the book is useful for all who have interest, as well as concern, in the food mutagens.

Hikoya Hayatsu
Okayama
October, 1990

THE EDITOR

Hikoya Hayatsu, Ph.D., is Professor of Bioorganic Chemistry in the Faculty of Pharmaceutical Sciences, Okayama University at Okayama, Japan

Dr. Hayatsu graduated from the University of Tokyo in 1957 and obtained his M.S. and Ph.D. degrees from the same university in 1959 and 1964, respectively. As a post doctoral fellow in the University of Wisconsin during the period of 1964 to 1967, he was in the team directed by Dr. H. G. Khorana, and was engaged in the chemical synthesis of genetic-code triplets. He became an Associated Professor of the University of Tokyo, Faculty of Pharmaceutical Sciences, in 1967. It was in 1978 that he assumed his present position.

Dr. Hayatsu is President of the Japan Environmental Mutagen Society. He is on the editorial board of *Mutation Research, Japanese Journal of Cancer Research*, and *Journal of Pharmacobio-Dynamics*. He received the Miyata Science Award and the Japan Environmental Mutagen Society Award. He has been the recepient of many research grants from the Ministry of Education, Science, and Culture and the Ministry of Health and Welfare.

Dr. Hayatsu is the author of more than 200 papers. His major research interest has been chemical modification of nucleic acids, and currently this subject has become related to mutagen actions on genetic materials.

CONTRIBUTORS

Hans-Ulrich Aeschbacher
Head
Department of Genetic Toxicology
Nestec Ltd.-Nestlé Research Centre
Lausanne, Switzerland

Jan Alexander
Deputy Chief
Department of Environmental
  Medicine
National Institute of Public Health
Oslo, Norway

Bruce N. Ames
Professor
Division of Biochemistry and Molecular
  Biology
University of California
Berkeley, California

A. W. Andrews
Laboratory Head
Microbial Mutagenesis Laboratory
Program Resources, Inc.
Frederick, Maryland

Sakae Arimoto
Assistant Researcher
Faculty of Pharmaceutical Sciences
Okayama University
Tsushima, Okayama, Japan

Helmut Bartsch
Unit Chief
Unit of Environmental Carcinogens and
  Host Factors
International Agency for Research on
  Cancer
Lyon, France

Silvio De Flora
Director
Institute of Hygiene and Preventive
  Medicine
University of Genoa
Genoa, Italy

Virginia C. Dunkel
Chief
Genetic Toxicology Branch
Center for Food Safety and Applied
  Nutrition
U.S. Food and Drug Administration
Washington, D.C.

James S. Felton
Section Leader, Molecular Biology
Biomedical Sciences Division
Lawrence Livermore National Laboratory
Livermore, California

Marlin Friesen
Scientist
Department of Environmental
  Carcinogens and Host Factors
International Agency for Research on
  Cancer
Lyon, France

David W. Gaylor
Director
Biometry Staff
National Center for Toxicological
  Research
U.S. Food and Drug Administration
Jefferson, Arkansas

Lois Swirsky Gold
Director, Carcinogenic Potency Program
Division of Biochemistry and Molecular
  Biology
University of California
Berkeley, California

Yuzo Hayashi
Chief
Division of Pathology
National Institute of Hygienic Sciences
Tokyo, Japan

Hikoya Hayatsu
Professor
Faculty of Pharmaceutical Sciences
Okayama University
Okayama, Japan

Kazuyuki Hiramoto
Research Associate
Faculty of Pharmaceutical Sciences
Okayama University
Okayama, Japan

Motoi Ishidate, Jr.
Head
Department of Genetics and Mutagenesis
Natioanl Institute of Hygienic Science
Tokyo, Japan

Alberto Izzotti
Fellowship Owner
Institute of Hygiene and Preventive
  Medicine
University of Genoa
Genoa, Italy

Fred F. Kadlubar
Associate Director for Research
National Center for Toxicological
  Research
U.S. Food and Drug Administration
Jefferson, Arkansas

Tetsuta Kato
Lecturer
Tokyo College of Pharmacy
Tokyo, Japan

Kiyomi Kikugawa
Professor
Tokyo College of Pharmacy
Tokyo, Japan

Mark G. Knize
Biomedical Scientist
Biomedical Sciences Division
Lawrence Livermore National Laboratory
Livermore, California

Christian Malaveille
Scientist
Department of Environmental
  Carcinogens and Host Factors
International Agency for Research on
  Cancer
Lyon, France

Minako Nagao
Division Chief
Carcinogenesis Division
National Cancer Center Research
  Institute
Tokyo, Japan

Kazuo Negishi
Associate Professor
Gene Research Center
Okayama Univeristy
Okayama, Japan

Tomoe Negishi
Lecturer
Faculty of Pharmaceutical
  Sciences
Okayama University
Okayama, Japan

Hiroko Ohgaki
Scientist
Laboratory of Neuropathology
Institute of Pathology
University of Zurich
Zurich, Switzerland

Hiroshi Ohshima
Senior Scientist
Department of Environmental
  Carcinogens and Host Factors
International Agency for Research on
  Cancer
Lyon, France

Shogo Ozawa
Assistant
Department of Pharmacology
Keio University
Tokyo, Japan

Brigitte Pignatelli
Scientist
Department of Environmental
  Carcinogens and Host Factors
International Agency for Research on
  Cancer
Lyon, France

Margie Profet
Staff Scientist
Division of Biochemistry and Molecular
  Biology
University of California
Berkeley, California

Mark H. Schiffman
Clinical Investigator
Environmental Epidemiology Branch
National Cancer Institute (U.S.A.)
Bethesda, Maryland

Takashi Sugimura
President
National Cancer Center
Tokyo, Japan

Shozo Takayama
Director
National Cancer Center Research Institute
Tokyo, Japan

Yoshio Ueno
Professor
Department of Toxicology and Microbial
  Chemistry
Faculty of Pharmaceutical Sciences
Science University of Tokyo
Tokyo, Japan

Roger Van Tassell
Research Associate
Department of Anaerobic Microbiology
Virginia Polytechnic Institute and State
  University
Blacksburg, Virginia

Keiji Wakabayashi
Section Head
Carcinogenesis Division
National Cancer Center Research Institute
Tokyo, Japan

Håkan Wallin
Research Fellow
Department of Environmental Medicine
National Institute of Public Health
Oslo, Norway

John H. Weisburger
Senior Member
Department of Biochemical Pharmacology
American Health Foundation
Valhalla, New York

Yasushi Yamazoe
Associate Professor
Department of Pharmacology
School of Medicine
Keio Univeristy
Tokyo, Japan

Patrizia Zanacchi
University Researcher
Institute of Hygiene and Preventive
  Medicine
University of Genoa
Genoa, Italy

Errol Zeiger
Head, Mutagenesis Group
Experimental Toxicology Branch
National Institute of Environmental
  Health Sciences
Research Triangle Park, North Carolina

TABLE OF CONTENTS

Chapter 1

INTRODUCTION

Hikoya Hayatsu

Epidemiological studies show that the major causes of human cancer are diet and tobacco.[1,2] While humans can avoid, in principle, coming into contact with cigarette smoke, they cannot avoid eating food. Therefore, the presence of carcinogens, either natural or artificial, in food has been of great concern to mankind. The recent development of mutagenicity assays has provided an efficient screening method for detecting potential carcinogens.[3] Predicting the carcinogenicity of a given agent on the basis of its mutagenicity alone is not always accurate, but knowledge of its chemical structure will add information to make the prediction more reliable.[4] A representative case is the finding of the presence of mutagenic heterocyclic amines in cooked food. These compounds were detected first by the bacterial assay for mutagenicity and later proved to be carcinogenic in rodents (see Chapter 2). Another example is AF2, a food preservative used in Japan until the early 1970s. This compound was found to be mutagenic and soon proved to be carcinogenic and as a result was banned from use as a food additive (see Chapters 11.4 and 12).

The presence of mutagens in food and its consequences are clearly important issues for scientists. A major portion of this book is devoted to reviews on the detection of food mutagens and the assessment of their possible risks. Naturally occurring, seminaturally occurring (for example, in the process of cooking), and artificially added mutagenic food components are dealt with in various chapters. Here it may be instructive to point out that there is great variance for individual mutagenic substances in their pattern of consumption by humans. This variance stems from the kind of food in which the mutagen in question is present (eaten widely or only locally, taken every day or only rarely, etc.) and from the nature of the mutagen (always or only at times present, highly or not carcinogenic, present in large or small amounts, etc.).

A goal in the study of food mutagens is to find preventive measures against human exposure to the mutagens. There could be various strategies toward this purpose: for example, (1) removal from food of the causes for generation of mutagens and (2) fortification of diet with food rich in mutagen inhibitors. These strategies are amply discussed in several chapters of this book.

The ultimate goal of this science is to provide the public with a recipe for a healthy diet; what kind of food should be chosen from edible materials for preparing the diet and how they should be cooked. As the contents of the book show, this field is undergoing rapid progress, and the proposition of a recipe for a healthy diet is still premature.

Water for drinking as well as for use in cooking food is an important constituent of the diet. Tap water and well water have been generally regarded as safe. However, current studies have indicated that chlorination of surface water can sometimes produce mutagens. Chlorination of humic acid, a naturally occurring contaminant of some surface waters, results in the formation of 3-chloro-4-(dichloromethyl)-5-hydroxy-2(5H)-furanone, a strong *Salmonella typhimurium* TA100 mutagen.[5,6] Furthermore, the source water from which tap water is derived may contain mutagens. Such a case is found in the river system in the Kyoto-Osaka area in Japan, where the source water for the population in Osaka comes from the Yodo River which flows through Kyoto and collects wastes. The water of the Yodo has been found to be continuously contaminated with strong mutagens that originate from a sewage plant of the City of Kyoto.[7] These recent findings provide an alarm for the notion of water safety. Although it would be premature to deal with this problem in this book, it is an important subject for future studies of dietary mutagens.

Overall, this book provides a list of mutagens in food and their detailed accounts. It also contains considerations on the consequences of the presence of such mutagens, particularly the possible carcinogenic hazard to humans. I hope that this publication can serve as a means to attract researchers of a wide range of disciplines into this exciting field of science.

REFERENCES

1. **Wynder, E. L. and Gori, G. B.,** Contribution of the environment to cancer incidence: an epidemiologic exercise, *J. Natl. Cancer Inst.,* 58, 825, 1977.
2. **Doll, R. and Peto, R.,** The causes of cancer: quantitative estimates of avoidable risks of cancer in the United States today, *J. Natl. Cancer Inst.,* 66, 1191, 1981.
3. **Ames, B. N.,** Identifying environmental chemicals causing mutations and cancer, *Science,* 204, 587, 1979.
4. **Ashby, J. and Tennant, R. W.,** Chemical structure, Salmonella mutagenicity and extent of carcinogenicity as indicators of genotoxic carcinogenesis among 222 chemicals tested in rodents by the U.S. NCI/NTP, *Mutat. Res.,* 204, 17, 1988.
5. **Meier, J. R., Knohl, R. B., Coleman, W. E., Ringhand, H. P., Munch, J. W., Kaylor, W. H., Streicher, R. P., and Kopfler, F. C.,** Studies on the potent bacterial mutagen, 3-chloro-4-(dichloromethyl)-5-hydroxy-2(5*H*)-furanose: aqueous stability, XAD recovery and analytical determination in drinking water and in chlorinated humic acid solutions, *Mutat. Res.,* 189, 363, 1987.
6. **Kronberg, L. and Vartiainen, T.,** Ames mutagenicity and concentration of the strong mutagen 3-chloro-4-(dichloromethyl)-5-hydroxy-2(5*H*)-furanone and its geometric isomer E-2-chloro-3-(dichloromethyl)-4-oxo-butenoic acid in chlorine-treated tap waters, *Mutat. Res.,* 206, 177, 1988.
7. **Sakamoto, H. and Hayatsu, H.,** A simple method for monitoring mutagenicity of river water. Mutagens in Yodo river system, Kyoto-Osaka, *Bull. Environ. Contam. Toxicol.,* 44, 521, 1990.

Chapter 2

MUTAGENS IN COOKED FOOD

Chapter 2.1

HISTORY OF THE STUDY OF COOKED FOOD MUTAGENS

Hikoya Hayatsu

By the end of the year 1989, 15 different mutagenic heterocyclic amines were isolated from cooked food, heated proteins, and heated amino acids (see Table 1). They are potent mutagens as tested by the Ames Salmonella assay. Moreover, ten of these compounds have been reported to be carcinogenic in rodents. It is no wonder that these findings have aroused ever-increasing concern and interest from scientists, and the general public alike, in the presence of these chemical entities in the diet.

In this chapter, the history, particularly the earliest findings, of this important discovery is recorded. A list of these compounds with brief descriptions for individual substances is given in the next chapter.

By the early 1970s, there was much interest in environmental genotoxins, and studies had been performed using assay systems, available at that time, on important issues such as mutagenic food additives (see Chapter 12). Since the creation by B. N. Ames of the *Salmonella typhimurium* strains that are extremely sensitive to a wide variety of mutagens,[1,2] studies aimed at detecting and identifying potential carcinogens in the environment have flourished. In an early report from Ames' laboratory, cigarette smoke condensate was shown to be strongly mutagenic.[3] This report aroused the interest of several scientists to more thoroughly investigate the mutagenic components of tobacco.

One of those scientists was T. Sugimura of the National Cancer Center Research Institute in Japan. He and his co-workers found in 1975 that the mutagenicity of cigarette smoke condensate could not be accounted for by its benzo(*a*)pyrene content.[4] According to Sugimura's classic paperback *Hatsugan Busshitsu* (Carcinogens), written in Japanese and published in 1982,[5] he was excited to find that tobacco tar had 20,000 times the mutagenicity expected from its benzo(*a*)pyrene content. Following this observation, he and his colleagues almost immediately stumbled on the discovery of cooked food mutagens. Sugimura writes:

Soon on a Sunday [in the fall of 1975] my wife broiled mackerel pike in the kitchen of our apartment, with smoke filling up the rooms. A thought suddenly took shape in me: Why couldn't this smoke be as nasty as cigarette smoke? Why don't we test the mutagenicity of the smoke from cooking fish?

However, the proposal for experiments to explore the mutagenicity of cooked-fish smoke was, quite understandably, not entirely welcomed by the research members of the institute. Scientists tend to regard themselves as the world's top-grade and are usually disgraceful to daily-life matters such as this. Indeed, when I discussed this proposal with several of our staff, their responses were not very enthusiastic. In spite of this generally reluctant atmosphere, Dr. Nagao and her associates agreed to explore the possibility and started some experiments.

The initial phase of major scientific discoveries may always be as humble as this: We bought an open gas flame appliance, and broiled lots of sun-dried fish on it. We had no authentic apparatus for collecting smoke, and had to devise our own way. The smoke was sucked through an inverted glass funnel placed above the fish and then through a filter membrane, the same type of membrane that had been used in the collection of cigarette smoke. The smoke particles thus collected on the filter showed, as I anticipated, a strong mutagenicity.

Having found out the mutagenicity in the fish smoke, I expected that the charred part of the fish surface would contain mutagens, as the smoke must simply be the particulates derived from the charred skin. Obviously the charred portion is a more important subject for study than the smoke, because the smoke is not eaten by people but the charred part is. Moreover, the smoke will go out through the kitchen ventilator into the atmosphere and

TABLE 1
Mutagenic Heterocyclic Amines: A Chronological Listing[a]

Year of report	Abbreviated name of heterocyclic amine	Ref.
1977	Trp-P-1	Sugimura et al.[8]
1977	Trp-P-2	Sugimura et al.[8]
1977	Phe-P-1	Sugimura et al.[8]
1978	AαC	Yoshida et al.[17]
1978	MeAαC	Yoshida et al.[17]
1978	Glu-P-1	Yamamoto et al.[19]
1978	Glu-P-2	Yamamoto et al.[19]
1980	IQ	Kasai et al.[20]
1980	MeIQ	Kasai et al.[21]
1981	MeIQx	Kasai et al.[22]
1981	Orn-P-1	Yokota et al.[23]
1984	7,8-DiMeIQx	Negishi et al.[24]
1985	4,8-DiMeIQx	Negishi et al.[25]
1986	PhIP	Felton et al.[26]
1988	IQx	Becher et al.[27]

[a]   This table is a modification of a table in Reference 12. See Chapter 2.2 for a full description of the compounds.

be diluted. Certainly, it is not a trivial matter to have mutagens in the edible part of cooked fish. We then discovered that the charred part of fish was mutagenic.*

A report of this observation was submitted to *Cancer Letters* in September 1976 and was published in 1977.[6] These researchers soon found that proteins and amino acids, but not starch, fat, or nucleic acids, can generate strong mutagenicities upon heating.[7] Among amino acids tested, tryptophan was the most active. The mutagenic pyrolysis products, named Trp-P-1 and Trp-P-2, were then successfully isolated from heated tryptophan and their structures were established within a short period of time. The memorable report on this achievement was submitted early in 1977 and published promptly as a communication in *Proceedings of the Japan Academy.*[8]

Subsequent isolations of many other heterocyclic amines from pyrolysates of food, proteins, and amino acids accomplished by Sugimura and his associates are documented in current reviews from their laboratory[9-12] (see also Table 1).

Yoshida, Mizusaki, and their co-workers in the Japan Tobacco Monopoly Company Research Center were the other group of scientists who had used the Salmonella assay for testing the mutagenesis of cigarette smoke condensates.[13] The mutagenicity found was very high and, interestingly, was correlated with the amount of nitrogen fertilizer that had been used for growing the tobacco. Consequently, they began to suspect that the burning of nitrogen-containing components in tobacco leaves, probably amino acids and proteins that were known to be present in the leaves, might have produced mutagens.[14] At the October 1976 annual meeting of the Environmental Mutagen Society of Japan, Yoshida et al. reported their observation on the generation of mutagenicity from heating of proteins and amino acids, notably tryptophan.[15,16] Their continued effort to characterize the mutagens culminated in the isolation and identification of AαC and MeAαC from the pyrolysate of globulin, a work published in 1978.[17]

Apart from the Japanese workers, several scientists in the U.S., where heated meat is

---

*   Translated from Sugimura, T., *Hatsugan Busshitsu*, Chuo-Koron Sha, Tokyo, 1982, 105—106. With permission.

the major diet, became aware of the importance of these findings. Commoner et al. showed in 1978 that heating meat extract at commonly used cooking temperatures can generate compounds that are highly mutagenic and yet distinct from benzo(a)pyrene.[18]

The ensuing discoveries of strongly mutagenic heterocyclic amines are summarized in Table 1. Extensive reviews have been published about these mutagens.[9-12,28-30]

REFERENCES

1. **Ames, B. N., Durston, W. E., Yamasaki, E., and Lee, F. D.,** Carcinogens are mutagens: a simple test system combining liver homogenates for activation and bacteria for detection, *Proc. Natl. Acad. Sci. U.S.A.,* 70, 2281, 1973.
2. **Ames, B. N., McCann, J., and Yamasaki, E.,** Methods for detecting carcinogens and mutagens with the Salmonella/mammalian-microsome mutagenicity test, *Mutat. Res.,* 31, 347, 1975.
3. **Kier, L. D., Yamasaki, E., and Ames, B. N.,** Detection of mutagenic activity in cigarette smoke condensates, *Proc. Natl. Acad. Sci. U.S.A.,* 71, 4159, 1974.
4. **Sugimura, T., Sato, S., Nagao, M., Yahagi, T., Matsushima, T., Seino, Y., Takeuchi, M., and Kawachi, T.,** Overlapping of cacinogens and mutagens, in *Fundamentals in Cancer Prevention. Proc. 6th Int. Symp., Princess Takamatsu Cancer Research Fund,* Magee, P. N., et al., Eds., University of Tokyo Press, Tokyo, 1976, 191.
5. **Sugimura, T.,** *Hatsugan Busshitsu,* Chuo-Koron Sha, Tokyo, 1982.
6. **Nagao, M., Honda, M., Seino, Y., Yahagi, T., and Sugimura, T.,** Mutagenicities of smoke condensates and the charred surface of fish and meat, *Cancer Lett.,* 2, 221, 1977.
7. **Nagao, M., Honda, M., Seino, Y., Yahagi, T., Kawachi, T., and Sugimura, T.,** Mutagenicities of protein pyrolysates, *Cancer Lett.,* 2, 335, 1977.
8. **Sugimura, T., Kawachi, T., Nagao, M., Yahagi, T., Seino, Y., Okamoto, T., Shudo, K., Kosuge, T., Tsuji, K., Wakabayashi, K., Iitaka, Y., and Itai, A.,** Mutagenic principle(s) in tryptophan and phenylalanine pyrolysis products, *Proc. Jpn. Acad.,* 52, 58, 1977.
9. **Sugimura, T.,** Carcinogenicity of mutagenic heterocyclic amines formed during the cooking process, *Mutat. Res.,* 150, 33, 1985.
10. **Sugimura, T.,** Studies on environmental chemical carcinogenesis in Japan, *Science,* 233, 312, 1986.
11. **Sugimura, T., Sato, S., and Wakabayashi, K.,** Mutagens/carcinogens in pyrolysates of amino acids and proteins and in cooked foods: heterocyclic aromatic amines, in *Chemical Induction of Cancer,* Vol. IIIC, Woo, Y.-T., Lai, D. Y., Arcos, J. C., and Argus, M. F., Eds., Academic Press, San Diego, 1988, 681.
12. **Sugimura, T., Wakabayashi, K., Nagao, M., and Ohgaki, H.,** Heterocyclic amines in cooked food, in *Food Toxicology; A Prespective on the Relative Risk,* Taylor, S. L. and Scanlan, R. A., Eds., Marcel Dekker, New York, 1989, 31.
13. **Mizusaki, S., Takashima, T., and Tomaru, K.,** Factors affecting mutagenic activity of cigarette smoke condensate in *Salmonella typhimurium* TA1538, *Mutat. Res.,* 48, 29, 1977.
14. **Mizusaki, S., Okamoto, H., Akiyama, A., and Fukuhara, Y.,** Relation between chemical consituents of tobacco and mutagenic activity of cigarette smoke condensate, *Mutat. Res.,* 48, 319, 1977.
15. **Matsumoto, I., Yoshida, D., Mizusaki, S., and Okamoto, H.,** Mutagenicity of pyrolysates of amino acids, peptides and proteins in *Salmonella typhimurium* TA98, in *Proc. 5th Annu. Meet. Environmental Mutagen Society of Japan,* The Environmental Mutagen Society of Japan, Tokyo, 1976, 6; *Mutat. Res.,* 51, 219 (abstr.), 1978.
16. **Matsumoto, T. Yoshida, D., Mizusaki, S., and Okamoto, H.,** Mutagenic activity of amino acid pyrolysates in *Salmonella typhimurium* TA98, *Mutat. Res.,* 48, 279, 1977.
17. **Yoshida, D., Matsumoto, T., Yoshimura, R., and Mizusaki, T.,** Mutagenicity of amino-α-carbolines in pyrolysis products of soybean globulin, *Biochem. Biophys. Res. Commun.,* 83, 915, 1978.
18. **Commoner, B., Vithayathil, A. J., Dolara, P., Nair, S., Madyastha, P., and Cuca, G. C.,** Formation of mutagens in beef and beef extract during cooking, *Science,* 201, 913, 1978.
19. **Yamamoto, T., Tsuji, K.., Kosuge, T., Okamoto, T., Shudo, K., Takeda, K., Iitaka, Y., Yamaguchi, K., Seino, Y., Yahagi, T., Nagao, M., and Sugimura, T.,** Isolation and structure determination of mutagenic substances in L-glutamic acid pyrolysate, *Proc. Jpn. Acad.,* 54B, 248, 1978.
20. **Kasai, H., Yamaizumi, Z., Wakabayashi, K., Nagao, M., Sugimura, T., Yokoyama, W., Miyazawa, T., Spingarn, N. E., Weisburger, J. H., and Nishimura, S.,** Potent novel mutagens produced by broiling fish under normal conditions, *Proc. Jpn. Acad.,* 56B, 278, 1980.

21. **Kasai, H., Yamaizumi, Z., Wakabayashi, K., Nagao, M., Sugimura, T., Yokoyama, S., Miyazawa, T., and Nishimura, S.,** Structure and chemical synthesis of Me-IQ, a potent mutagen isolated from broiled fish, *Chem. Lett.,* 1391, 1980.

22. **Kasai, J., Yamaizumi, S., Shiomi, T., Yokoyama, S., Miyazawa, T., Wakabayashi, K., Nagao, M., Sugimura, T., and Nishimura, S.,** Structure of a potent mutagen isolated from fried beef, *Chem. Lett.,* 485, 1981.

23. **Yokota, M., Narita, K., Kosuge, T., Wakabayashi, K., Nagao, M., Sugimura, T., Yamaguchi, K., Shudo, K., Iitaka, Y., and Okamoto, T.,** A potent mutagen isolated from a pyrolysate of L-ornithine, *Chem. Pharm. Bull.,* 29, 1473, 1981.

24. **Negishi, C., Wakabayashi, K., Tsuda, M., Sato, S., Sugimura, T., Saito, H., Maeda, M., and Jägerstad, M.,** Formatin of 2-amino-3,7,8-trimethylimidazo[4,5-*f*]quinoxaline, a new mutagen, by heating a mixture of creatinine, glucose and glycine, *Mutat. Res.,* 140, 55, 1984.

25. **Negishi, C., Wakabayashi, K., Yamaizumi, Z., Saito, H., Sato, S., Sugimura, T., and Jägerstad, M.,** Identification of 4,8-DiMeIQx, a new mutagen. Selected abstracts of papers presented at the 13th annual meeting of the Environmental Mutagen Society of Japan, 12-13 Oct. 1984, Tokyo (Japan), *Mutat. Res.,* 147, 267, 1985.

26. **Felton, J. S., Knize, M. G., Shen, N. H., Lewis, P. R., Andresen, B. D., Happe, J., and Hatch, F. T.,** The isolation and identification of a new mutagen from fried ground beef: 2-amino-1-methyl-6-phenylimidazo[4,5-*b*]pyridine (PhIP), *Carcinogenesis,* 7, 1081, 1986.

27. **Becher, G., Knize, M. G., Nes, I. F., and Felton, J. S.,** Isolation and identification of mutagens from a fried Norwegian meat product, *Carcinogenesis,* 9, 247, 1988.

28. **Hatch, F. T., Felton, J. S., Stuermer, D. H., and Bjeldanes, L. F.,** Identification of mutagens from the cooking of food, in *Chemical Mutagens, Principles and Methods for Their Detection,* Vol. 9, de Serres, F. J., Ed., Plenum Press, New York, 1984, 111.

29. **de Meester, C.,** Bacterial mutagenicity of heterocyclic amines found in heat-processed food, *Mutat. Res.,* 221, 235, 1989.

30. **Hatch, F. T., Knize, M. G., Healy, S. K., Slezak, T., and Felton, J. S.,** Cooked-food mutagen reference list and index, *Environ. Mol. Mutagen.,* 12 (Suppl. 14), 1, 1988.

Chapter 2.2

A LIST OF MUTAGENIC HETEROCYCLIC AMINES

Tomoe Negishi, Minako Nagao, Kazuyuki Hiramoto, and Hikoya Hayatsu

TABLE OF CONTENTS

NOTES FOR THE DATA

The ultraviolet absorption spectra and the pKa values were measured in the Faculty of Pharmaceutical Sciences, Okayama University by Dr. Sakae Arimoto and Miss Makiko Akashi using heterocyclic amine samples of high purity (>99%). The pKa determinations were done spectroscopically (see Chapter 6.1 for details). The authors thank them for their assistance.

The data for the mutagenicity in *Salmonella typhimurium* are taken for References 1 and 67. All of the heterocyclic amines are promutagens, requiring metabolic activation for causing mutagenesis in Salmonella.

Harman and norharman, which can arise from cooking of food, are not mutagenic themselves, but because they are comutagens they are included in this list.

Trp-P-1

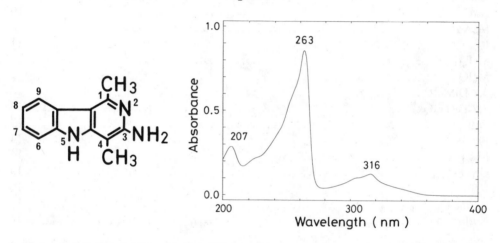

3-Amino-1,4-dimethyl-5*H*-pyrido[4,3-*b*]indole; CAS reg. no.: 62450-06-0; $C_{13}H_{13}N_3$; mol wt 211.3; pKa 8.55. Occurrence: broiled beef,[1] broiled chicken,[1] fried ground beef,[1] broiled sardine.[2] Synthesis: from 2,5-lutidine in 5 steps,[3] from 4-nitro-3,6-dichloro-picolinic acid in 4 steps.[4] Derivatives: 1-demethyl,[5] 3-acetyl,[5] 2-NHOH.[6] Genotoxicity data: *Salmonella typhimurium* TA98 3.9 × 10^4 rev/μg, mammalian cultured cells[7] (human, Chinese hamster, Syrian golden hamster), Drosophila,[8,9] mouse,[10] rat.[11] Carcinogenicity (+). UV spectrum (above): concentration 0.017 m*M* (a stock aqueous solution was diluted with water).

Trp-P-2

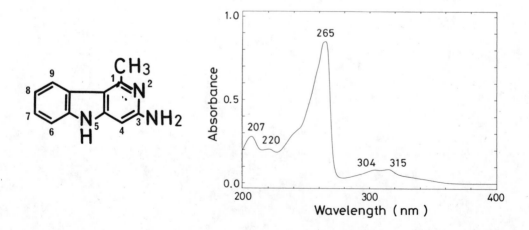

3-Amino-1-methyl-5H-pyrido[4,3-b]indole; CAS reg. no.: 62450-07-1; $C_{12}H_{11}N_3$; mol wt 197.2; pKa 8.5 (lit. 8.2[12]). Occurrence: broiled beef,[1] fried ground beef,[1] broiled chicken,[1] broiled mutton,[1] broiled sardine,[2] cigarette smoke condensate.[1] Synthesis: from indole-2-carboxylic acid in 3 steps or 9 steps,[3,4] from cyanomethylindole and acetonitrile in 1 step.[13] Derivatives: 2-NO$_2$,[14,15] 2-NHOH,[14,15] 2-NO,[15] 1-ethyl,[5] demethyl.[5] Genotoxicity data: *Salmonella typhimurium* TA98 1.04×10^5 rev/µg, mammalian cultured cells[7] (human, Chinese hamster, golden hamster), Drosophila.[8,9] Carcinogenicity (+). UV spectrum (above): concentration 0.028 mM (a stock aqueous solution was diluted with water).

Glu-P-1

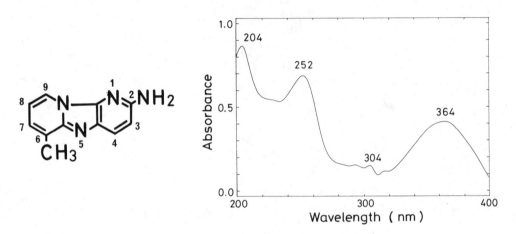

2-Amino-6-methyldipyrido[1,2-a:3′,2′-d]imidazole; CAS reg. no.: 67730-11-4; $C_{11}H_{10}N_4$; mol wt 198.2; pKa 6.0. Occurrence: casein pyrolysate.[16] Synthesis: from 3-amino-8-methylimidazo[1,2-a]pyridine (prepared from 2-amino-4-methylpyridine) and 2-chloroacrylonitrile in 1 step.[17] Derivatives: 2-NO$_2$,[15,18] 2-NHOH,[15,18] 2-NO.[15] Genotoxicity data: *Salmonella typhimurium* TA98 4.9×10^4 rev/µg, mammalian cultured cells[7] (human, Chinese hamster, Syrian golden hamster), Drosophila,[8] mouse.[10] Carcinogenicity (+). Oncogene activation: (N-OAc-Glu-P-1).[19,20] UV spectrum (above): concentration 0.04 mM (a stock methanol solution was diluted with water).

Glu-P-2

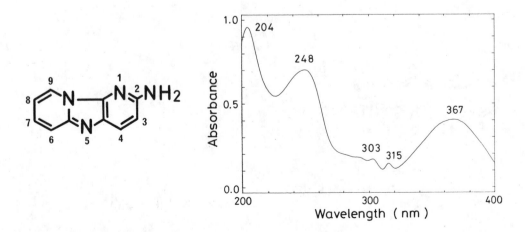

2-Aminodipyrido[1,2-*a*:3′,2′-*d*]imidazole; CAS reg. no.: 67730-10-3; $C_{10}H_8N_4$; mol wt 184.2; pKa 5.85. Occurrence: broiled cuttle fish,[21] casein pyrolysate.[16] Synthesis: from 3-aminoimidazo[1,2-*a*]pyridine (prepared from 2-aminopyridine) in 1 step.[17] Derivatives: 3-methyl,[22] 4-methyl,[22] 7-methyl,[22] 8-methyl,[22] 9-methyl.[22] Genotoxicity data: *Salmonella typhimurium* TA98 1.9×10^3 rev/μg, mammalian cultured cells[7] (human, Chinese hamster), Drosophila.[8] Carcinogenicity (+). Oncogene activation.[23] UV spectrum (above): concentration 0.03 m*M* (a stock methanol solution was diluted with water).

AαC

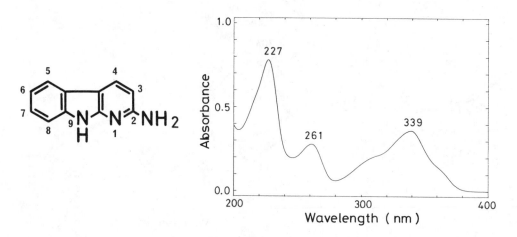

2-Amino-9*H*-pyrido[2,3-*b*]indole; CAS reg. no.: 26148-68-5; $C_{11}H_9N_3$, mol wt 183.2; pKa 4.6. Occurrence: broiled beef,[1] broiled chicken,[1] broiled mutton,[1] cigarette smoke condensate,[1] grilled onion,[24] grilled mushroom.[24] Synthesis: from 6-bromo-2-picolinic acid and *o*-phenylenediamine in 4 steps.[25] Genotoxicity data: *Salmonella typhimurium* TA98 300 rev/μg, mammalian cultured cells[7] (human, Chinese hamster), Drosophila.[8] Carcinogenicity (+). UV spectrum (above): concentration 0.03m*M* (a stock methanol solution was diluted with water).

MeAαC

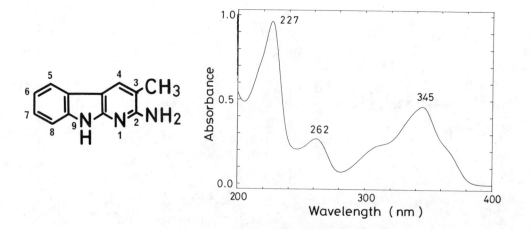

2-Amino-3-methyl-9*H*-pyrido[2,3-*b*]indole; CAS reg. no. 68006-83-7; $C_{12}H_{11}N_3$; mol wt 197.2; pKa 4.9. Occurrence: broiled mutton,[1] cigarette smoke condensate,[1] broiled beef,[24] broiled chicken,[24] grilled mushroom.[24] Synthesis: from 2-aminoindole and 3-amino-2-methylacrylonitrile in 1 step.[26] Genotoxicity data: *Salmonella typhimurium* TA98 200 rev/μg, Drosophila.[8] Carcinogenicity (+). UV spectrum (above): concentration 0.03 m*M* (a stock solution was diluted with water).

IQ

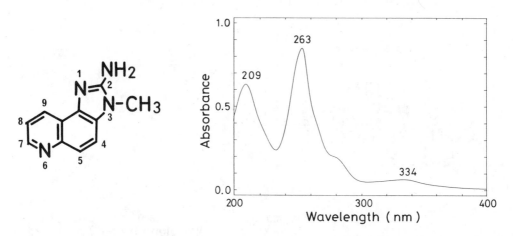

2-Amino-3-methylimidazo[4,5-*f*]quinoline; CAS reg. no. 76180-96-6; $C_{11}H_{10}N_4$; mol wt 198.2; pKa 3.8, 6.6 (lit. 3.5, 6.6[27]). Occurrence; broiled beef,[1] fried beef,[28] broiled sardine,[28] broiled salmon,[28] cigarette smoke condensate.[29] Synthesis: from 5,6-diaminoquinoline in 2 steps,[30] from 6-aminoquinoline in 5 steps,[31] from 5-amino-6-nitroquinoline in 4 steps,[32] from 6-chloroquinoline in 4 steps.[33] Derivatives: 2-NO$_2$,[34,35] 2-NHOH,[35,36] 2-N-aminopropyl,[37] 3-ethyl,[38] 7-OH,[39] 1-methyl-2-NO$_2$,[34] 6-deaza-2-NO$_2$,[34] 1-methyl-6-deaza-2-NO$_2$,[34] demethyl-6-deaza-2-NO$_2$,[34] 3-demethyl,[40] 3,5-dimethyl.[40] Genotoxicity data: *Salmonella typhimurium* TA98 4.3×10^5 rev/μg, mammalian cultured cells[7] (Chinese hamster), Drosophila.[8] Carcinogenicity (+). Oncogene activation.[41-43] UV spectrum (above): concentration 0.025 m*M* (a stock aqueous solution was diluted with water).

MeIQ

2-Amino-3,4-dimethylimidazo[4,5-*f*]quinoline; CAS reg. no. 77094-11-2; $C_{12}H_{12}N_4$; mol wt 212.3; pKa 3.95, 6.4. Occurrence: broiled sardine,[28] broiled salmon.[28] Synthesis: from 6-amino-7-methylquinoline (prepared from *m*-acetyltoluidine) in 5 steps,[31,44] from 6-chloro-7-methylquinoline in 4 steps.[33] Derivatives: 3-ethyl,[38] 7-OH.[45] Genotoxicity data: *Salmonella typhimurium* TA98 6.6×10^5 rev/μg, mammalian cultured cells[7] (Chinese hamster), Drosophila.[8] Carcinogenicity (+). UV spectrum (above): concentration 0.02 m*M* (a stock methanol solution was diluted with water).

MeIQx

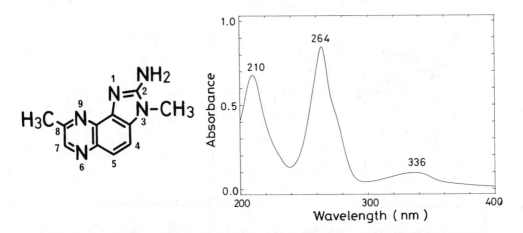

2-Amino-3,8-dimethylimidazo[4,5-*f*]quinoxaline; CAS reg. no.: 77500-04-0; $C_{11}H_{11}N_5$, mol wt 213.2; pKa: <2, 6.25 (lit. 6.4[46]). Occurrence: broiled beef,[1] fried ground beef,[1] broiled chicken,[1] broiled mutton,[1] food-grade beef extract,[47] smoked, dried bonito products.[48,49] Synthesis: from 6-amino-3-methylquinoxaline (prepared from 1,2,4-triaminobenzene and methylglyoxal) in 6 steps,[50] from 4'-fluorobenzenesulfonanilide in 7 steps,[51] from 7-chloro-2-methylquinoxaline in 4 steps,[33] from 5-chloro-2,1,3-benzoselenadiazole in 6 steps,[33] from 1-chloro-2,4-dinitrobenzene in 5 steps.[52] Derivatives: 2-NHOH,[6] 2-NO$_2$,[53] 3-ethyl,[38] 2-N-aminopropyl.[37] Genotoxicity data: *Salmonella typhimurium* TA98 1.45×10^5 rev/μg, mammaliam cultured cells[7] (Chinese hamster), Drosophila.[8] Carcinogenicity (+). UV spectrum (above): concentration 0.02 m*M* (a stock methanol solution was diluted with water).

4,8-DiMeIQx

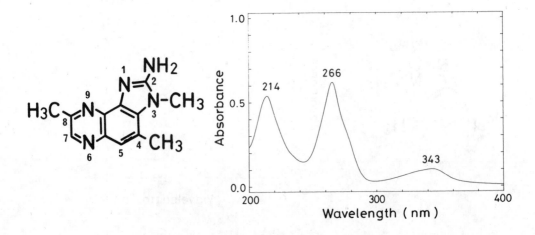

2-Amino-3,4,8-trimethylimidazo[4,5-*f*]quinoxaline; CAS reg. no.: 5896-78-9; $C_{12}H_{13}N_5$; mol wt 227.3; pKa<2, 6.25. Occurrence: fried ground beef,[1] broiled chicken,[1] broiled mutton,[1] smoked, dried fish products,[49] fried fish.[54] Synthesis: from 5-amino-2,4-dinitrotoluene in 8 steps,[52] from 2-fluoro-5-nitrotoluene in 10 steps.[55] Genotoxicity data: *Salmonella typhimurium* TA98 1.83 × 10^5 rev/μg. UV spectrum (above): concentration 0.015 m*M* (a stock methanol solution was diluted with water).

7,8-DiMeIQx

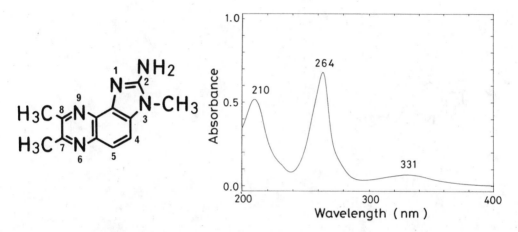

2-Amino-3,7,8-trimethylimidazo[4,5-*f*]quinoxaline; $C_{12}H_{13}N_5$; mol wt 227.3; pKa <2, 6.45. Occurrence: heated mixture of creatinine, glucose, and glycine.[56] Synthesis: from 7-chloro-2,3-dimethylquinoxaline in 4 steps,[33] from 5-chloro-2,1,3-benzoselenadiazole in 6 steps,[33] from 4-nitro-1,2-phenylenediamine and biacetyl in 7 steps.[57] Genotoxicity data: *Salmonella typhimurium* TA98 1.63 × 10^5 rev/μg. UV spectrum (above): concentration 0.015 m*M* (a stock methanol solution was diluted with water).

PhIP

2-Amino-1-methyl-6-phenylimidazo[4,5-*b*]-pyridine; CAS reg. no: 105650-23-5; $C_{13}H_{12}N_4$; mol wt 224.3; pKa 5.7. Occurrence: fried ground beef,[58,59] fried fish.[54] Synthesis: from 3-phenylpyridine in 4 steps.[60] Derivatives: 2-NO$_2$,[61] 2-NHOH,[62] 2-N-aminopropyl.[37] Genotoxicity data: *Salmonella typhimurium* TA98 1.8×10^3 rev/μg, mammalian cultured cells[63] (Chinese hamster). Carcinogenicity (+). UV spectrum (above): concentration 0.03 mM (a stock methanol solution was diluted with water).

Phe-P-1

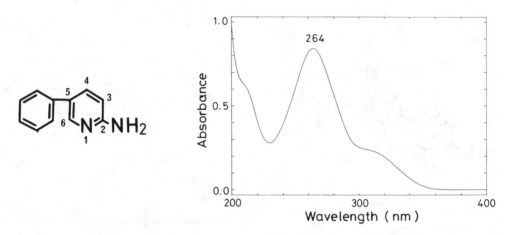

2-Amino-5-phenylpyridine: CAS reg. no.: 33421-40-8; $C_{11}H_{10}N_2$; mol wt 170.2; pKa 6.5. Occurrence: broiled sardine.[2] Synthesis: from imidazole and dichloromethylbenzene in 2 steps.[64] Genotoxicity data: *Salmonella typhimurium* TA98 41 rev/μg. UV spectrum (above): concentration 0.03 mM (a stock methanol solution was diluted with water).

Lys-P-1

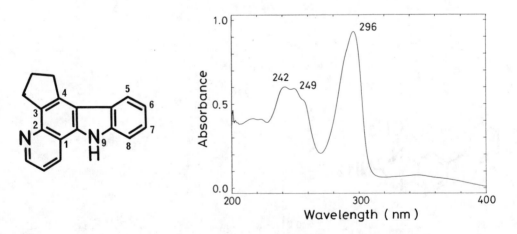

3,4-Cyclopentenopyrido[3,2-*a*]carbazole; $C_{18}H_{14}N_2$; mol wt 258.3; pKa 5.4 (in 50% methanol solution). Occurrence: pyrolytic tar of L-lysine[65] but no reports of detection in foods. Synthesis: from indane in 7 steps.[65] Genotoxicity data: *Salmonella typhimurium* TA98 90 rev/μg. UV spectrum (above): concentration 0.017 mM (a stock methanol solution was diluted with methanol).

Orn-P-1

4-Amino-6-methyl-1H-2,5,10,10b-tetraazafluoranthene; $C_{13}H_{11}N_5$; mol wt 237.3. Occurrence: pyrolytic tar of L-ornithine[66] but no reports of detection in foods. Genotoxicity data: *Salmonella typhimurium* TA98 5.7 × 10⁴ rev/μg.

IQx

2-Amino-3-methylimidazo[4,5-f]quinoxaline; $C_{10}H_9N_5$; mol wt 199.2. Occurrence: fried ground beef.[59] Synthesis: from 4-methylamino-3-nitro-1,2-benzene diamine.[33] Genotoxicity data: *Salmonella typhimurium* TA98 7.5 × 10⁴ rev/μg.

Harman

1-Methyl-9H-pyrido[3,4-b]indole; CAS reg. no.:486-84-0; $C_{12}H_{10}N_2$; mol wt 182.2. Synthesis: described in *Merck Index*; Occurrence: pyrolytic tar of L-tryptophan, egg yolk and soybean,[68] cigarette smoke condensate,[69] toasted bread,[70] broiled beef,[70] broiled sardine,[70] mushroom,[71] Japanese sake.[72] Genotoxicity data: comutagenic with 2-aminofluorene[73] (enhancement), mutagenic on mammalian cultured cells (Chinese hamster).[74]

Norharman

9*H*-Pyrido[3,4-*b*]indole; CAS reg. no.: 244-63-3; $C_{11}H_8N_2$; mol wt 168.2. Occurrence: pyrolytic tar of egg yolk, soybean, and tryptophan,[68] cigarette smoke condensate,[69] toasted bread,[70] broiled beef,[70] broiled sardine.[70] Genotoxicity data: comutagenic with aniline, *o*-toluidine, yellow OB (1-[(2-methylphenyl)azo]-2-naphthalenamine), *N,N*-dimethyl-4-aminoazobenzene, 3-aminopyridine, 2-amino-3-methylpyridine, and *N*-nitrosodiphenylamine;[75,76] mutagenic on mammalian cultured cells (Chinese hamster).[74]

REFERENCES

1. **Sugimura, T., Sato, S., and Wakabayashi, K.,** Mutagens/carcinogens in pyrolysates of amino acids and proteins and in cooked food: heterocyclic aromatic amines, in *Chemical Induction of Cancer,* Vol. IIIC, Woo, Y.-T, Lai, D. Y., Arcos, J. C., and Argus, M. F., Eds., Academic Press, San Diego, 1988, 681.
2. **Yamaizumi, Z., Shiomi, T., Kasai, H., Nishimura, S., Takahashi, Y., Nagao, M., and Sugimura, T.,** Detection of potent mutagens, Trp-P-1 and Trp-P-2, in broiled fish, *Cancer Lett.,* 9, 75, 1980.
3. **Akimoto, H., Kawai, A., Nomura, H., Nagao, M., Kawachi, T., and Sugimura, T.,** Synthesis of potent mutagens in tryptophan pyrolysates, *Chem. Lett.,* p. 1061, 1977.
4. **Akimoto, H., Kawai, A., and Nomura, H.,** Synthesis of 3-amino-5*H*-pyrido[4,3-*b*]indoles, carcinogenic γ-carbolines, *Bull. Chem. Soc. Jpn.,* 58, 123, 1985.
5. **Nagao, M., Takahashi, Y., Yahagi, T., Sugimura, T., Takeda, K., Shudo, K., and Okamoto, T.,** Mutagenicities of γ-carboline derivatives related to potent mutagens found in tryptophan pyrolysates, *Carcinogenesis,* 1, 451, 1980.
6. **Yamashita, K., Umemoto, A., Grivas, S., Kato, S., and Sugimura, T.,** *In vitro* reaction of hydroxyamino derivatives of MeIQx, Glu-P-1 and Trp-P-1 with DNA: ^{32}P-postlabelling analysis of DNA adducts formed *in vivo* by the parent amines and *in vitro* by their hydroxyamino derivatives, *Mutagenesis,* 3, 515, 1988.
7. **Furihata, C. and Matsushima, T.,** Mutagens and carcinogens in food, *Annu. Rev. Nutr.,* 6, 67, 1986.
8. **Yoo, M.-A, Ryo, H., Todo, T., and Kondo, S.,** Mutagenic potency of heterocyclic amines in the Drosophila wing spot test and its correlation to carcinogenic potency, *Jpn. J. Cancer Res. (Gann),* 76, 468, 1985.
9. **Fujikawa, K., Inagaki, E., Uchibori, M., and Kondo, S,.** Comparative induction of somatic eye-color mutations and sex-linked recessive lethals in *Drosophila melanogaster* by tryptophan pyrolysates, *Mutat. Res.,* 122, 315, 1983.
10. **Jensen, N. J.,** Pyrolytic products from tryptophan and glutamic acid are positive in the mammalian spot test, *Cancer Lett.,* 20, 241, 1983.
11. **Ishikawa, T., Takayama, S., Kitagawa, T., Kawachi, T., and Sugimura, T.,** Induction of enzyme-altered islands in rat liver by tryptophan pyrolysis products, *J. Cancer Res. Clin. Oncol.,* 95, 221, 1979.
12. **Kimura, T., Nakayama, T., Kurosaki, Y., Suzuki, Y., Arimoto, S., and Hayatsu, H.,** Absorption of 3-amino-1-methyl-5*H*-pyrido[4,3-*b*]indole, a mutagen-carcinogen present in tryptophan pyrolysate, from the gastrointestinal tract in the rat, *Jpn. J. Cancer Res. (Gann),* 76, 272, 1985.
13. **Takeda, K., Ohta, T., Shudo, K., Okamoto, T., Tsuji, K., and Kosuge, T.,** Synthesis of a mutagenic principle isolated from tryptophan pyrolysate, *Chem. Pharm. Bull.,* 25, 2145, 1977.
14. **Hashimoto, Y., Shudo, K., and Okamoto, T.,** Activation of a mutagen, 3-amino-1-methyl-5*H*-pyrido[4,3-*b*] indole. Identification of 3-hydroxyamino-1-methyl-5*H*-pyrido[4,3-*b*]indole and its reaction with DNA, *Biochem. Biophys. Res. Commun.,* 96, 355, 1980.
15. **Saito, K., Yamazoe, Y., Kamataki, T., and Kato, R.,** Synthesis of hydroxyamino, nitroso and nitro derivatives of Trp-P-2 and Glu-P-1, amino acid pyrolysate mutagens, and their direct mutagenicities towards *Salmonella typhimurium* TA98 and TA98NR, *Carcinogenesis,* 4, 1547, 1983.
16. **Yamaguchi, K., Zenda, H., Shudo, K., Kosuge, T., Okamoto, T., and Sugimura, T.,** Presence of 2-aminodipyrido[1,2-*a*:3′,2′-*d*]imidazole in casein pyrolysate, *Gann,* 70, 849, 1979.
17. **Takeda, K., Shudo, K., Okamaoto, T., and Kosuge, T.,** Synthesis of mutagenic principles isolated from L-glutamic acid pyrolysate, *Chem. Pharm. Bull.,* 26, 2924, 1978.
18. **Hashimoto, Y., Shudo, K., and Okamoto, T.,** Modification of DNA with potent mutacarcinogenic 2-amino-6-methyldipyrido[1,2-*a*:3′,2′-*d*]imidazole isolated from a glutamic acid pyrolysate: structure of the modified nucleic acid base and initial chemical event caused by the mutagen, *J. Am. Chem. Soc.,* 104, 7636, 1982.

19. **Hashimoto, Y., Kawachi, E., Shudo, K., and Sekiya, T.,** Activation of c-Ha-*ras* proto-oncogene by *in vitro* chemical modification with 2-amino-6-methyldipyrido[1,2-a:3',2'-*d*]imidazole (Glu-P-1) and 4-nitroquinoline N-oxide (4NQO), *Nucleic Acids Res. Symp. Ser.,* 17, 135, 1986.

20. **Hashimoto, Y., Kawachi, E., Shudo, K., Sekiya, T., and Sugimura, T.,** Transforming activity of human c-Ha-*ras*-1 proto-oncogene generated by the binding of 2-amino-6-methyldipyrido[1,2- a:3',2'-*d*]imidazole and 4-nitroquinoline N-oxide: direct evidence of cellular transformation by chemically modified DNA, *Jpn. J. Cancer Res. (Gann),* 78, 211, 1987.

21. **Yamaguchi, K., Shudo, K., Okamoto, T., Sugimura, T., and Kosuge, T.,** Presence of 2-aminodipyrido-[1,2-a:3',2'-*d*]imidazole in broiled cuttlefish, *Gann,* 71, 743, 1980.

22. **Takeda, K., Shudo, K., Okamoto, T., Nagao, M., Wakabayashi, K., and Sugimura, T.,** Effect of methyl substitution on mutagenicity of 2-aminodipyrido[1,2-a:3',2'-*d*]imidazole, *Carcinogenesis,* 1, 889, 1980.

23. **Ishizaka, Y., Ochiai, M., Ishikawa, F., Sato, S., Miura, Y., Nagao, M., and Sugimura, T.,** Activated N-*ras* oncogene in a transformant derived from a rat small intestinal adenocarcinoma induced by 2-aminodipyrido[1,2-a:3',2'-*d*]imidazole, *Carcinogenesis,* 8, 1575, 1987.

24. **Matsumoto, T., Yoshida, D., and Tomita, H.,** Determination of mutagens, amino-α-carbolines in grilled foods and cigarette smoke condensate, *Cancer Lett.,* 12, 105, 1981.

25. **Matsumoto, T., Yoshida, D., and Tomita, H.,** Synthesis of 2-amino-9*H*-pyrido[2,3-*b*]indole isolated as a mutagenic principle from pyrolytic products of protein, *Agric. Biol. Chem.,* 43, 675, 1979.

26. **Matsumoto, T., Yoshida, D., and Tomita, H.,** Synthesis and mutagenic activity of alkyl derivatives of 2-amino-9*H*-pyrido[2,3-*b*]indole, *Argic. Biol. Chem.,* 45, 2031, 1981.

27. **Kasai, H., Yamaizumi, Z., Nishimura, S., Wakabayashi., K, Nagao, M., Spingarn, N. E., Weisburger, J. H., Yokoyama, S., and Miyazawa, T.,** A potent mutagen in broiled fish. I. 2-Amino-3-methyl-3*H*-imidazo[4,5-*f*]quinoline, *J. Chem. Soc. Perkin Trans. 1,* p. 2290, 1981.

28. **Yamaizumi, Z., Kasai, H., Nishimura, S., Edmonds, C. G., and McCloskey, J. A.,** Stable isotope dilution quantification of mutagens in cooked foods by combined liquid chromatography-thermospray mass spectrometry, *Mutat. Res.,* 173, 1, 1986.

29. **Yamashita, M., Wakabayashi, K., Nagao, M., Sato, S., Yamaizumi, Z., Takahashi, M., Kinae, N., Tomita, I., and Sugimura, T.,** Detection of 2-amino-3-methylimidazo[4,5-*f*]quinoline in cigarette smoke condensate, *Jpn. J. Cancer Res. (Gann),* 77, 419, 1986.

30. **Kasai, H., Nishimura, S., Wakabayashi, K., Nagao, M., and Sugimura, T.,** Chemical synthesis of 2-amino-3-methylimidazo[4,5-*f*]quinolin (IQ), a potent mutagen isolated from broiled fish, *Proc. Jpn. Acad.,* 56 (Ser. B), 382, 1980.

31. **Lee, C.-S, Hashimoto, Y., Shudo, K., and Okamoto, T.,** Synthesis of mutagenic heteroaromatics: 2-aminoimidazo[4,5-*f*]quinolines, *Chem. Pharm. Bull.,* 30, 1857, 1982.

32. **Ziv, J., Knapp, S., and Rosen, J. D.,** Convenient synthesis of food mutagen 2-amino-3-methylimidazo[4,5-*f*]quinoline (IQ) and IQ-d$_3$, *Synthet. Commun.,* 18, 973, 1988.

33. **Grivas, S.,** Efficient synthesis of mutagenic imidazo[4,5-*f*]quinoxalin-2-amines via readily accessible 2,1,3-benzoselenadiazoles, *Acta Chem. Scand. Ser. B.,* 40, 404, 1986.

34. **Dirr, A. and Wild, D.,** Synthesis and mutagenic activity of nitro-imidazoarenes. A study on the mechanism of the genotoxicity of heterocyclic arylamines and nitroarenes, *Mutagenesis,* 3, 147, 1988.

35. **Okamoto, T., Shudo, K., Hashimoto, Y., Kosuge, T., Sugimura, T., and Nishimura, S.,** Identification of a reactive metabolite of the mutagen, 2-amino-3-methylimidazolo[4,5-*f*]quinoline, *Chem. Pharm. Bull.,* 29, 590, 1981.

36. **Snyderwine, E. G., Roller, P. P., Wirth, P. J., Adamson, R. H., Sato, S., and Thorgeirsson, S. S.,** Synthesis, purification and mutagenicity of 2-hydroxyamino-3-methylimidazolo[4,5-*f*]quinoline, *Carcinogensis,* 8, 1017, 1980.

37. **Watkins, B. E., Knize, M. G., Morris, C. J., Andresen, B. D., Happe, J., Vanderlaan, M., and Felton, J. S.,** The synthesis of haptenic derivatives of aminoimidazoazaarene cooked food mutagens, *Heterocycles,* 26, 2069, 1987.

38. **Jägerstad, M. and Grivas, S.,** The synthesis and mutagenicity of the 3-ethyl analogues of the potent mutagens IQ, MeIQ, MeIQx and its 3,7-dimethyl isomer, *Mutat. Res.,* 144, 131, 1985.

39. **Bashir, M. and Kingston, D. G. I.,** Biological formation and chemical synthesis of 2-amino-3,6-dihydro-3-methyl-7*H*-imidazolo[4,5-*f*]quinolin-7-one, the major metabolite of the dietary carcinogen 2-amino-3-methyl-3*H*-imidazolo[4,5-*f*]quinoline(IQ) by normal intestinal bacteria, *Heterocycles,* 26, 2877, 1987.

40. **Nagao, M., Wakabayashi, K., Kasai, H., Nishimura, S., and Sugimura, T.,** Effect of methyl substitution on mutagenesis of 2-amino-3-methylimidazo[4,5-*f*]quinoline, isolated from broiled sardine, *Carcinogenesis,* 2, 1147, 1981.

41. **Ishikawa, F., Takaku, F., Nagao, M., Ochiai, M., Hayashi, K., Takayama, S., and Sugimura, T.,** Activated oncogenes in a rat hepatocellular carcinoma induced by 2-amino-3-methylimidazo[4,5-*f*]quinoline, *Jpn. J. Cancer Res. (Gann),* 76, 425, 1985.

42. **Ishikawa, F., Takaku, F., Ochiai, M., Hayashi, K., Hirohashi, S., Terada, M., Takayama, S., Nagao, M., and Sugimura, T.,** Activated c-*raf* gene in a rat hepatocellular carcinoma induced by 2-amino-3-methylimidazo[4,5-*f*]quinoline, *Biochem. Biophys. Res. Commun.,* 132, 186, 1985.

43. **Ishikawa, F., Takaku, F., Hayashi, K., Nagao., M., and Sugimura, T.,** Activation of rat c-*raf* during transfection of hepatocellular carcinoma DNA, *Proc. Natl. Acad. Sci. U.S.A.,* 83, 3209, 1986.

44. **Kasai, H., Yamaizumi, Z., Wakabayashi, K., Nagao, M., Sugimura, T., Yokoyama, S., Miyazawa, T., and Nishimura, S.,** Structure and chemical synthesis of Me-IQ, a potent mutagen isolated form broiled fish, *Chem. Lett.,* p. 1391, 1980.

45. **Bashir, M. and Kingston, D. G. I.,** Isolation, structure elucidation, and synthesis of the major anaerobic bacterial metabolite of the dietary carcinogen 2-amino-3,4-dimethyl-3*H*-imidazo[4,5-*f*]quinolin (MeIQ), *Heterocycles,* 29, 1127, 1989.

46. **Hayatsu, H., Matsui, Y., Ohara, Y., Oka, T., and Hayatsu, T.,** Characterization of mutagenic fractions in beef extract and in cooked ground beef. Use of blue-cotton for efficient extraction, *Gann,* 74, 472, 1983.

47. **Takahashi, M., Wakabayashi, K., Nagao, M., Yamamoto, M., Masui, T., Goto, T., Kinae, N., Tomita, I., and Sugimura, T.,** Quantification of 2-amino-3-methylimidazo[4,5-*f*]quinoline (IQ) and 2-amino-3,8-dimethylimidazo[4,5-*f*]quinoxaline (MeIQx) in beef extracts by liquid chromatography with electrochemical detection (LCEC), *Carcinogenesis,* 6, 1195, 1985.

48. **Kikugawa, K., Kato, T., and Hayatsu, H.,** The presence of 2-amino-3,8-dimethylimidazo[4,5-*f*]quinoxaline in smoked dry bonito (katsuobushi), *Jpn. J. Cancer Res. (Gann),* 77, 99, 1986.

49. **Kato, T., Kikugawa, K., and Hayatsu, H.,** Occurrence of the mutagens, 2-amino-3,8-dimethylimidazo[4,5-*f*]quinoxaline (MeIQx) and 2-amino-3,4,8-trimethylimidazo[4,5-*f*]quinoxaline (4,8-Me$_2$IQx) in some Japanese smoked, dried fish products, *J. Agric. Food Chem.,* 34, 810, 1986.

50. **Kasai, H., Shiomi, T., Sugimura, T., and Nishimura, S.,** Synthesis of 2-amino-3,8-dimethylimidazo [4,5-*f*]quinoxaline (MeIQx), a potent mutagen isolated from fried beef, *Chem. Lett.,* p. 675, 1981.

51. **Grivas, S. and Olsson, K.,** An improved synthesis of 3,8-dimethyl-3*H*-imidazo[4,5-*f*]quinoxaline ("MeIQx") and its 2-^{14}C-labelled analogue, *Acta Chem. Scand. Ser. B.,* 39, 31, 1985.

52. **Knapp, S., Ziv, J., and Rosen, J. D.,** Synthesis of the food mutagens MeIQx and 4,8-DiMeIQx by copper (I) promoted quinoxaline formation, *Tetrahedron,* 45, 1293, 1989.

53. **Grivas, S.,** Synthesis of 3,8-dimethyl-2-nitro-3*H*-imidazo[4,5-*f*]quinoxaline, the 2-nitro analogue of the food carcinogen MeIQx, *J. Chem. Res. (S),* p. 84, 1988.

54. **Zhang, X.-M., Wakabayashi, K., Liu, Z.-C., Sugimura, T., and Nagao, M.,** Mutagenic and carcinogenic heterocyclic amines in Chinese cooked foods, *Mutat. Res.,* 201, 181, 1988.

55. **Grivas, S.,** A convenient synthesis of the potent mutagen 3,4,8-trimethyl-3*H*-imidazo[4,5-*f*]quinoxalin-2-amine, *Acta Chem. Scand. Ser. B.,* 39, 213, 1985.

56. **Negishi, C., Wakabayashi, K., Tsuda, M., Sato, S., Sugimura, T., Saito, H., Maeda, M., and Jägerstad, M.,** Formation of 2-amino-3,7,8-trimethylimidazo[4,5-*f*]quinoxaline, a new mutagen, by heating a mixture of creatinine, glucose and glycine, *Mutat. Res.,* 140, 55, 1984.

57. **Lovelette, C., Barnes, W. S., Weisburger, J. H., and Williams, G. M.,** Improved synthesis of the food mutagen 2-amino-3,7,8-trimethyl-3*H*-imidazo[4,5-*f*]quinoxaline and activity in a mammalian DNA repair system, *J. Agric. Food Chem.,* 35, 912, 1987.

58. **Felton, J. S., Knize, M. G., Shen, N. H., Lewis, P. R., Andresen, B. D., Happe, J., and Hatch, F. T.,** The isolation and identification of a new mutagen from fried ground beef: 2-amino-1-methyl-6-phenylimidazo[4,5-*b*]pyridine (PhIP), *Carcinogenesis,* 7, 1081, 1986.

59. **Becher, G., Knize, M. G., Nes, I. F., and Felton, J. S.,** Isolation and identification of mutagens from a fried Norwegian meat product, *Carcinogenesis,* 9, 247, 1988.

60. **Knize, M. G. and Felton, J. S.,** The synthesis of the cooked beef mutagen 2-amino-1-methyl-6-phenylimidazo[4,5-*b*]pyridine and its 3-methyl isomer, *Heterocycles,* 24, 1815, 1986.

61. **Adolfsson, L. and Olsson, K.,** A convenient synthesis of mutagenic ^{3}H-imidazo[4,5-*f*]quinolin-2-amines and of their 2-^{14}C-labelled analogues, *Acta Chem. Scand. Ser. B,* 37, 157, 1983.

62. **Holme, J. A., Wallin, H., Brunborg, G., Søderlund, E. J., Hongslo, J. K., and Alexander, J.,** Genotoxicity of the food mutagen 2-amino-1-methyl-6-phenylimidazo[4,5-*b*]pyridine (PhIP): formation of 2-hydroxamino-PhIP, a directly acting genotoxic metabolite, *Carcinogenesis,* 10, 1389, 1989.

63. **Thompson, L. H., Tucker, J. D., Stewart, S. A., Christensen, M. L., Salazar, E. P., Carrano, A. V., and Felton, J. S.,** Genotoxicity of compounds from cooked beef in repair-deficient CHO cells versus *Salmonella* mutagenicity, *Mutagenesis,* 2, 483, 1987.

64. **Tsuji, K., Yamamoto, T., Zenda, H., and Kosuge, T.,** Studies on active principles of tar. VII. Production of biological active substances in pyrolysis of amino acids. II. Antifungal constituents in pyrolysis products of phenylalanine, *Yakugaku Zasshi (J. Pharm. Soc. Jpn.),* 98, 910, 1978.

65. **Wakabayashi, K., Tsuji, K., Kosuge, T., Yamaguchi, K., Shudo, K., Takeda, K., Iitaka, Y., Okamoto, T., Yahagi, T., Nagao, M., and Sugimura, T.,** Isolation and structure determination of a mutagenic substance in L-lysine pyrolysate, *Proc. Jpn. Acad.,* 54 (Ser. B), 569, 1978.

66. **Yokota, M., Narita, K., Kosuge, T., Wakabayashi, K., Nagao, M., Sugimura, T., Yamaguchi, K., Shudo, K., Iitaka, Y., and Okamoto, T.,** a potent mutagen isolated from a pyrolysate of L-ornithine, *Chem. Pharm. Bull.,* 29, 1473, 1981.

67. **Sugimura, T., Wakabayashi, K., Nagao, M., and Ohgaki, H.,** Heterocyclic amines in cooked food, in *Food Toxicology: A Perspective on the Relative Risks,* Taylor, S. L. and Scanlan, R. A., Eds., Marcel Dekker, New York, 1989, 31.

68. **Kosuge, T., Tsuji, K., Wakabayashi, K., Okamoto, T., Shudo, K., Iitaka, Y., Itai, A., Sugimura, T., Kawachi, T., Nagao, M., Yahagi, T., and Seino, T.,** Isolation and structure studies of mutagenic principles in amino acid pyrolysates, *Chem. Pharm. Bull.,* 26, 611, 1978.

69. **Poindexter, E. H., Jr. and Carpenter, R. O.,** The isolation of harman and norharman from tobacco and cigarette smoke, *Phytochemistry,* 1, 215, 1962.

70. **Yasuda, T., Yamaizumi, Z., Nishimura, S., Nagao, M., Takahashi, M., Fujiki, H., Sugimura, T., and Tsuji, K.,** Detection of comutagenic compounds, harman and norharman in pyrolysis product of protein and food by gas chromatography-mass spectrometry, *Proc. 3rd Annu. Meet. Med. GCMS Soc.,* p. 97, 1978.

71. **Takeuchi, T., Ogawa, K., Iinuma, H., Suda, H., Ukita, K., Nagatsu, T., Kato, M., Umezawa, H., and Tanabe, O.,** Monoamine oxidase inhibitors isolated from fermented broths, *J. Antibiot.,* 26, 162, 1973.

72. **Takase, S. and Murakami, H.,** Studies on the fluorescence of sake and identification of harman, *Agric. Biol. Chem.,* 30, 869, 1966.

73. **Matsumoto, T., Yoshida, D., and Mizusaki, S.,** Enhancing effect of harman on mutagenicity in Salmonella, *Mutat. Res.,* 56, 85, 1977.

74. **Nakayasu, M., Nakasato, F., Sakamoto, H., Terada, M., and Sugimura, T.,** Mutagenic activity of norharman and harman in Chinese hamster lung cells in assays with diphtheria toxin resistance as a marker, *Cancer Lett.,* 17, 249, 1983.

75. **Sugimura, T. and Nagao, M.,** Mutagenic factors in cooked foods., *CRC Crit. Rev. Toxicol.,* 6, 189, 1979.

76. **Sugimura, T., Nagao, M., and Wakabayashi, K.,** Metabolic aspects of the comutagenic action of norharman, in *Biological Reactive Intermediates,* Vol. 2, Snyder, R., Parke, D. V., Kocsis, J., Jollow, D. J., and Gibson, G. G., Eds., Plenum Press, New York, 1982, 1011.

Chapter 2.3

DNA MODIFICATION *IN VITRO* AND *IN VIVO* WITH HETEROCYCLIC AMINES

Kazuo Negishi, Minako Nagao, Kazuyuki Hiramoto, and Hikoya Hayatsu

TABLE OF CONTENTS

I. NONCOVALENT INTERACTIONS OF HETEROCYCLIC AMINES WITH DNA

Physicochemical affinity of mutagens to DNA is thought to facilitate covalent bindings with DNA, because molecules which interact with DNA should covalently bind to DNA more easily than free molecules. Pezzuto et al. synthesized Trp-P-2, Trp-P-1, and their derivatives, which showed various degrees of mutagenicity toward *Salmonella typhimurium* in the presence of the microsomal activation system. All these compounds had affinity to calf thymus DNA, with association constants ranging 0.4 to 4 $\times$ 10^4 M^{-1}. A reasonable correlation was found with these compounds between the mutagenicity and the DNA affinity.[1] Imamura et al. reported that Glu-P-2 and its methylated derivatives including Glu-P-1 had a DNA affinity comparable to that of the Trp-P derivatives.[2] The association constants for calf thymus DNA are 0.3 to 1.7 $\times$ 10^4 M^{-1}. Glu-P-1 has a greater affinity than Glu-P-2 and is more mutagenic. IQ, MeIQ, and MeIQx were shown to have high affinity to DNA.[3] The association constants for calf thymus DNA are about 1 $\times$ 10^6 M^{-1}. The order of the association constants is coincident with that of the mutagenic potential in *S. typhimurium*. All the results suggest that the DNA affinity of the mutagens in a closely related group correlates with their mutagenicity. Trp-P-1, however, has a higher affinity than Trp-P-2 but is less mutagenic. Among methylated Glu-P-2 derivatives, 9-methyl-Glu-P-2 has the highest affinity to DNA but is much less mutagenic than 4-methyl-Glu-P-2 (Glu-P-1). This finding indicates that in some cases factors other than the affinity may be more important for the mutagenicity.

The nature of the DNA affinity of heterocyclic amines has been studied with the use of a closed circular DNA. Nonmutagenic heterocyclic amines, harman and norharman, were the first that were shown to unwind superhelical plasmid DNA.[4] Glu-P-1, Glu-P-2,[5] and IQ derivatives[6] also unwind DNA. Clearly, these heterocyclic amines can intercalate into DNA. The unwinding angles for norharman, Glu-P-1, and Glu-P-2 were estimated to be 17 $\pm$ 3°, 20 $\pm$ 3°, and 18 $\pm$ 3°, respectively.

II. REACTIONS OF ACTIVATED HETEROCYCLIC AMINES WITH DNA *IN VITRO*

Hetrocyclic amines themselves have no ability of covalent interaction with biomolecules. The hydroxyamino form of the hetrocyclic amines, or further acylated hydroxyamino derivatives, can bind covalently to DNA and other molecules.[6] Hashimoto et al.[7,8] have studied the reaction of an acetoxy derivative of Glu-P-1 (Glu-P-1-NHOAc) with DNA and deoxyribonucleosides. They have shown that (guanin-8-yl)-Glu-P-1 is a major product of the reaction. The structure of this product was confirmed with comparison to the sample of (guanin-8-yl)-Glu-P-1 synthesized by another route. The yield of this adduct in the reaction with GpC was 100-fold greater than that in the reaction with guanosine, guanylic acid, or poly(G). Glu-P-1-NHOAc was supposed to intercalate into a pair of GpC first and then bind covalently. Hashimoto and Shudo studied the sequence specificity of the modification with Glu-P-1-NHOAc using a ^{32}P-labeled DNA fragment as a target.[9] They reported a preferential modification of the guanine residues in the region with GC clusters, a result suggesting a reaction pathway via intercalation. The same adduct, (guanin-8-yl)-Glu-P-1, was obtained when Glu-P-1 was incubated with DNA in the presence of microsomal enzymes.[8] They also demonstrated the formation of (guanin-8-yl)-Trp-P-2 in DNA treated with either Trp-P-2 in the presence of microsomal enzymes or with acetoxy-Trp-P-2 (Trp-P-2-NHOAc).[10] More recently, IQ-NHOH has been shown to bind to DNA, forming (guanin-9-yl)-IQ. The rates of the reaction of synthetic homopolynucleotides with IQ-NHOH are in the order poly(G) $>>$ ploy(A) $>$ poly(C) = poly(U). The binding of IQ-NHOH to DNA is enhanced sixfold when acetic anhydride is present in the reaction mixture, a fact suggesting that acetoxy-IQ

(IQ-NHOAc) is more reactive than IQ-NHOH. (Guanin-8-yl)-IQ is a major product from DNA modified with IQ-NHOH or IQ-NHOAc.[11] Loukakou et al. reported the DNA binding of Glu-P-3, a synthetic mutagen similar to and more mutagenic than Glu-P-1. An activated form of Glu-P-3, 3-*N,N*-acetoxyacetylamino-4,6-dimethyldipyrido[1,2-*a*:3′,2′-*d*]imidazole (acetyl-Glu-P-3-NOAc), binds covalently to guanine in DNA. A major adduct was identified as (guanin-8-yl)-acetyl-Glu-P-3.[12] This binding induces large conformational changes in the DNA molecule, as detected by a decrease in Tm and by an increase in the susceptibility of DNA to nuclease S1, a single-strand-specific nuclease. Poly(dG-dC)·poly(dG-dC) and poly(dA-dC)·poly(dG-dT) modified with acetyl-Glu-P-3-NOAc can undergo the change in conformation from B-form to left-handed Z-form more easily than the unmodified molecules.[13]

Major reactive species derived *in vivo* from administered heterocyclic amines in living cells are believed to be the acetate or the sulfate of the hydroxyamino form. Other reaction pathways, however, might also be possible. Seryl-tRNA synthetase can activate Trp-P-2-NHOH[14] and proline-dependent activation of Trp-P-2-NHOH was reported.[15] These data indicate the possibility that amino acid esters of the hydroxyamino form of heterocyclic amines may attack DNA if they are formed in the cells. Trp-P-2-NHOH itself seems to have some reactivity to DNA.[16,17] The reactivity of Trp-P-2-NHOH to DNA is enhanced by the presence of ascorbate.[18] Cysteamine also enhances the binding of Trp-2-NHOH to DNA by protecting Trp-P-2-NHOH from its oxidative degradation.[17]

Trp-P-2-NHOH has been reported to show another type of genotoxicity. It induces DNA single-strand breaks on treatment of DNA in solutions as well as on treatment of mammalian cells in culture.[17] Active oxygens generated during the oxidative degradation of Trp-P-2-NHOH have been shown to be responsible for the breaks of DNA in solution. This mechanism may be operating in the breaks of the cellular DNA as well, because simultaneous production of superoxide anion inside cells has been observed.[19] Hayashi et al. reported that breaks of cellular DNA were induced on treatment of hepatocytes with Trp-P-1, Trp-P-2, Glu-P-1, Glu-P-2, and IQ.[20] To obtain this effect, microsomal activation of these compounds into hydroxyamino derivatives seemed to be obligatory.

III. ADDUCT FORMATIONS *IN VIVO*

The analysis of DNA from rats injected with Trp-P-2 has shown that (guanin-8-yl)-Trp-P-2 is formed. (Guanin-8-yl)-Glu-P-1 was demonstrated in DNA from rats injected with Glu-P-1.[21] Adducts were found in DNA from female BALB/c mice fed with [2-^{14}C]MeIQx. Among various organs examined, the liver showed the highest adduct level.[22]

^{32}P-Postlabeling was used for the detection of modified bases in DNA. This technique allows detection of adducts with a high sensitivity, although chemical structures of the adducts are generally difficult to identify.[23] With this technique, DNA modifications induced *in vivo* by MeAαC and AαC were detected.[24] The DNA extracted from various tissues of rats fed a diet containing either MeAαC or AαC at the concentration of 0.08% for 1 week was enzymatically digested, labeled with ^{32}P, and then the modified bases were quantitated with chromatographic analysis. The MeAαC treatment induced DNA-adduct formation in the salivary glands, the pancreas, and the liver in a preferential manner. The levels of total adducts formed were 1 per 10^5 nucleotides in the liver and 0.2 to 0.5 per 10^5 nucleotides in the other two organs. In contrast, AαC-induced DNA-adduct formation in the liver was at the level of 0.3 per 10^5 nucleotides and only a very few adducts were formed in the salivary glands and in the pancreas. These results are consistent with the induction by MeAαC but not by AαC of atrophy in the salivary glands and in the pancreas. Snyderwine et al. have analyzed adduct formations in the DNA from the livers of Cynomolgus monkeys fed IQ (20 mg/kg). In the chromatogram of ^{32}P-postlabeled nucleotides from the digest of this DNA, they obtained eight different spots of modified nucleotides. One of them was identified as (guanin-8-yl)-IQ nucleotide.[25] The content of modified bases in the liver was about 1 in

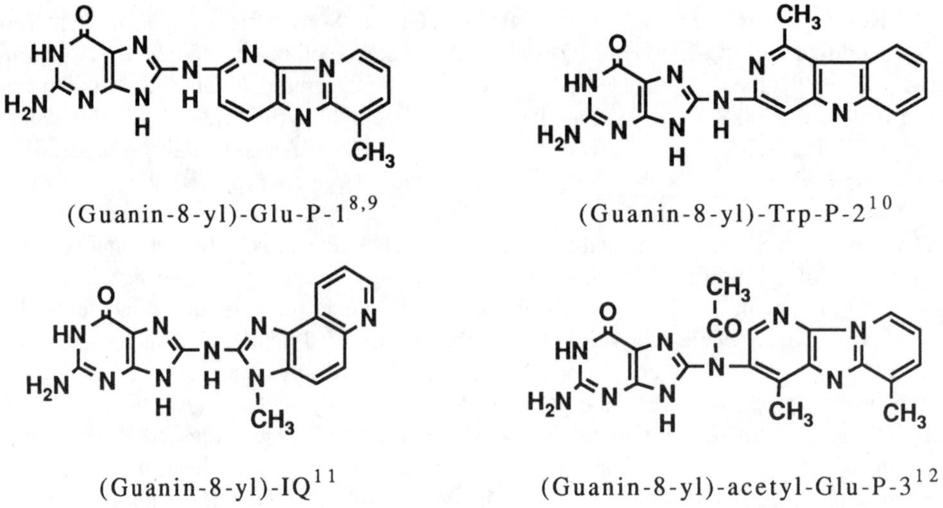

(Guanin-8-yl)-Glu-P-1[8,9]

(Guanin-8-yl)-Trp-P-2[10]

(Guanin-8-yl)-IQ[11]

(Guanin-8-yl)-acetyl-Glu-P-3[12]

FIGURE 1.   Structures of DNA adducts induced with heterocyclic amines.

10^5 nucleotides. A slightly lower amount of adducts was found in the kidney. In the colon, stomach, and bladder, the levels of adducts were much less. Schut et al. also reported that administration of IQ to F-344 rats induced DNA adducts in the liver and the small and large intestines, the three organs which are target tissues in the rat by the carcinogenic action of IQ.[26]

The DNA adducts formed in rats *in vivo* on treatment intragastrically with 50 mg/kg of MeIQx, Glu-P-1, or Trp-P-1 were also analyzed by the postlabeling method. The level of total adducts formed was about 1 in 10^6 nucleotides. The products were compared to the adducts obtained from *in vitro* modification of DNA with MeIQx-NHOH, Glu-P-2-NHOH, or Trp-P-1-NHOH in the presence or absence of acetic anhydride.[27] In the case of MeIQx and Glu-P-1, almost all the adducts formed *in vitro* with the hydroxylamines were the same as those formed in the rat liver. Addition of acetic anhydride to the *in vitro* reaction did not affect the relative ratio of the products. In contrast, the chromatogram of adducts formed by an *in vitro* treatment of DNA with Trp-P-1-NHOH gave a pattern different from those formed *in vivo*. Addition of acetic anhydride to Trp-P-1-NHOH and DNA affected the spot pattern of the chromatogram. These results indicate a complex nature of Trp-P-1 metabolism. Comutagens harman and norharman also cause DNA lesions. Adducts were identified in DNA extracted from the organs of mice fed food containing 0.1% of harman. The levels of harman-DNA adducts in the liver and the kidney were 1.6 and 0.57 per 10^7 nucleotides, respectively. Similar treatment of mice with norharman causes DNA adducts in the kidney, glandular stomach, and large intestine, but not in the liver or the brain.[28] Harman and norharman are only comutagenic to *S. typhimurium,* but they are mutagenic to Chinese hamster lung cells.[29] Mammalian cells may have enzymes that convert harman and norharman to species reactive with DNA.

In almost all these *in vivo* binding experiments with heterocyclic amines, DNA extracted from the liver showed the highest frequency of binding among DNAs from various organs. The exceptions are PhIP and norharman. PhIP-DNA adducts were relatively high in the lung, the pancreas, and the heart (around 2 per 10^6 nucleotides) and lowest in the liver (0.2 per 10^6 nucleotides) when rats fed with diet containing 0.05% PhIP for 4 weeks were analyzed.[30] There is a report showing that DNA adducts induced *in vivo* with Trp-P-2, Glu-P-2, MeIQ, 4,8-DiMeIQx, and 7,8-DiMeIQx can be detected by the ^{32}P-postlabeling method.[31]

DNA binding of the hydroxyamino derivatives of heterocyclic amines is dependent on their esterification in cells.[6,32,33] The extent of the dependence seems to vary among heterocyclic amines. The activated molecules might yield nitrenium ions,[33,34] which react with DNA.

Analysis of the mutation spectrum induced by Trp-P-2 in human cultured cells reveals that most of the mutations can be explained in terms of the modification of guanine. Some other modifications, presumably damages in thymine, may also be involved.[35]

IV. ONCOGENE ACTIVATION

Carcinogenicity of heterocyclic amines implies their ability to induce mutations in whole animals in the initiation step of carcinogenesis. A proto-oncogene, c-Ha-*ras*-1, was shown to be activated on an *in vitro* treatment of plasmids containing the proto-oncogene with an activated heterocyclic amine, Glu-P-1-NHOAc. The treated plasmid can transform NIH3T3 cells. In the transformants, the CCGG sequence in the region which codes for the 11th and the 12th amino acids of the oncogene product was mutated.[36] These results indicate that (guanin-8-yl)-Glu-P-1 residues, which are known to be formed exclusively in the DNA treated *in vitro* with Glu-P-1-NHOAc, cause the mutation, thereby inducing the activation of the oncogene. *In vivo* situations are still unclear. DNA extracted from malignant tissues of rats induced by heterocyclic amines can transform NIH3T3 cells. The activation of H-*ras, raf,* and N-*ras* proto-oncogenes was detected in the transformants.[37-39] However, the mutations responsible for the oncogene activation in the malignant tissues of rats have not yet been identified. Further analysis of the activated oncogenes will clarify the molecular events taking place in the heterocyclic amine-mediated oncogene activation.

REFERENCES

1. **Pezzuto, J. M., Lau, P. P., Luh, Y., Moore, P. D., Wogan, G. N., and Hecht, S. M.,** There is a correlation between the DNA affinity and mutagenicity of several 3-amino-1-methyl-5*H*-pyrido[4,3-*b*]indoles, *Proc. Natl. Acad. Sci. U.S.A.,* 77, 1427, 1980.
2. **Imamura, M., Takeda, K., Shudo, K., Okamoto, T., Nagata, C., and Kodama, M.,** Non-covalent interaction with DNA of the mutagens 2-amino-dipyrido[1,2-*a*:3′,2′-*d*]imidazole and methyl-substituted isomers, *Biochem. Biophys. Res. Commun.,* 96, 611, 1980.
3. **Watanabe, T., Yokoyama, S., Hayashi, K., Kasai, H., Nishimura, S., and Miyazawa, T.,** DNA-binding of IQ, Me-IQ and Me-IQx, strong mutagens found in broiled foods, *FEBS Lett.,* 150, 434, 1982.
4. **Hayashi, K., Nagao, M., and Sugimura, T.,** Interactions of norharman and harman with DNA, *Nucleic Acids Res.,* 4, 3679, 1977.
5. **Imamura, M., Shudo, K., Okamoto, T., and Andoh, T.,** Interaction of mutagens isolated from L-glutamic acid pyrolysate with DNA, *Biochem. Biophys. Res. Commun.,* 97, 968, 1980.
6. **Kato, R. and Yamazoe, Y.,** Metabolic activation and covalent binding to nucleic acids of carcinogenic heterocyclic amines from cooked foods and amino acid pyrolysates, *Jpn. J. Cancer Res. (Gann),* 78, 297, 1987.
7. **Hashimoto, Y., Shudo, K., and Okamoto, T.,** Metabolic activation of a mutagen, 2-amino-6-methyl-dipyrido[1,2-*a*:3′,2′-*d*]imidazole. Identification of 2-hydroxyamino-6-methyldipyrido[1,2-*a*:3′,2′-*d*]imidazole and its reaction with DNA, *Biochem. Biophys. Res. Commun.,* 92, 971, 1980.
8. **Hashimoto, Y., Shudo, K., and Okamoto, T.,** Modification of DNA with potent mutacarcinogenic 2-amino-6-methyldipyrido[1,2-*a*:3′,2′-*d*]imidazole isolated from a glutamic acid pyrolysate: structure of the modified nucleic acid base and initial chemical event caused by the mutagen, *J. Am. Chem. Soc.,* 104, 7636, 1982.
9. **Hashimoto, Y. and Shudo, K.,** Sequence selective modification of DNA with muta-carcinogenic 2-amino-6-methyldipyrido[1,2-*a*:3′,2′-*d*]imidazole, *Biochem. Biophys. Res. Commun.,* 116, 1100, 1983.
10. **Hashimoto, Y., Shudo, K., and Okamoto, T.,** Structural identification of a modified base in DNA covalently bound with mutagenic 3-amino-1-methyl-5*H*-pyrido[4,3-*b*]indole, *Chem. Pharm. Bull.,* 27, 1058, 1979.

11. **Snyderwine, E. G., Roller, P. P., Adamson, R. H., Sato, S., and Thorgeirsson, S. S.,** Reaction of N-hydroxylamine and N-acetoxy derivatives of 2-amino-3-methylimidazolo[4,5-*f*]quinoline with DNA. Synthesis and identification of N-(deoxyguanosin-8-yl)-IQ, *Carcinogenesis*, 9, 1061, 1988.

12. **Loukakou, B., Hébert, E., Saint-Ruf, G., and Leng, M.,** Reaction of DNA with a mutagenic 3-N,N-acetoxyacetylamino-4,6-dimethyldipyrido[1,2-*a*:3′,2′-*d*]imidazole (N-AcO-AGlu-P-3) related to glutamic acid pyrolysates, *Carcinogenesis*, 6, 377, 1985.

13. **Hébert, E., Loukakou, B., Saint-Ruf, G., and Leng, M.,** Conformational changes induced in DNA by the *in vitro* reaction with the mutagenic amine: 3-N,N-acetoxyacetylamino-4,6-dimethyldipyrido(1,2-*a*:3′,2′-*d*)imidazole, *Nucleic Acids Res.*, 12, 8553, 1984.

14. **Yamazoe, Y., Tada, M., Kamataki, T., and Kato, R.,** Enhancement of binding of N-hydroxy-Trp-P-2 to DNA by seryl-tRNA synthetase, *Biochem. Biophys. Res. Commun.*, 102, 432, 1981.

15. **Yamazoe, Y., Shimada, M., Kamataki, T., and Kato, R.,** Covalent binding of N-hydroxy-Trp-P-2 to DNA by cytosolic proline-dependent system, *Biochem. Biophys. Res. Commun.*, 107, 165, 1982.

16. **Mita, S., Ishii, K., Yamazoe, Y., Kamataki, T., Kato, R., and Sugimura, T.,** Evidence for the involvement of N-hydroxylation of 3-amino-1-methyl-5*H*-pyrido[4,3-*b*]indole by cytochrome P-450 in the covalent binding to DNA, *Cancer Res.*, 41, 3610, 1981.

17. **Wakata, A., Oka, N., Hiramoto, K., Yoshioka, A., Negishi, K., Wataya, Y., and Hayatsu, H.,** DNA strand cleavage *in vitro* by 3-hydroxyamino-1-methyl-5*H*-pyrido[4,3-*b*]indole, a direct-acting mutagen formed in the metabolism of carcinogenic 3-amino-1-methyl-5*H*-pyrido[4,3-*b*]indole, *Cancer Res.*, 45, 5867, 1985.

18. **Mita, S., Yamazoe, Y., Kamataki, T., and Kato, R.,** Effects of ascorbic acid on the nonenzymatic binding to DNA and the mutagenicity of N-hydroxylated metabolite of a tryptophan-pyrolysis product, *Biochem. Biophys. Res. Commun.*, 105, 1396, 1982.

19. **Wataya, Y., Yamane, K., Hiramoto, K., Ohtsuka, Y., Okubata, Y., Negishi, K., and Hayatsu, H.,** Generation of intracellular active oxygens in mouse FM3A cells by 3-hydroxyamino-1-methyl-5*H*-pyrido[4,3-*b*]indole, the activated Trp-P-2, *Jpn. J. Cancer Res. (Gann)*, 79, 576, 1988.

20. **Hayashi, S., Møller, M. E., and Thorgeirsson, S. S.,** Genotoxicity of heterocyclic amines in the *Salmonella*/hepatocyte system, *Jpn. J. Cancer Res. (Gann)*, 76, 835, 1985.

21. **Hashimoto, Y., Shudo, K., and Okamoto, T.,** Modification of nucleic acids with muta-carcinogenic heteroaromatic amines *in vivo*. Identification of modified bases in DNA extracted from rats injected with 3-amino-1-methyl-5*H*-pyrido[4,3-*b*]indole and 2-amino-6-methyldipyrido[1,2-*a*:3′,2′-*d*]imidazole, *Mutat. Res.*, 105, 9, 1982.

22. **Alldrick, A. J. and Lutz, W. K.,** Covalent binding of [2-^{14}C]2-amino-3,8-dimethylimidazo[4,5-*f*]-quinoxaline (MeIQx) to mouse DNA *in vivo*, *Carcinogenesis*, 10, 1419, 1989.

23. **Watson, W. P.,** Post-radiolabelling for detecting DNA damage, *Mutagenesis*, 2, 319, 1987.

24. **Yamashita, K., Takayama, S., Nagao, M., Sato, S., and Sugimura, T.,** Amino-methyl-α-carboline-induced DNA modification in rat salivary glands and pancreas detected by ^{32}P-postlabeling method, *Proc. Jpn. Acad.*, 62 (Ser. B), 45, 1986.

25. **Snyderwine, E. G., Yamashita, K., Adamson, R. H., Sato, S., Nagao, M., Sugimura, T., and Thorgeirsson, S. S.,** Use of the ^{32}P-postlabeling method to detect DNA adducts of 2-amino-3-methylimidazolo[4,5-*f*]quinoline (IQ) in monkeys fed IQ: identification of the N-(deoxyguanosin-8-yl)-IQ adduct, *Carcinogenesis*, 9, 1739, 1988.

26. **Schut, H. A. J., Putman, K. L., and Randerath, K.,** DNA adduct formation of the carcinogen 2-amino-3-methylimidazo[4,5-*f*]quinoline in target tissues of the F-344 rat, *Cancer Lett.*, 41, 345, 1988.

27. **Yamashita, K., Umemoto, A., Grivas, S., Kato, S., and Sugimura, T.,** *In vitro* reaction of hydroxyamino derivatives of MeIQx, Glu-P-1 and Trp-P-1 with DNA: ^{32}P-postlabelling analysis of DNA adducts formed *in vivo* by the parent amines and *in vitro* by their hydroxyamino derivatives, *Mutagenesis*, 3, 515, 1988.

28. **Yamashita, K., Ohgaki, H., Wakabayashi, K., Nagao, M., and Sugimura, T.,** DNA adducts formed by the comutagens harman and norharman in various tissues of mice. *Cancer Lett.*, 42, 179, 1988.

29. **Nakayasu, M., Nakasato, F., Sakamoto, H., Terada, M., and Sugimura, T.,** Mutagenic activity of norharman and harman in Chinese hamster lung cells in assay with diphtheria toxin resistance as a marker, *Cancer Lett.*, 17, 249, 1983.

30. **Takayama, K., Yamashita, K., Wakabayashi, K., Sugimura, T., and Nagao, M.,** DNA modification by 2-amino-1-methyl-6-phenylimidazo[4,5-*b*]pyridine in rats, *Jpn. J. Cancer Res.*, 80, 1145, 1989.

31. **Yamashita, K., Umemoto, A., Grivas, S., Kato, S., Sato, S., and Sugimura, T.,** Heterocyclic amine-DNA adducts analyzed by ^{32}P-postlabeling method, *Nucleic Acids Symp. Ser.*, 19, 111, 1988.

32. **Snyderwine, E. G., Wirth, P. J., Roller, P. P., Adamson, R. H., Sato, S., and Thorgeirsson, S. S.,** Mutagenicity and *in vitro* covalent DNA binding of 2-hydroxyamino-3-methylimidazolo[4,5-*f*]quinoline, *Carcinogenesis*, 9, 411, 1988.

33. **Turteltaub, K. W., Watkins, B. E., Vanderlaan, M., and Felton, J. S.,** Role of metabolism on the DNA binding of MeIQx in mice and bacteria, *Carcinogenesis*, 11, 43, 1990.

34. **Wild, D., Dirr, A., Fasshauer, I., and Henschler, D.,** Photolysis of arylazides and generation of highly electrophilic DNA-binding and mutagenic intermediates, *Carcinogenesis*, 10, 335, 1989.

35. **Akagi, T., Morota, K., Iyehara-Ogawa, H., Kimura, H., and Kato, T.,** Mutational specificity of the carcinogen 3-amino-1-methyl-5*H*-pyrido[4,3-*b*]indole in mammalian cells, *Carcinogenesis,* 11, 841, 1990.
36. **Hashimoto, Y., Kawachi, E., Shudo, K., Sekiya, T., and Sugimura, T.,** Transforming activity of human c-Ha-*ras*-1 proto-oncogene generated by the binding of 2-amino-6-methyldipyrido[1,2-*a*:3',2'-*d*]imidazole and 4-nitroquinoline N-oxide: direct evidence of cellular transformation by chemically modified DNA, *Jpn. J. Cancer Res.(Gann),* 78, 211, 1987.
37. **Ishikawa, F., Takaku, F., Nagao, M., Ochiai, M., Hayashi, K., Takayama, S., and Sugimura, T.,** Activated oncogenes in a rat hepatocellular carcinoma induced by 2-amino-3-methylimidazo[4,5-*f*]quinoline, *Jpn. J. Cancer Res. (Gann),* 76, 425, 1985.
38. **Ishikawa, F., Takaku, F., Hayashi, K., Nagao, M., and Sugimura, T.,** Activation of rat c-*raf* during transfection of hepatocellular carcinoma DNA, *Proc. Natl. Acad. Sci. U.S.A.,* 83, 3209, 1986.
39. **Ishizaka, Y., Ochiai, M., Ishikawa, F., Sato, S., Miura, Y., Nagao, M., and Sugimura, T.,** Activated N-*ras* oncogene in a transformant derived from a rat small intestinal adenocarcinoma induced by 2-aminodipyrido[1,2-*a*:3',2'-*d*]imidazole, *Carcinogenesis,* 8, 1575, 1987.

Chapter 3

DIETARY CARCINOGENS AND MUTAGENS FROM PLANTS

Bruce N. Ames, Margie Profet, and Lois Swirsky Gold

TABLE OF CONTENTS

I. INTRODUCTION

The pesticides in our diet are 99.99% natural. Plants produce an enormous variety of toxins to protect themselves against fungi, insects, and animal predators. These natural pesticides in plants are present in very much greater variety and at levels thousands of times higher than synthetic pesticides. To protect crops from pests, there is also a tradeoff between nature's pesticides and synthetic pesticides.

Although only 47 natural pesticides have been tested in animal cancer tests, about half are carcinogens. Similarly, about half of all chemicals tested in animal cancer tests at the maximum tolerated dose (MTD) are positive. The proportion of natural pesticides that are clastogenic (break chromosomes in tissue culture) is the same as for synthetic chemicals.

Despite the wide array of natural carcinogens in vegetables, epidemiological evidence suggests that vegetable intake in humans reduces cancer. Recent advances in understanding of the role of cell proliferation in the mechanisms of carcinogenesis indicate that the risk to humans from low doses of rodent carcinogens has been markedly overestimated.

We argue that a high percentage of all chemicals, natural or synthetic, will be rodent carcinogens because testing chemicals—whether mutagens or nonmutagens—at the MTD induces chronic cell proliferation. Cell proliferation is itself mutagenic in several ways. A dividing cell is much more susceptible to mutation than a quiescent cell. Cell proliferation is by far the most effective way to convert a heterozygous mutation (e.g., in a tumor suppressor gene) to homozygosity or hemizygosity through nondisjunction and induction of mitotic recombination. Therefore, in the presence of normal spontaneous mutation rates, chronic induction of cell proliferation, such as occurs in testing chemicals at the MTD, is carcinogenic. Cell proliferation due to toxicity is not observed at low doses (i.e., it shows a threshold). Therefore, the cancer risk of nongenotoxic carcinogens at low doses is likely to be negligible; the risk from genotoxic carcinogens is likely to be lower than is commonly assumed.

II. CARCINOGENS AND TERATOGENS ARE COMMON IN RODENT TESTS

More than half of the chemicals tested to date in both rats and mice have been found to be carcinogens at the high doses administered,[1,2] the MTD. Synthetic industrial chemicals account for almost all (82%) of the 427 chemicals tested in both species. However, although more than 99.9% of the chemicals humans eat are natural, only 75 *natural* chemicals have been tested in both rats and mice; about *half* (47%) are carcinogens.[2-6]

The high proportion of positives is not simply due to selection of suspicious chemical structures. While some synthetic or natural chemicals were selected for testing precisely because of structure, many were selected simply because they were widely used, e.g., they were high-volume industrial compounds, pesticides, natural or synthetic food additives, dyes or food colors, or drugs. The natural world of chemicals has never been looked at systematically.[2] We explain in Section VI why current understanding of the mechanisms of carcinogenesis justifies the prediction that a high proportion of all chemicals—natural and synthetic—will prove to be carcinogenic at the MTD.[1]

A chemical is classified as a carcinogen in our analysis if the authors of the study give it a positive evaluation in at least one experiment. Clearly carcinogens are not all the same: some have been tested many times in several species while others have been examined at only one site in one species; some, such as safrole, are positive in two species and form DNA adducts in animals; some, like 5- and 8-methoxypsoralen, are clearly genotoxic, whereas others, like D-limonene, are not.

A high proportion of positives is also reported for teratogenicity tests (tests to determine the potential to cause reproductive damage). One-third of the 2800 chemicals tested in laboratory animals have been shown to cause reproductive damage at MTDs.[7] Thus, it seems likely that a sizable percentage of both natural and man-made chemicals will be reproductive toxins—when tested at the MTD.

Since such a high proportion of test agents is positive in animal studies, it is important to try to rank possible carcinogenic hazards to humans from exposures to various chemicals. Natural chemicals can be used as a reference for evaluating carcinogenic hazards from synthetic chemicals. A chemical pollutant should not be treated as a significant carcinogenic hazard if its possible hazard seems far below that of many common food items. In recent years, we have attempted to use the animal data to find out how human exposures to rodent carcinogens compare to one another in terms of possible hazard. To evaluate possible hazard we determine how close the human exposure level is to the dose that induces tumors in rodents.[1,8] The emphasis in this work is upon comparing and ranking human exposures from a variety of sources rather than upon risk assessment.

III. NATURE'S PESTICIDES: MUTAGENICITY AND CARCINOGENICITY

"Plants are not just food …The world is not green. It is colored lectin, tannin, cyanide, caffeine, aflatoxin, and canavanine".[9]

A. DIETARY PESTICIDES ARE 99.99% ALL NATURAL

Nature's pesticides are one important group of natural chemicals that we have investigated. All plants produce toxins to protect themselves against fungi, insects, and animal predators such as man.[10-12] Tens of thousands of these natural pesticides have been discovered, and every species of plant contains its own set of different toxins, usually a few dozen. In addition, when plants are stressed or damaged, such as during a pest attack, they increase their natural pesticide levels manyfold, occasionally to levels that are acutely toxic to humans. We estimate that Americans eat about 1,500 mg/day of natural pesticides, 10,000 times more than man-made pesticide residues. Concentrations of natural pesticides are usually measured in parts per thousand or million[11] rather than parts per billion (ppb), the usual concentration of synthetic pesticide residues or of water pollutants.[1,13] We estimate that a human ingests roughly 5,000 to 10,000 different natural pesticides and their breakdown products. Table 1 shows 49 natural pesticides (and breakdown products) that are ingested when eating cabbage and indicates how few have been tested for carcinogenicity or clastogenicity. Lima beans contain a different array of 33 natural toxins that, in stressed plants, range in concentration from 0.2 to 33 parts per thousand fresh weight; none appears to have been tested yet for carcinogenicity or teratogenicity.[32] A large literature has examined the toxicity of many of these compounds to plant predators.[12,32]

Alfalfa sprouts, for example, contain canavanine, a highly toxic arginine analog that is incorporated into protein in place of arginine when eaten by animals. Canavanine, which occurs in alfalfa sprouts at about 1.5% of their dry weight,[33] is the active agent in causing a lupus erythematosus-like syndrome in monkeys that are fed alfalfa sprouts or canavanine.[33] Lupus in man is characterized by a defect in the immune system that is associated with autoimmunity, antinuclear antibodies, chromosome breaks and various types of pathology.[33] Chromosome breaks in lupus appear to be caused by oxygen radicals since they are prevented by superoxide dismutase.[34] The effects of canavanine/alfalfa sprouts could be due in part to the production of oxygen radicals during phagocytization of antibody complexes with canavanine-containing protein.

Surprisingly few plant toxins have been tested in animal cancer bioassays,[35] but among those tested, again about half (25/47)[180] are carcinogenic. Even though only a tiny proportion of plant toxins in our diet has been tested, the 25 natural pesticide carcinogens identified so far are present in the following foods: anise, apples, apricots, bananas, basil, broccoli,

TABLE 1
49 Natural Pesticides (and Metabolites) in Cabbage

Glucosinolates
2-Propenyl glucosinolate (sinigrin)[a]
3-Methyl-thio-propyl glucosinolate
3-Methyl-sulfinyl-propyl glucosinolate
3-Butenyl glucosinolate
2-Hydroxy-3-butenyl glucosinolate
4-Methyl-thio-butyl glucosinolate
4-Methyl-sulfinyl-butyl glucosinolate
4-Methylsulfonyl-butyl glucosinolate
Benzyl glucosinolate
2-Phenyl-ethyl glucosinolate
Propyl glucosinolate
Butyl glucosinolate

Indole glucosinolates and related indoles
3-Indolyl-methyl glucosinolate (glucobrassicin)
1-Methoxy-3-indolylmethyl (neoglucobrassicin)
Indole-3-carbinol (I3C)[a]
Indole-3-acetonitrile (IAN)[a]
3,3′-Diindolylmethane (I33′)[a]

Isothiocyanates and goitrin
Allyl isothiocyanate[a]
3-Methyl-thio-propyl isothiocyanate
3-Methyl-sulfinyl-propyl isothiocyanate
3-Butenyl isothiocyanate
5-Vinyloxazolidine-2-thione (goitrin)
4-Methylthiobutyl isothiocyanate
4-Methylsulfinylbutyl isothiocyanate
4-Methylsulfonylbutyl isothiocyanate
4-Pentenyl isothiocyanate
Benzyl isothiocyanate
Phenylethyl isothiocyanate

Nitriles
1-Cyano-2,3-epthiopropane
1-Cyano-3,4-epithiobutane
1-Cyano-3,4-epithiopentane
threo-1-Cyano-2-hydroxy-3,4-epithiobutane
erthro-1-Cyano-2-hydroxy-3,4-epithiobutane
2-Phenylpropionitrile
Allyl cyanide[a]
1-Cyano-2-hydroxy-3-butene
1-Cyano-3-methylsulfinylpropane
1-Cyano-4-methylsulfinylbutane

Alcohols
Menthol
Neomenthol
Isomenthol

Ketones
Carvone[a]

Phenols and tannins
2-Methoxyphenol
3-Caffoylquinic acid (chlorogenic acid)[a]
4-Caffoylquinic acid[a]
5-Caffoylquinic acid (neochlorogenic acid)[a]
4-*p*-Coumaroylquinic acid
5-*p*-Coumaroylquinic acid
5-Feruloylquinic acid

TABLE 1 (continued)
49 Natural Pesticides (and Metabolites) in Cabbage

a   Discussed below; all others untested. *Clastogenicity:* Chlorogenic acid[14] and allyl isothiocyanate are positive.[15] Chlorogenic acid and its metabolite caffeic acid are also mutagens,[16-18] as is allyl isothiocyanate.[19] *Carcinogenicity:* Allyl isothiocyanate induced papillomas of the bladder in male rates (a neoplasm that is unusually rare in control rats) and was classified by NTP as carcinogenic. There was no evidence of carcinogenicity in mice; however, NTP indicated "the mice probably did not receive the MTD."[20,21] Sinigrin (the glucosinolate, i.e., thioglycoside of allyl isothiocyanate) is cocarcinogenic for the rat pancreas.[22] Carvone is negative in mice.[23] Indole acetonitrile has been shown to form a carcinogen, nitroso indole acetonitrile, in the presence of nitrite.[24] Caffeic acid is a carcinogen [25,26] and clastogen[14] and is a metabolite of its esters 3-, 4-, and 5-caffoylquinic acid (chlorogenic and neochlorogenic acid). *Metabolities:* Sinigrin gives rise to allyl isothiocyanate on eating raw cabbage (e.g., coleslaw); in cooked cabbage it also is metabolized to allyl cyanide, which is untested. Indole carbinol forms dimers and trimrs on ingestion, which mimic dioxin (TCDD) (see text). *Occurrence.*[27,28] *Toxicology:* The mitogenic effects of goitrin (which is goitrogenic) and various organic cyanides from cabbage suggest that they may be potential carcinogens.[29,30] Aromatic cyanides related to those from cabbage have been shown to be mutagens and are metabolized to hydrogen cyanide and potentially mutagenic aldehydes.[31]

Brussels sprouts, cabbage, cantaloupe, caraway, carrots, cauliflower, celery, cherries, cinnamon, cloves, cocoa, coffee, collard greens, comfrey tea, dill, eggplant, endive, fennel, grapes, grapefruit juice, honey, honeydew melon, horseradish, kale, lettuce, mangoes, mushrooms, mustard, nutmeg, orange juice, parsley, parsnips, peaches, pears, black pepper, pineapples, plums, potatoes, radishes, raspberries, rosemary, tarragon, and turnips. Thus it is probable that almost every plant product in the supermarket contains natural carcinogens. The levels of the known natural carcinogens in the above plants are commonly thousands of times higher than the levels of man-made pesticides. Table 2 shows a variety of carcinogens in the parts per million (ppm) range in plant foods. The catechol-type phenolics such as tannins, and caffeic acid and its esters, are more widespread in plant species than other natural pesticides (e.g., Tables 1 and 2). It may be that these phenolics have an antimicrobial role analogous to the respiratory burst of oxygen radicals from mammalian phagocytic cells. The phenolics autooxidize when a plant is wounded, e.g., the browning when an apple is cut, yielding a burst of mutagenic oxygen radicals.

1. Residues of Man-Made Pesticides

The U.S. Food and Drug Administration (FDA) has assayed food for man-made pesticide residues of the 200 compounds thought to be of greatest importance,[13] including the residues of a few industrial chemicals such as polychlorinated biphenyls (PCBs). The FDA found residues for 105% of these chemicals and estimates that U.S. daily per capita intake of the sum of these 105 chemicals averages about 0.09 mg/day, which we calculate to be an average concentration of about 0.10 ppm in foods derived from plants.

About half (0.04 mg) of this intake of synthetic pesticides is composed of four chemicals (ethylhexyl diphenyl phosphate, malathion, dicloran, and chlorpropham), which were not carcinogenic in rodent tests.[1] Thus, the intake of carcinogens from residues (0.05 mg a day, if one assumes that all the other residues are carcinogenic, which is unlikely) is extremely tiny (averaging about 0.05 ppm in plant foods) relative to the background of natural substances, such as shown in Table 2.[1,10]

The latest figures from the FDA about actual exposures do not include every known man-made pesticide, but they constitute a reasonable attempt at doing so. A 1987 National Academy of Sciences report, *Regulating Pesticides in Food*,[70] suggested that some of the pesticides not sampled by the FDA, particularly those used on tomatoes, should have lower allowable limits and should be added to the FDA sampling program. Nevertheless, the estimate of 0.05 mg of possibly carcinogenic pesticide residues consumed in a day seems to be a reasonable rough estimate.

In comparison, the consumption of naturally occurring carcinogens is enormous.[1] There

TABLE 2
Concentrations of Natural Pesticide Carcinogens

Plant	Carcinogen	Concentration (ppm)
Parsley	5- and 8-methoxypsoralen	14
Parsnip, cooked	5- and 8-methoxypsoralen	32
Celery	5- and 8-methoxypsoralen	0.8
New cultivar	5- and 8-methoxypsoralen	6.2
Stressed	5- and 8-methoxypsoralen	25
Mushroom, commercial	p-Hydrazinobenzoate	11
	Glutamyl-p-hydrazinobenzoate	42
Cabbage	Sinigrin[a] (allyl isothiocyanate)	35—590
Collard greens	Sinigrin[a] (allyl isothiocyanate)	250—788
Cauliflower	Sinigrin[a] (allyl isothiocyanate)	12—66
Brussels sprouts	Sinigrin[a] (allyl isothiocyanate)	110—1,560
Mustard (black)	Sinigrin[a] (allyl isothiocyanate)	16,000—72,000
Horseradish	Sinigrin[a] (allyl isothiocyanate)	4,500
Orange juice	Limonene	31
Mango	Limonene	40
Pepper, black	Limonene	8,000
Basil	Estragole	3,800
Fennel	Estragole	3,000
Nutmeg	Safrole	3,000
Mace	Safrole	10,000
Pepper, black	Safrole	100
Pineapple	Ethyl Acrylate	0.07
Basil	Benzyl acetate	82
Jasmine tea	Benzyl acetate	230
Honey	Benzyl acetate	15
Apple, pear, plum, cherry, carrot, celery, lettuce, potato, endive, coffee (brewed), grapes, eggplant	Caffeic acid	50—200
Thyme, basil, anise, caraway, rosemary, tarragon, marjoram, savory, sage, dill, absinthe	Caffeic acid	>1,000
Apricot, cherry, plum, peach	Chlorogenic acid[b] (caffeic acid)	50—500
Coffee (brewed)	Chlorogenic acid[b] (caffeic acid)	>1,000
Apple, pear, peach, apricot, plum, cherry, Brussels sprouts, kale, cabbage, broccoli, coffee (brewed)	Neochlorogenic acid[b] (caffeic acid)	50—500

Note: Carcinogen references: References 3 to 6 and 5-methoxypsoralen (light-activated),[36] 8-methoxypsoralen,[37] p-hydrazinobenzoate and glutamyl-p-hydrazinobenzoate,[38,39] allyl isothiocyanate,[20,21] D-limonene,[40] estragole and safrole,[35,41] ethyl acrylate,[39] benzyl acetate,[42] caffeic acid (N. Ito, personal communication).[25,26,43] *Concentration references:* 5- and 8-methoxypsoralen,[11,44-48] p-hydrazinobenzoates,[38,39] sinigrin,[27,28,49] D-limonene,[50-52] estragole and safrole,[53-56] ethyl acrylate,[57] benzyl acetate,[58-60] caffeic acid, chlorogenic acid, neochlorogenic acid.[61-69] *Mutagenicity and clastogenicity references:* see text.

a   Sinigrin is a cocarcinogen[22] and is metabolized to the carcinogen allyl isothiocyanate, although no adequate test has been done on sinigrin itself. The proportion converted to allyl isothiocyanate or to allyl cyanide depends on food preparation.[27,28]

b   Chlorogenic and neochlorogenic acid are metabolized to the carcinogen caffeic acid but have not been tested for carcinogenicity themselves. The clastogenicity and mutagenicity of the above compounds are referenced in Table 1.

are at least 10 mg of rodent carcinogens in a cup of coffee (caffeic acid, catechol, hydrogen peroxide, furfural, and methylglyoxal) and 0.8 mg of carcinogenic estragole in a basil leaf. In addition we eat about 1500 mg of mostly untested natural pesticides. Cooking our daily food produces about 2000 mg of mostly untested burnt material which contains many known carcinogens such as the nitrosamines formed by the use of gas ovens.[1,10] Thus, the possible carcinogenic hazards from man-made pesticide residues appear to be minimal compared to the background of naturally occuring chemicals.

2. TCDD (Dioxin) Compared with Alcohol and Broccoli

TCDD is a substance of great public concern, because it is a carcinogen and teratogen in rodents at extremely low doses. However, the doses humans ingest are very low relative to the doses that cause cancer and reproductive damage in rodents under experimental conditions.

TCDD can be compared with alcohol. Although the dose of alcohol known to cause cancer or birth defects in rodents is extremely high, the doses that humans are exposed to are also very high. Alcoholic beverages are the most important human teratogen. In contrast, there is no persuasive evidence that TCDD is carcinogenic or teratogenic in man, although it is at high doses in rodents.

The Environmental Protection Agency's (EPA) human "reference dose" (formerly "acceptable dose limit") of TCDD is 6 femtograms (fg) per kilogram per day. If one compares the teratogenic potential of TCDD to that of alcohol for causing birth defects (after adjusting for their respective potencies as determined in rodent tests), then a daily consumption of the reference dose of TCDD is equivalent in teratogenic potential to a daily consumption of alcohol from 1/3,000,000 of a beer. That is equivalent to drinking a single beer (15 g ethyl alcohol) over a period of 8000 years. A daily slice of bread, or a daily glass of orange juice, contains much more natural alcohol than that.

Alcoholic beverages in man are clearly carcinogenic as well as teratogenic.[71] A comparison of the carcinogenic potential of TCDD with that of alcohol, ajdusting for the potency in rodents, shows that ingesting the TCDD reference dose of 6 fg/kg/day is equivalent to ingesting one beer every 345 years. Since the average consumption of alcohol in the U.S. is equivalent to more than one beer per day per capita, and since five drinks a day are a carcinogenic risk in man, the great concern over TCDD at levels in the range of the reference dose seems unreasonable.

TCDD binds to the Ah receptor in mammalian cells; evidence strongly suggests that all of the harmfull effects of TCDD are through this binding.[72] A wide variety of natural substances also bind the Ah receptor, and insofar as they have been examined, they have similar properties to TCDD. A cooked steak, for instance, contains polycyclic hydrocarbons, which bind to the Ah receptor and mimic TCDD. In addition, our diet contains a variety of flavones and other plant substances that bind to the Ah receptor. For example, indole carbinol (IC), a breakdown product of glucobrassiein, a glucosinolate that is present in large amounts in broccoli (500 mg/kg), cabbage, cauliflower, and other members of the *Brassica* family,[3] makes dimers and trimers at the pH of the stomach that induce the same set of enzymes as does TCDD.[74] When given before aflatoxin or other carcinogens, IC protects against carcinogenesis, as does TCDD.[75] However, when given *after* aflatoxin or other carcinogens, IC is a strong promoter of carcinogenesis, as is TCDD.[76] This stimulation of carcinogenesis has also been shown for cabbage itself.[77]

The EPA reference dose of 6 fg/kg of TCDD per day should be compared with 5 mg of IC per 100 g of broccoli (one portion) (see also cabbage in Table I). Although the affinity of the indole derivatives in binding to Ah receptors is less than TCDD by a factor of about 8000, the effective dose to the Ah receptor from a helping of broccoli appears to be roughly 1500 times higher than the TCDD reference dose, taking into account the very long lifetime of TCDD in the body (several years). Although these IC derivatives appear to be much more of a potential hazard than TCDD, it is not clear whether at the low doses of human exposure *either* is a hazard. Another study[78] shows that when sunlight oxidizes tryptophan, a normal amino acid, it converts it to a variety of indoles that bind to the Ah receptor and mimic the action of TCDD. It seems likely that many more of these "natural dioxins" will be discovered in the future.

B. CLASTOGENICITY/MUTAGENICITY STUDIES

In order to identify chemicals that may be important risk factors for human cancer, it may be useful to look at other types of studies as well. For example, Ishidate et al.[15] reviewed experiments on the clastogenicity (chromosome breakage) of 951 chemicals in mammalian cell cultures. Of these 951 chemicals, we identified 72 as natural plant pesticides: 35 (48%) were positive for clastogenicity in some or all tests. This is similar to the results of the remaining chemicals: 467/879 (53%) were positive in some or all tests.

Of particular interest are the levels at which some of these plant toxins were clastogenic:

1.   Allyl isothiocyanate was clastogenic at a concentration of 0.0005 ppm, which is approximately 200,000 times less than the concentration of its thioglycoside in cabbage. It was among the most potent chemicals in the compendium, and it is also positive at unusually low levels in transforming[79] and mutating animal cells.[19] (See discussion of allyl isothiocyanate cancer tests in Table 1).
2.   Safrole was clastogenic at a concentration of about 100 ppm, which is 30 times less than the concentration in nutmeg, 100 times less than the concentration in mace, and roughly equal to the concentration in black pepper. The carcinogens safrole and estragole, and a number of related dietary natural pesticides that have not been tested in animal cancer tests, have been shown to give DNA adducts in mice.[80]
3.   Caffeic acid was clastogenic at a concentration of 260 to 500 ppm, which is 5 to 10 times less than the concentration in roasted coffee beans and many spices, and is roughly equal to the concentration in apples, lettuce, endive, and potato skin. The genotoxic activity of coffee to mammalian cells has been demonstrated.[81] Chlorogenic acid, a precursor of caffeic acid, was clastogenic at a concentration of 150 ppm, which is 100 times less than its concentration in roasted coffee beans and is roughly within the range of its concentration in apples, pears, plums, peaches, cherries, and apricots. Chlorogenic acid and its metabolite caffeic acid are mutagens (Table 1).

Benzyl acetate and ethyl acrylate mutate mouse lymphoma cells.[19] Plant phenolics such as caffeic acid, chlorogenic acid, and tannins (esters of gallic acid) have been reviewed for their mutagenicity, clastogenicity, and carcinogenicity.[82] The carcinogenicity and genetic toxicology of plant carcinogens has been recently reviewed.[35]

IV. THE TOXICOLOGY OF MAN-MADE TOXINS IS NO DIFFERENT FROM THAT OF NATURAL TOXINS

It is often assumed that, because plants are part of human evolutionary history, natural selection has equipped humans with mechanisms to cope with natural toxic chemicals, but that these mechanisms are insufficient to cope with synthetic chemicals.[83] We find this assumption flawed[84] for several reasons:

(1) Humans, like other mammals, have developed many types of *general* defenses against the large amounts and enormous variety of nature's pesticides in plants. These defenses are effective against both natural and synthetic toxins and include the following:

(A) The continuous shedding of cells exposed to toxins: the surface layers of the mouth, esophagus, stomach, intestine, colon, skin, and lungs are shed every few days.
(B) The induction of a wide variety of general detoxifiying enzymes, such as antioxidant defenses or the glutathione transferases for detoxifiying alkylating agents.[85] Human cells that are exposed to small doses of an oxidant, such as radiation or hydrogen peroxide, induce antioxidant defenses and become much more resistant to higher doses.[86-92] The defenses against oxidant carcinogens are induced by both synthetic (e.g., the herbicide paraquat) and natural oxidants and are effective against both.
(C) The active excretion of planar hydrophobic molecules (natural or synthetic) out of liver and intestinal cells.[93]

(D) DNA repair, which is effective against DNA adducts formed from both synthetic and natural chemicals, and is inducible in response to DNA damage.

(E) Animals' olfactory and gustatory perception of bitter, acrid, astringent, and pungent chemicals, which are commonly associated with toxicity, are presumably most effective against some natural toxins. Whether these stimuli are monitoring toxicity itself or particular chemicals is unclear; however, many synthetic toxic compounds are pungent, acrid, or astringent. Even though mustard (allyl isothiocyanate), pepper, garlic, onions, etc. have some of these attributes, humans are ignoring the warnings.

(F) Humans can circumvent some natural pesticides (and also some synthetic residues) by peeling or cooking food.

Most of these defense systems are inducible in response to toxic stress, to make us well buffered against toxins. Experimental evidence indicates that these general defenses are effective against both natural and synthetic compounds,[94] since the basic mechanisms of carcinogenesis are not unique to either.

The fact that defenses are usually general, rather than specific for each chemical, makes good evolutionary sense. The reason that predators of plants evolved general defenses against toxins is presumably to be prepared to counter a diverse and ever-changing array of plant toxins in a co-evolutionary world; an herbivore that had defenses only against a set of specific toxins would be at a great disadvantage in obtaining new foods when favored foods became scarce.

(2) Various toxins that have been present throughout vertebrate evolutionary history nevertheless cause cancer in vertebrates. For example, mold aflatoxins have been shown to cause cancer in trout, rats, mice, monkeys, and probably humans;[95,96] many of the common elements (e.g., salts of lead, cadmium, beryllium, nickel, chromium, selenium, and arsenic) are carcinogenic, despite their presence throughout all of evolution. Furthermore, epidemiological studies from various parts of the world show that certain natural chemicals appear to be of considerable carcinogenic risk to humans: in Polynesia, the ingestion of flour made from the cycad plant (which contains the carcinogen cycasin) has been correlated with high incidences of liver cancer;[35] the chewing of betel nuts around the world has been correlated with high incidences of esophageal cancer.[35] The phorbol esters present in the Euphorbiacea, some of which are used as folk remedies or herb teas, are potent promoters of carcinogenesis and may have been a cause of nasopharyngeal cancer in China and esophageal cancer in Curacao.[97,98]

Furthermore, there is no reason to think that natural selection in human beings would have eliminated the hazard of carcinogenicity from plant toxins past the reproductive age, which is when most cancers occur.

Plants have been evolving and refining their chemical weapons for at least 500 million years and incur large fitness costs in producing these chemicals. If these chemicals were not effective in deterring predation, i.e., in harming predators like humans, plants would not be naturally selected to produce them.

(3) Many natural toxins have the same mechanisms of toxicity that certain synthetic toxins do. As discussed above, cabbage and broccoli contain the chemical indole-3-carbinol, whose breakdown products bind to the body's Ah receptor, induce enzymes, and possibly cause cell proliferation—just as does dioxin (TCDD), one of the most potent and feared industrial chemicals (see Section III). Thus, natural and synthetic chemicals can be toxic in similar ways.

(4) The human diet has changed drastically in the last few thousand years, and most of us are eating recently introduced plants (such as coffee, cocoa, potatoes, tomatoes, corn, avocados, mangoes, olives, and kiwi fruit) that our ancestors did not. Natural selection works far too slowly for humans to have evolved specific resistance to the food toxins that are comparatively new to their diets. Cruciferous vegetables such as cabbage, broccoli, kale,

cauliflower, and mustard were used in ancient times primarily for medicinal purposes and spread as foods across Europe in the Middle Ages.[28]

(5) Humans in nonindustrial, agricultural societies ingest cow's or goat's milk and other dairy products contaminated by the natural toxins from plants that were eaten by foraging animals, because toxins that are absorbed through the animal's gut are often secreted in the milk. Since the plants foraged by cows vary from place to place and are usually inedible for human consumption, the plant toxins that are secreted in the milk are, in general, not toxins to which humans could have easily adapted. Poisoning from the milk of foraging animals was quite common in previous centuries. Abraham Lincoln's mother, for example, died from drinking cow's milk that had been contaminated with toxins from the snakeroot plant.[99] When cows and goats forage on lupine, their offspring may have severe teratogenic abnormalitites, such as "crooked calf" syndrome caused by the anagyrine in lupine.[100-102] Such significant amounts of these teratogens can be transferred to the animals' milk, so that drinking the milk during pregnancy is a teratogenic hazard for humans.[100-102] In one rural California family, a baby boy, a litter of puppies, and goat kids all had "crooked" bone birth-defect abnormalities. The pregnant woman and the pregnant dog had both been drinking milk obtained from the family goats, which had been foraging on lupine (the main forage in winter).[100-102]

(6) It has been shown that plants contain anticarcinogenic chemicals that protect us against carcinogens.[103] However, these anticarcinogens, e.g., plant antioxidants, *do not distinguish* whether carcinogens are synthetic or natural in origin. Thus, they help to protect us against both.

(7) It has been argued that synthetic carcinogens can be synergistic with each other. However, this is also true of natural chemicals, which are by far the major source of chemicals in our diet.

(8) DDT, the first major synthetic insecticide, is often viewed as the typically dangerous synthetic pesticide. It bioconcentrates in the food chain due to its unusual lipophilicity and persists for years. Many natural pesticides, however, also bioconcentrate: the teratogens solanine and chaconine, for example, are found in tissues of all potato eaters.[104,105] Although DDT was unusual with respect to bioconcentration, it was remarkably nontoxic to mammals, saved millions of lives, and was never shown to cause harm to humans.[106] To a large extent DDT replaced lead arsenate, a major pesticide used before the modern era. When the undesirable bioconcentration and persistence of DDT were realized, less persistent chemicals were developed to replace it. Examples are the synthetic pyrethroids, which, although like DDT they inhibit the same sodium channel of insects,[107] are degraded rapidly in the environment and can often be used at a concentration as low as a few grams per acre.

V. TRADEOFFS BETWEEN NATURAL AND SYNTHETIC PESTICIDES

"It has been suggested that one consequence of crop plant domestication is the deliberate or inadvertent selection for reduced levels of secondary compounds that are distasteful or toxic. Insofar as many of these chemicals are involved in the defense of plants against their enemies, the reduction due to artificial selection in these defenses may account at least in part for the increased susceptibility of crop plants to herbivores and pathogens..."[108]

Since no plot of land is immune to attack by insects, plants need chemical defenses—either natural or synthetic—in order to survive pest attack. Therefore, there is a fundamental tradeoff between nature's pesticides and man-made pesticides. For example, during ripening, fruits drastically lower their production of natural toxins in order to entice birds, deer, and certain other animals to eat them and disperse their mature seeds; ripe fruit, as opposed to unripe fruit, is therefore especially vulnerable to insect and disease organisms.[109]

Through selective breeding and the use of synthetic pesticides, farmers have been able to drastically reduce the levels of natural toxins in crop plants. Cultivated plant foods commonly contain on average fewer natural toxins than do their wild counterparts. For example, the wild potato *Solanum acaule*, the progenitor of the cultivated strains of potato, has a glycoalkaloid content about 3 times that of the cultivated strains *Solanum* × *Curtilobum* and *Solanum* × *Juzepczuki*.[110] The leaves of the wild cabbage *Brassica oleracea* (the progenitor of cabbage, broccoli, and cauliflower) contain about twice as many glucosinolates as cultivated cabbage.[111] (One of these glucosinolates, sinigrin, breaks down by enzymatic action to the carcinogen allyl isothiocyanate.) The wild bean *Phaseolus lunatus* contains about 3 times as many cyanogenic glucosides as does the cultivate bean.[112] Similar reductions in toxicity through agriculture have been reported as well in lettuce, lima bean, mango, and cassava.[113]

This trend may be reversing, however. One consequence of disproportionate concern about tiny traces of synthetic pesticide residues is that plant breeders are developing plants for the organic food market that are highly insect-resistant — high in natural toxins. Two recent cases illustrate the potential hazards of this approach to pest control. (1) When a major grower introduced a new variety of highly insect-resistant celery into commerce, a flurry of complaints to the Centers for Disease Control from all over the country followed because people who handled the celery developed severe rashes when they were subsequently exposed to sunlight. Some detective work found that the pest-resistant celery contained 6200 ppb of carcinogenic (and mutagenic) psoralens instead of the 800 ppb present in normal celery (Table 2).[11,47,48] It is not known whether other natural pesticides in the celery were increased as well. The celery is still on the market. (2) An insect-resistant potato, at a cost of millions of dollars, had to be withdrawn from the market because of its acute toxicity to humans—a consequence of higher levels of the alkaloids solanine and chaconine. Solanine and chaconine block nerve transmission by inhibiting cholinesterase and are known teratogens. They were widely introduced into the human diet about 400 years ago with the dissemination of the potato from the Andes. Total alkaloids are present in normal potatoes at a level of 15 mg per 200-g potato (75 ppm), which is less than a tenfold safety margin from the toxic level for humans.[114] Neither solanine nor chaconine has been tested for carcinogenicity. In contrast, the cholinesterase inhibitor malathion, the main synthetic organophosphate pesticide present in our diet (0.02 mg/day), has been thoroughly tested and is not a carcinogen in rodents.

There is a tendency for nonscientists to think of *chemicals* as being only synthetic and to characterize synthetic chemicals as toxic, as if every natural chemical were not also toxic at some dose. Event the recent National Academy of Sciences report cited above[70] states: ''Advances in classical plant breeding... offer some promise for nonchemical pest control in the future. Nonchemical approaches will be encouraged by tolerance revocations...'' The report was particularly concerned with some pesticides used on tomatoes. Of course, tomatine, one of the natural toxins in tomatoes, is a chemical, too, and was introduced from Peru 400 years ago. Neither tomatine nor its aglycone tomatidine, an antifungal steroid-like molecule, has been tested in rodent cancer bioassays. Tomatine is present at 36 mg per 100-g tomato (360 ppm), a concentration that is much closer to the acutely toxic level than are man-made pesticide residues.

Certain cultivated crops have become popular in developing countries because they thrive without requiring synthetic pesticides. However, the tradeoffs of cultivating these naturally pest-resistant crops are that they are highly toxic and require extensive processing to detoxify them. For example, cassava root, which is a major food crop in Africa, is quite resistant to pests and disease; however, it contains cyanide at such high levels that only a laborious process of washing, grinding, fermenting, and heating can make it edible.[115] In India, the pest-resistant grain *Lathyrus sativus* is cultivated to make some types of dahl. Its seeds contain the neurotoxin beta-*N*-oxalyl aminoalanine, which causes a crippling condition of the nervous system (neurolathyrism) in many people who eat it.[116]

As an alternative to synthetic pesticides, many "organic" farmers use the natural pesticides from one plant species against pests that attack a different plant species. It is legal for farmers who advertise their produce as "organic" to spray their crops with large amounts of pesticides synthesized by other plants—such as the pyrethrins from chrysanthemum plants.[109] These naturally derived pesticides have usually not been tested for carcinogenicity, mutagenicity, or teratogenicity, as have synthetic pesticides; therefore, the safety compared to synthetically derived pesticides should not be prematurely assumed.

VI. CARCINOGENS AND MUTAGENS MAY BE LESS HAZARDOUS AT LOW DOSES THAN IS COMMONLY ASSUMED

It is prudent to assume that if a chemical is a carcinogen in rats and mice at the MTD, it is also likely to be a carcinogen in humans *at the MTD*. However, the understanding of the mechanisms of carcinogenesis is critical to the attempt to predict risk to humans at low doses that are often hundreds of thousands of times below the dose at which an effect is observed in rodents. *Quantitative carcinogenic risk assessment in the absence of an understanding of the mechanisms of carcinogenesis is not scientifically justifiable.*[1,117]

A. MECHANISMS OF CARCINOGENESIS
The study of the mechanisms of carcinogenesis is a rapidly developing field that is essential for evaluating both the role of mutagenicity tests and the methods for risk assessment in regulatory policy. Both DNA damage and cell proliferation (i.e., promotion) are important aspects of carcinogenesis and agents that increase either are proper carcinogens.[1,118-120]

1. Endogenous Rates of DNA Damage Are Enormous
From oxidative damage alone there are 10^4 hits/cell/day in man and 10 times higher rates in rodents.[121] This high rate has been proposed as a major factor contributing to aging and the degenerative diseases of aging such as cancer.[122]

2. Cell Proliferation Is Itself Mutagenic in Numerous Ways
A.   A dividing cell is much more at risk for mutation than a quiescent cell. The time interval for DNA repair during cell division is short, and adducts are converted to gaps during replication. Single-strand DNA is also more sensitive than double-strand DNA. Endogenous or exogenous damage is therefore increased if cells are proliferating.
B.   Cell division triggers mitotic recombination, the conversion of adducts to gaps, gene conversion, and nondisjunction, which together are orders of magnitude more effective than an independent second mutation[123-126] in converting a heterozygous recessive tumor suppressor mutation to homozygosity.[127-129]
C.   Cell division allows gene duplication, which can cause expression of oncogenes that are otherwise not expressed.[130]
D.   Cell division allows adducts in the noncoding strand to convert to mutations.
E.   Cell division allows 5-methyl C in DNA to be lost, which can result in dedifferentiation.[131,132]

3. Cell Proliferation Can Overcome Growth Inhibition
In nondividing tissues, such as the liver (the major target site for carcinogenesis in rodents)[2], mutation is not sufficient for carcinogenesis because cells are communicating and inhibiting growth of neighboring cells.[133-135] Unless there is a cluster of proliferating cells, an essential step for carcinogenesis may be lacking.

4. Exogenous Factors Such as Natural and Synthetic Toxins Can Cause Cell Proliferation

A.   Toxicity can cause injury to tissues, resulting in cell proliferation. In an experimental cancer model, the surgical removal of part of the liver causes neighboring cells to proliferate.[136,137] The incidence of liver cancer is very low in humans (but not in some strains of mice) *unless* the liver is chronically damaged. Alcohol excess, for example, causes cirrhosis of the liver, which is a risk factor for cancer. Salt is a major risk factor in human stomach cancer because it causes cell proliferation.[138-145] Chronic toxicity can also cause an inflammatory reaction, since phagocytic cells unleash a barrage of oxidants in destroying dead cells at a wound. The oxidants produced are the same as in ionizing radiation, so chronic inflammation is the equivalent of irradiating the tissue.[146] Oxidants produced as a result of inflammation could stimulate oncogenes and cell proliferation.[147-150] Chronic inflammation is, as expected, a risk factor for cancer;[151-154] asbestos carcinogenesis is one of many examples of this,[155] and asbestos and the NO_x in cigarette smoke may be primarily promotional carcinogens.

B.   Viruses, particularly those associated with chronic infection, cause cell killing and consequent cell proliferation and are thus risk factors for cancer. Two examples are the human virus hepatitis B, a major cause of liver cancer in the world,[156,157] and human papilloma virus 16 (HPV16), a major risk factor for cervical cancer whose main effect on cells is to increase cell proliferation.[158] Some oncogenic viruses also cause cancer by direct genetic mechanisms.

C.   Hormones can also cause cell proliferation and are major risk factors for a number of human cancers such as breast cancer.[159]

D.   Some chemicals interfere with cell-cell communication, thereby causing cell proliferation and carcinogenesis, as has been emphasized by Trosko and his associates.[134,135]

Thus, agents causing cell proliferation are proper carcinogens and appear to be the most numerous and important class of human carcinogen. The classical tumor promoters such as phenobarbital and tetradecanoyl phorbol acetate cause cell proliferation and are in fact complete carcinogens in animals when tested thoroughly.[160] Evidence that the cell proliferation induced by certain mitogens is less potentially carcinogenic than that induced by toxicity suggests that there are other modifying factors.[161] These factors could be the death of hyperplastic tissue after the stimulus is removed or differential toxicity of inititated cells.

B. ANIMAL CANCER TESTS AND CELL PROLIFERATION

The high proportion of carcinogens among chemicals tested at the MTD emphasizes the importance of understanding cancer mechanisms in order to determine the relevance of rodent cancer test results for humans. A list of carcinogens is not enough. The main rule in toxicology is that "the dose makes the poison": at some level, every chemical becomes toxic, but there are safe levels below that. However, the precedent of radiation, which is both a mutagen and a carcinogen, gave credence to the idea that there could be effects of chemicals even at low doses. A scientific consensus evolved in the 1970s that we should treat carcinogens differently, that we should assume that even low doses might cause cancer, even though we lacked methods for measuring effects at low levels. This idea evolved because most carcinogens appeared to be mutagens and because it was expected that only a small proportion of chemicals would be carcinogenic. However, it seems time to take account of new information. Because administering chemicals at the MTD in animal cancer tests commonly causes cell proliferation and inflammatory reactions,[1,120,162,163] it seems likely that a high percentage of all chemicals, both man-made and natural, will cause cell proliferation at the MTD and thereby increase tumor incidence.

1. Mutagenicity

Analyses of animal cancer tests to date indicate that a high proportion (~40%) of chemicals that are carcinogenic under bioassay conditions are not mutagenic,[6,164,165] although mutagens (in contrast to nonmutagens) are (a) more likely to be carcinogenic, (b) more likely to be positive in both rats and mice, (c) toxic at lower doses, and (d) more likely to cause tumors at multiple sites.[6] Since cell proliferation is itself mutagenic, nonmutagens at the MTD are likely to be acting by this mechanism. Since the MTD approaches the level that kills the animal because of toxicity, and cytotoxicity is a threshold process, for nongenotoxic chemicals the exact dose is clearly critical for tumor induction.

2. The Dose-Response Relation

Both theory and experimental observations suggest an upward-curving (quadratic-type) dose-response relation in carcinogenesis and multiplicative interactions.[1,84,95,117,120,166-168] Carcinogenesis is a multihit and multistage process. Several mutations are necessary for carcinogenesis, and tumor suppressor genes are recessive. There are many layers of defense against carcinogens, and most of these defenses are inducible. There is much evidence to show that agents causing cell proliferation may have thresholds, i.e., that no proliferative effect is observed below a certain dose.[118-120] Inducing both mutation and cell proliferation should give a multiplicative effect in an animal cancer test. Multiplicative interactions are common in human cancer causation.

Even such a well-characterized mutagen as diethylnitrosamine may be carcinogenic at the MTD primarily by inducing cell proliferation. At doses near the MTD, the induced ethylated adducts show a linear dose response, and the induced cell proliferation shows a threshold; the tumors induced at high doses, however, show a clearly upward-curving dose response.[168] A similar case is seen with the mutagen formaldehyde.[168] Thus, when chemicals are tested at the MTD, cell proliferation appears to be the primary aspect of carcinogenesis, even for mutagens. Mutagens, because they damage DNA, are very effective at killing cells and thus are also very effective at causing cell proliferation and inflammatory reactions.

Recent work suggests that cell killing is also an important factor in radiation carcinogenesis.[169,170] In addition, low doses of radiation induce antioxidant defenses which protect against the mutagenic and killing effects of larger doses of radiation or other oxidizing agents.[86-92]

If a chemical is nonmutagenic and its carcinogenicity is due to cell proliferation that results from near-toxic doses, one might commonly expect a threshold in the dose response.[1,118,119,162,163] An analysis of the shape of the dose-response curves in 344 National Cancer Institute/National Toxicology Program (NCI/NTP) animal cancer tests indicates that even at the high doses used, a quadratic dose response is compatible with more of the data than a linear one.[166] Another analysis of 52 NTP cancer tests indicates that more than two thirds of the carcinogenic effects would not have been detected if the high dose had been reduced from the estimated MTD to one half the MTD.[171]

If toxicity were not a factor in animal cancer tests at the MTD and one half the MTD, the dose response might commonly plateau because chemicals would give all of the animals tumors at doses well below the toxic dose. From our database of NTP rodent cancer tests, we observed that approximately 10% of the dose-response functions indicated a possible plateau. For the compounds in which this was observed, the result was generally not replicated in other target sites in the same experiment, in the other sex of the same species, or in other species.[172] Our explanation for the observation that a plateau in the dose response is uncommon is that the MTD of a carcinogen causes cell death, which allows neighboring cells to proliferate; it also causes constant oxygen radical production from phagocytosis, which stimulates cell proliferation and induces chronic inflammation.[172] Thus, even though mutagens can cause cancer at low doses in the absence of cell proliferation, the primary contribution to carcinogenicity for mutagens and nonmutagens at the MTD is cell proliferation. This is consistent with the theoretical analysis of the mutagenicity of cell proliferation given above.

These considerations suggest that at doses close to the toxic dose, any chemical, whether synthetic or natural, is a potential rodent and human carcinogen. Some chemicals that might be close to the toxic dose are those in an occupational setting[8] or certain natural pesticides. The chemicals of some concern at doses far below their toxic dose are likely to be the natural mutagens that we get from food, and cooking of our food, because there are so many of them present at high levels. However, much more important for human cancer are likely to be the lifestyle factors that cause cell proliferation or increase endogenous mutation.

VII. HUMAN CANCER

The major preventable risk factors for cancer appear to be tobacco, dietary imbalances,[71,173-178] hormones,[159] and viruses,[156-158] as has been discussed extensively in the literature. Epidemiologists are constantly finding clues as to the dietary, viral, and hormonal risk factors for the different types of human cancer, and these hypotheses are then refined by animal and metabolic studies. It seems likely that this approach will lead to the elucidation of the causal factors of the major human cancers in the next decade. Epidemiology suggests dietary imbalances as a major area of interest, with insufficient vegetable consumption being a major deficiency.[179] Discovering the protective factors in vegetables is an area of major interest. The important issue is not to identify new chemicals that are carcinogenic in rodents at massive doses but to discover and eliminate the important causes of human cancer.

ACKNOWLEDGMENTS

This work was supported by National Cancer Institute Outstanding Investigator Grant CA39910, by National Institute of Environmental Health Sciences Center Grant ES01896, and by National Institute of Environmental Health Sciences/Department of Energy Interagency Agreement 222-YO1-ES-10066 through the Lawrence Berkeley Laboratory. This paper has been adapted from B. N. Ames and L. S. Gold, I. Chemical carcinogenesis: too many rodent carcinogens, *Proc. Natl. Acad. Sci. U.S.A.*, 87; B. N. Ames, M. Profet, and L. S. Gold, II. Dietary pesticides (99.99% all natural), *Proc. Natl. Acad. Sci. U.S.A.*; and B. W. Ames, M. Profet, and L. S. Gold, III. Nature's chemicals and synthetic chemicals: comparative toxicology, *Proc. Natl. Acad. Sci. U.S.A.*, 87, in press.

REFERENCES

1. **Ames, B. N., Magaw, R., and Gold, L. S.,** Ranking possible carcinogenic hazards, *Science,* 236, 271, 1987.
2. **Gold, L. S., Bernstein, L., Magaw, R., and Stone, T. H.,** Interspecies extrapolation in carcinogenesis: prediction between rats and mice, *Environ. Health Perspect.,* 81, 211—219, 1989.
3. **Gold, L. S., Sawyer, C. B., Magaw, R., Backman, G. M., de Veciana, M., Levinson, R., Hooper, N. K., Havender, W. R., Bernstein, L., Peto, R., Pike, M. C., and Ames, B. N.,** A carcinogenic potency database of the standardized results of animal bioassays, *Environ. Health Perspect.,* 58, 9, 1984.
4. **Gold, L. S., de Veciana, M., Backman, G. M., Magaw, R., Lopipero, P., Smith, M., Blumenthal, M., Levinson, R., Bernstein, L., and Ames, B. N.,** Chronological supplement to the Carcinogenic Potency Database: standardized results of animal bioassays published through December 1982, *Environ. Health Perspect.,* 67, 161, 1986.

5. **Gold, L. S.., Slone, T. H., Backman, G. M., Magaw, R., Da Costa, M. Lopipero, P., Blumenthal, M., and Ames, B. N.,** Second chronological supplement to the Carcinogenic Potency Database: standardized results of animal bioassays published through December 1984 and by the National Toxicology Program through May 1986, *Environ. Health Perspect.*, 74, 237, 1987.

6. **Gold, L. S., Slone, T. H., Backman, G. M., Eisenberg, S., Da Costa, M., Wong, M., Manley, N. B., Rohrbach, L., and Ames, B. N.,** Third chronological supplement to the Carcinogenic Potency Database: standardized results of animal bioassays published through December 1986 and by the National Toxicology Program through June 1987, *Environ. Health Perspect.*, 84, 215—285, 1990.

7. **Schardein, J. L., Schwetz, B. A., and Kenal, M. F.,** Species sensitivities and prediction of teratogenic potential, *Environ. Health Perspect.*, 61, 55, 1985.

8. **Gold, L. S., Backman, G. M., Hooper, N. K., and Peto, R.,** Ranking the potential carcinogenic hazards to workers from exposures to chemicals that are tumorigenic in rodents, *Environ. Health Perspect.*, 76, 211, 1987.

9. **Janzen, D. H.,** Promising directions of study in tropical animal plant interactions, *Ann. Mo. Bot. Gard.*, 64, 706, 1977.

10. **Ames, B. N.,** Dietary carcinogens and anticarcinogens: oxygen radicals and degenerative diseases, *Science*, 221, 1256, 1983.

11. **Beier, R. C.,** Natural pesticides and bioactive components in foods, in *Reviews of Environmental Contamination and Toxicology*, Ware, G. W., Ed., Springer-Verlag, New York, 47—137, 1990.

12. **Rosenthal, G. A. and Janzen, D. H., Eds.,** *Herbivores: Their Interaction with Secondary Plant Metabolites*, Academic Press, New York, 1979.

13. **Gartrell, M. H., Craun, J. C., Podrebarac, D. S., and Gunderson, E. L.,** Dietary intakes of pesticides, selected elements, and other chemicals, *J. Assoc. Off. Anal. Chem.*, 69, 146, 1986.

14. **Stich, H. F., Rosin, M. P., Wu, C. H., and Powrie, W. D.,** A comparative genotoxicity study of chlorogenic acid (3-O-caffeoylquinic acid), *Mutat. Res.*, 90, 201, 1981.

15. **Ishidate, M., Jr., Harnois, M. C., and Sofuni, T.,** A comparative analysis of data on the clastogenicity of 951 chemical substances tested in mammalian cell cultures, *Mutat. Res.*, 195, 151, 1988.

16. **Ariza, R. R., Dorado, G., Barbancho, M., and Pueyo, C.,** Study of the causes of direct-acting mutagenicity in coffee and tea using the Ara test in *Salmonella typhimurium, Mutat. Res.*, 201, 89, 1988.

17. **Fung, V. A., Cameron, T. P., Hughes, T. J., Kirby, P. E., and Dunkel, V. C.,** Mutagenic activity of some coffee flavor ingredients, *Mutat. Res.*, 204, 219, 1988.

18. **Hanham, A. F., Dunn, B. P., and Stich, H. F.,** Clastogenic activity of caffeic acid and its relationship to hydrogen peroxide generated during autooxidation, *Mutat. Res.*, 116, 333, 1983.

19. **McGregor, D. B., Brown, A., Cattanach, P., Edwards., I., McBride, D., Riach, C., and Caspary, W. J.,** Responses of the L5178Y tk+/tk− mouse lymphoma cell forward mutation assay. III. 72 coded chemicals, *Environ. Mol. Mutagen.*, 12, 85, 1988.

20. **National Toxicology Program,** Carcinogenesis Bioassay of Allyl Isothiocyanate (CAS No. 57-06-7) in F344/N Rats and B6C3F$_1$ Mice (Gavage Study), Tech. Rep. 234, NIH Publ. No. 83-1790, National Toxicology Program, NIH, Research Triangle Park, NC, 1982.

21. **Huff, J. E., Eustis, S. L., and Haseman, J. K.,** Occurrence and relevance of chemically induced benign neoplasms in long-term carcinogenicity studies, *Cancer Metastasis Rev.*, 8, 1, 1989.

22. **Morse, M. A., Wang, C.-X., Amin, S. G., Hecht, S. S., and Chung F.-L.,** Effects of dietary sinigrin or indole-3-carbinol on O^6-methylguanine—DNA-transmethylase activity and 4-(methylnitrosamino)-1-(3-pyridyl)-1-butanone-induced DNA methylation and tumorigenicity in F344 rats, *Carcinogenesis*, 9, 1891, 1988.

23. **National Toxicology Program,** Draft Technical Report: Toxicology and Carcinogenesis Studies of d-Carvone in B6C3F$_1$ Mice and Toxicology Studies in F344/N Rats, Tech. Rep. 381, NIH Publ. No. 90-2836, National Toxicology Program, NIH, Research Triangle Park, NC, 1989.

24. **Wakabayashi, K., Suzuki, M., Sugimura, T., and Nagao, M.,** Induction of tumors by 1-nitrosoindole-3-acetonitrile, in Proc. 48th Annu. Meet. Japanese Cancer Association, Nagoya, Japan, October 1989, Abst. No. 284.

25. **Ito, N. and Hirose, M.,** The role of antioxidants in chemical carcinogenesis, *Jpn. J. Cancer Res. (Gann)*, 78, 1011, 1987.

26. **Hirose, M., Fukushima, S., Shirai, T., Hasegawa, R., Kato, T., Tanaka, H., Asakawa, E., and Ito, N.,** Stomach carcinogenicity of caffeic acid, sesamol and catechol in rats and mice, *Jpn. J. Cancer Res.*, 81, 207—212, 1990.

27. **VanEtten, C. H. and Tookey, H. L.,** Chemistry and biological effects of glucosinolates, in *Herbivores: Their Interaction with Secondary Plant Metabolites*, Rosenthal, G. A. and Janzen, D. H., Eds., Academic Press, New York, 1979, 471.

28. **Fenwick, G. R., Heaney, R. K., and Mullin, W. J.,** Glucosinolates and their breakdown products in food and food plants, *CRC Crit. Rev. Food Sci. Nutr.*, 18, 123, 1983.

29. **Nishie, K. and Daxenbichler, M. E.,** Toxicology of glucosinolates, related compounds (nitriles, R-goitrin, isothiocyanates) and vitamin U found in Cruciferae., *Food Cosmet. Toxicol.*, 18, 159, 1980.

30. **Nishie, K. and Daxenbichler, M. E.,** Hepatic effects of *R*-goitrin in Sprague-Dawley rats, *Food Chem. Toxicol.*, 20, 279, 1982.

31. **Villasenor, I. M., Lim-Sylianco, C. Y., and Dayrit, F.,** Mutagens from roasted seeds of *Moringa oleifera, Mutat. Res.*, 224, 209, 1989.

32. **Harborne, J. B.,** The role of phytoalexins in natural plant resistance, in *Natural Resistance of Plants to Pests: Roles of Allelochemicals,* Green, M. B. and Hedin, P. A., Eds., ACS Symp. 296, American Chemical Society, Washington, D.C., 1986, 22.

33. **Malinow, M. R., Bardana, E. J., Jr., Pirofsky, B., Craig, S. and McLaughlin, P.,** Systemic lupus erythematosus-like syndrome in monkeys fed alfalfa sprouts: role of a nonprotein amino acid, *Science,* 216, 415, 1982.

34. **Emerit, I., Michelson, A. M., Levy, A., Camus, J. P., and Emerit, J.,** Chromosome-breaking agent of low molecular weight in human systemic *Lupus erythematosus.* Protector effect of superoxide dismutase, *Human Genet.,* 55, 341, 1980.

35. **Hirono, I., Ed.,** *Naturally Occurring Carcinogens of Plant Origin: Toxicology, Pathology and Biochemistry, Bioactive Molecules,* Vol. 2, Kodansha/Elsevier Science, Tokyo/Amsterdam, 1987.

36. *IARC Monographs on the Evaluation of Carcinogenic Risks to Humans: Some Naturally Occurring and Synthetic Food Components, Furocoumarins and Ultraviolet Radiation,* Vol. 40. International Agency for Research on Cancer, Lyon, France, 1986.

37. Toxicology and Carcinogenesis Studies of 8-Methoxypsoralen (CAS No. 298-81-7) in F344/N Rats (Gavage Studies), Tech. Rep. 359, NIH Publ. No. 89-2814, National Toxicology Program, NIH, Research Triangle Park, NC, 1989.

38. **McManus, B. M., Toth, B., and Patil, K. D.,** Aortic rupture and aortic smooth muscle tumors in mice: induction by *p*-hydrazinobenzoic acid hydrochloride of the cultivated mushroom *Agaricus bisporus, Lab. Invest.,* 57, 78, 1987.

39. **Toth, B.,** Carcinogenesis by N^2-[γL(+)-glutamyl]-4-carboxyphenylhydrazine of *Agaricus bisporus* in mice, *Anticancer Res.,* 6, 917, 1986.

40. Toxicology and Carcinogenesis Studies of *d*-Limonene (CAS No. 5989-27-5) in F344/N Rats and B6C3F$_1$ Mice, Gavage Studies, Tech. Rep. 347, Peer Review Draft, April 1988. National Toxicology Program, NIH, Research Triangle Park, NC, 1988.

41. **Miller, E. C., Swanson, A. B., Phillips, D. H., Fletcher, T. L., Liem, A., and Miller, J. A.,** Structure-activity studies of the carcinogenicities in the mouse and rat of some naturally occurring and synthetic alkenylbenzene derivatives related to safrole and estragole, *Cancer Res.,* 43, 1124, 1983.

42. Toxicology and Carcinogenesis Studies of Benzyl Acetate (CAS No. 140-11-4) in F344/N Rats and B6C3F$_1$ Mice (Gavage Studies), Tech. Rep. 250, NIH Publ. No. 86-2506, National Toxicology Program, NIH, Research Triangle Park, NC, 1986.

43. **Hirose, M., Masuda, A., Imaida, K., Kagawa, M., Tsuda, H., and Ito, N.,** Induction of forestomach lesions in rats by oral administrations of naturally occurring antioxidants for 4 weeks, *Jpn. J. Cancer Res. (Gann),* 78, 317, 1987.

44. **Beier, R. C., Ivie, G. W., Oertli, E. H., and Holt, D. L.,** HPLC analysis of linear furocoumarins (psoralens) in healthy celery *(Apium graveolens), Food Chem. Toxicol.,* 21, 163, 1983.

45. **Chaudhary, S. K., Ceska, O., Têtu, C., Warrington, P. J., Ashwood-Smith, M. J., and Poulton, G. A.,** Oxypeucedanin, a major furocoumarin in parsley, *Petroselinum crispum, Planta, Med.,* 462, 1986.

46. **Ivie, G. W., Holt, D. L., and Ivey, M. C.,** Natural toxicants in human foods: psoralens in raw and cooked parsnip root, *Science,* 213, 909, 1981.

47. **Berkley, S. F., Hightower, A. W., Beier, R. C., Fleming, D. W., Brokopp, C. D., Ivie, G. W., and Broome, C. V.,** Dermatitis in grocery workers associated with high natural concentrations of furanocoumarins in celery, *Ann. Intern. Med.,* 105, 351, 1986.

48. **Seligman, P. J., Mathias, C. G. T., O'Malley, M. A., Beier, R. C., Fehrs, L. J., Serrill, W. S., and Halperin, W. E.,** Phytophotodematitis from celery among grocery store workers, *Arch. Dermatol.,* 123, 1478, 1987.

49. **Carlson, D. G., Daxenbichler, M. E., VanEtten, C. H., Kwolek, W. F., and Williams, P. H.,** Glucosinolates in crucifer vegetables: broccoli, Brussel sprouts, cauliflower, collards, kale, mustard greens, and kohlrabi, *J. Am. Soc. Hortic. Sci.,* 112, 173, 1987.

50. **Schreier, P., Drawert, F., and Heindze, I.,** Ueber die quantitative Zusammensetzung natuerlicher und technologisch veraenderter pflanzlicher Aromen. VII. Verhalten der Aromastoffe bei der Gefrierkonzentrierung von Orangensaft, *Chem. Mikrobiol. Technol. Lebensm.,* 6, 78, 1979.

51. **Engel, K. H. and Tressl, R.,** Studies on the volatile components of two mango varieties, *J. Agric. Food Chem.,* 31, 796, 1983.

52. **Hasslestrom, T., Hewitt, E. J., Konigsbacher, K. S., and Ritter, J. J.,** Composition of volatile oil of black pepper, *Piper nigrum, Agric. Food Chem.,* 5, 53, 1957.

53. **Hecker, E.,** Cocarcinogenesis and tumor promoters of the diterpene ester type as possible carcinogenic risk factors., *J. Cancer Res. Clin. Oncol.,* 99, 103, 1981.

54. **Miura, Y., Ogawa, K., and Tabata, M.,** Changes in the essential oil components during the development of fennel plants from somatic embryoids, *Planta Med.,* 53, 95, 1987.

55. **Archer, A. W.,** Determination of safrole and myristicin in nutmeg and mace by high-performance liquid chromatography, *J. Chromatogr.,* 438, 117, 1988.

56. **Concon, J. M., Swerczek, T. W., and Newburg, D. S.,** Black pepper (*Piper nigrum*): evidence of carcinogenicity, *Nutr. Cancer,* 1 (Spring), 22, 1979.

57. **Ohta, H., Kinjo, S., and Osajima, Y.,** Glass capillary gas chromatographic analysis of volatile components of canned Philippine pineapple juice, *J. Chromatogr.,* 409, 409, 1987.

58. **Wootton, M., Edwards, R. A., Faraji-Haremi, R., and Williams, P. J.,** Effect of accelerated storage conditions on the chemical composition and properties of Australian honeys. III. Changes in volatile components, *J. Apicult. Res.,* 17, 167, 1978.

59. **Luo, S. J., Gue, W. F., and Fu, H. J.,** Correlation between aroma and quality grade of Chinese jasmine tea, *Dev. Food Sci.,* 17, 191, 1988.

60. **Karawya, M. S., Hashim, F. M., and Hifnawy, M. S.,** Oils of *Ocimum basilicum* L. and *Ocimum rubrum* L. grown in Egypt. *J. Agric. Food Chem.,* 22, 520, 1974.

61. **Risch, B. and Herrmann, K.,** Die Gehalte and Hydroxyzimtsaure-Verbindungen und Catechinen in Kern- and Steinobst, *Z. Lebensm. Unters. Forsch.,* 186, 225, 1988.

62. **Schmidtlein, H. and Herrmann, K.,** Über die Phenolsäuren des Gemuses. IV. Hydroxyzimtsäuren und Hydroxybenzoesäuren weiterer Gemüsearten und der Kartoffeln, *Z. Lebensm. Unters. Forsch.,* 159, 255, 1975.

63. **Moller, B. and Herrmann, K.,** Quinic acid esters of hydroxycinnamic acids in stone and pome fruit, *Phytochemistry,* 22, 477, 1983.

64. **Mosel, H. D. and Herrmann, K.,** The phenolics of fruits, III. The contents of catechins and hydroxycinnamic acids in pome and stone fruits, *Z. Lebensm. Unters. Forsch.,* 154, 6, 1974.

65. **Schäfers, F. I. and Herrmann, K.,** Über das Vorkommen von Methyl- und Ethylestern der Hydroxyzimtsäuren und Hydroxybenzoesäuren im Gemüse, *Z. Lebensm. Unters. Forsch.,* 175, 117, 1982.

66. **Winter, M., Brandl, W., and Herrmann, K.,** Determination of hydroxycinnamic acid derivatives in vegetable, *Z. Lebensm. Unters. Forsch.,* 184, 11, 1987.

67. **Herrmann, K.,** Übersicht über nichtessentielle Inhaltsstoffe der Gemüsearten. III. Möhren, Sellerie, Pastinaken, Rote Rüben, Spinat, Salat, Endivien, Treibzichorie, Rhabarber und Artischocken, *Z. Lebensm. Unters. Forsch.,* 167, 262, 1978.

68. **Stöhr, H. and Herrmann, K.,** Über die Phenolsäuren des Gemüses. III. Hydroxyzimtsäuren und Hydroxybenzoesäuren des Wurzelgemüses, *Z. Lebensm. Unters. Forsch.,* 159, 219, 1975.

69. **Schuster, B., Winter, M., and Hermann, K.,** 4-O-β-D-Glucosides of hydroxybenzoic and hydroxycinnamic acids—their synthesis and determination in berry fruit and vegetable, *Z. Naturforsch.,* 41c, 511, 1986.

70. **National Research Council, Board of Agriculture,** *Regulating Pesticides in Food.,* National Academy Press, Washington, D.C., 1987.

71. *IARC Monographs on the Evaluation of Carcinogenic Risks to Humans: Alcohol Drinking,* Vol. 44, International Agency for Research on Cancer, Lyon, France, 1988.

72. **Knutson, J. C. and Poland, A.,** Response of murine epidermis to 2,3,7,8-tetrachlorodibenzo-*p*-dioxin: interaction of the *Ah* and *hr* loci, *Cell,* 30, 225, 1982.

73. **Bradfield, C. A. and Bjeldanes, L. F.,** High-performance liquid chromatographic analysis of anticarcinogenic indoles in *Brassica oleracea, J. Agric. Food Chem.,* 35, 46, 1987.

74. **Bradfield, C. A. and Bjeldanes, L. F.,** Structure-activity relationships of dietary indoles: a proposed mechanism of action as modifiers of xenobiotic metabolism, *J. Toxicol. Environ. Health,* 21, 311, 1987.

75. **Dashwood, R. H., Arbogast, D. N., Fong, A. T., Hendricks, J. D., and Bailey, G. S.,** Mechanisms of anti-carcinogenesis by indole-3-carbinol: detailed *in vivo* DNA binding dose-response studies after dietary administration with aflatoxin B1, *Carcinogenesis,* 9, 427, 1988.

76. **Bailey, G. S., Hendricks, J. D., Shelton, D. W., Nixon, J. E., and Pawlowski, N. E.,** Enhancement of carcinogenesis by the natural anticarcinogen indole-3-carbinol, *J. Natl. Cancer Inst.,* 78, 931, 1987.

77. **Birt, D. F., Pelling, J. C., Pour, P. M., Tibbels, M. G., Schweickert, L., and Bresnick, E.,** Enhanced pancreatic and skin tumorigenesis in cabbage-fed hamsters and mice, *Carcinogenesis,* 8, 913, 1987.

78. **Rannug, A., Rannug, U., Rosenkranz, H. S., Winqvist, L., Westerholm, R., Agurell, E., and Grafstrom, A.-K.,** Certain photooxidized derivatives of tryptophan bind with very high affinity to the Ah receptor and are likely to be endogenous signal substances, *J. Biol. Chem.,* 262, 15422, 1987.

79. **Kasamaki, A., Yasuhara, T., and Urasawa, S.,** Neoplastic transformation of Chinese hamster cells *in vitro* after treatment with flavoring agents, *J. Toxicol. Sci.,* 12, 383, 1987.

80. **Randerath, K., Randerath, E., Agrawal, H. P., Gupta, R. C., Schurdak, M. E., and Reddy, V.,** Postlabeling methods for carcinogen-DNA adduct analysis, *Environ. Health Perspect.,* 62, 57, 1985.

81. **Tucker, J. D., Taylor, R. T., Christensen, M. L., Strout, C. L., and Hanna, M. L.,** Cytogenetic response to coffee in Chinese hamster ovary AUXB1 cells and human peripheral lymphocytes, *Mutagenesis,* 4, 343, 1989.

82. **Stich, H. F. and Powrie, W. D.,** Plant phenolics as genotoxic agents and as modulators for the mutagenicity of other food components, in *Carcinogens and Mutagens in the Environment,* Vol. I, Stich, H., Ed., CRC Press, Boca Raton, FL, 1982, 135.

83. **Davis, D. L.,** Paleolithic diet, evolution, and carcinogens, *Science,* 238, 1633, 1987.

84. **Ames, B. N. and Gold, L. S.,** Response to letter: paleolithic diet, evolution, and carcinogens, *Science,* 238, 1634, 1987.

85. **Mannervik, B. and Danielson, U. H.,** Glutathione transferases—structure and catalytic activity, *CRC Crit. Rev. Biochem.,* 23, 283, 1988.

86. **Kondo, S.,** Mutation and cancer in relation to the atomic-bomb radiation effects, *Jpn. J. Cancer Res. (Gann),* 79, 785, 1988.

87. **Ootsuyama, A. and Tanooka, H.,** One hundred percent tumor induction in mouse skin after repeated β irradiation in a limited dose range, *Radiat. Res.,* 115, 488, 1988.

88. **Wolff, S., Afzal, V., Wiencke, J. K., Olivieri, G., and Michaeli, A.,** Human lymphocytes exposed to low doses of ionizing radiations become refractory to high doses of radiation as well as to chemical mutagens that induce double-strand breaks in DNA, *Int. J. Radiat. Biol.,* 53, 39, 1988.

89. **Yalow, R. S.,** Biologic effects of low-level radiation, in *Low-Level Radioactive Waste Regulation: Science, Politics, and Fear,* Burns, M. E., Ed., Lewis Publishers, Chelsea, MI, 1988, 239.

90. **Wolff, S., Olivieri, G., and Afzal, V.,** Adaptation of human lymphocytes to radiation or chemical mutagens: differences in cytogenetic repair, in *Chromosomal Aberrations: Basic and Applied Aspects,* Natarajan, A. T. and Obe, G., Eds., Springer-Verlag, New York, 1990.

91. **Cai, L. and Liu, S.,** Induction of cytogenetic adaptive response of somatic and germ cells *in vivo* and *in vitro* by low-dose X-irradiation, *Int. J. Radiat. Biol.,* 58, 187—194, 1990.

92. **Wolff, S., Wiencke, J. K., Afzal, V., Youngblom, J., and Cortés, F.,** The adaptive response of human lymphocytes to very low doses of ionizing radiation: a case of induced chromosomal repair with the induction of specific proteins, in *Low Dose Radiation: Biological Bases of Risk Assessment,* Baverstock, K. F. and Stather, J. W., Eds., Taylor & Francis, London, 1989.

93. **Klohs, W. D. and Steinkampf, R. W.,** Possible link between the intrinsic drug resistance of colon tumors and a detoxification mechanism of intestinal cells, *Cancer Res.,* 48, 3025, 1988.

94. **Jakoby, W. B., Ed.,** *Enzymatic Basis of Detoxification,* Vols. I and II, Academic Press, New York, 1980.

95. **Ames, B. N., Magaw, R., and Gold, L. S.,** Response to letter: carcinogenicity of aflatoxins, *Science,* 237, 1283, 1987c.

96. **IARC Monographs on the Evaluation of Carcinogenic Risks to Humans: Overall Evaluations of Carcinogenicity: An Updating of IARC Monographs Volumes 1—42,** Suppl. 7, International Agency for Research on Cancer, Lyon, France, 1987.

97. **Hirayama, T. and Ito, Y.,** A new view of the etiology of nasopharyngeal carcinoma, *Prev. Med.,* 10, 614, 1981.

98. **Hecker, E.,** Cocarcinogenesis and tumor promoters of the diterpene ester type as possible carcinogenic risk factors, *J. Cancer Res. Clin. Oncol.,* 99, 103, 1981.

99. **Beier, R. C. and Norman, J. O.,** The toxic factor in white snakeroot: identity, analysis and prevention, in *Public Health Significance of Natural Food Toxicants in Animal Feeds,* Keller, W. C., Beasley, V. R., and Robens, J. F., Eds., in press.

100. **Kilgore, W. W., Crosby, D. G., Craigmill, A. L., and Poppen, N. K.,** Toxic plants as possible human teratogens, *Calif. Agric.,* 35, 6, 1981.

101. **Crosby, D. G.,** Alkaloids in milk may cause birth defects, *Chem. Eng. News.,* 61, (April 11), 37, 1983.

102. **Warren, C. D.,** Toxic alkaloids from lupines, *Chem. Eng. News.,* 61 (June 13), 3, 1983.

103. **Wattenberg, L. W.,** Inhibition of carcinogenesis by minor anutrient constituents of the diet, Nutrition Society, in press.

104. **Matthew, J. A., Morgan, M. R. A., McNerney, R., Chan, H. W.-S., and Coxon, D. T.,** Determination of solanidine in human plasma by radioimmunoassay, *Food Chem. Toxicol.,* 21, 637, 1983.

105. **Harvey, M. H., Morris, B. A., McMillan, M., and Marks, V.,** Measurement of potato steroidal alkaloids in human serum and saliva by radioimmunoassay, *Human Toxicol.,* 4, 503, 1985.

106. **Jukes, T. H.,** Insecticides in health, agriculture and the environment, *Naturwissenschaften,* 61, 6, 1974.

107. **Miller, T. A. and Salgado, V. L.,** The mode of action of pyrethroids on insects, in *The Pyrethroid Insecticides,* Leahey, J. P., Ed., Taylor & Francis, London, 1985, 43.

108. **Berenbaum, M. R., Zangerl, A. R., and Nitao, J. K.,** Furanocoumarins in seeds of wild and cultivated parsnip, *Phytochemistry,* 23, 1809, 1984.

109. **Prokopy, R. J.,** Organic Compared with Non-organically Grown Apples, University of Amherst, MA, 1989.

110. **Schmiediche, P. E., Hawkes, J. G., and Ochoa, C. M.,** Breeding of the cultivated potato species Solanum × Juzepczukii Buk. and Solanum × Cutilobum Juz. et Buk. I. A study of the natural variation of S. × Juzepczukii, S. × Cutilobum and their wild progenitor, S. Acaule Bitt, *Euphytica,* 29, 685, 1980.

111. **Mithen, R. F., Lewis, B. G., Heaney, R. K., and Fenwick, G. R.,** Glucosinolates of wild and cultivated *Brassica* species, *Phytochemistry,* 26, 1969, 1987.

112. **Lucas, B. and Sotelo, A.,** A simplified test for the quantitation of cyanogenic glucosides in wild and cultivated seeds, *Nutr. Rep. Int.,* 29, 711, 1984.

113. **Rhoades, D. F.,** Evolution of plant chemical defense against herbivores, in *Herbivores: Their Interaction with Secondary Plant Metabolites,* Rosenthal, G. A. and Janzen, D. H., Eds., Academic Press, New York, 1979, 3.

114. **Jadhav, S. J., Sharma, R. P., and Salunkhe, D. K.,** Naturally occurring toxic alkaloids in foods, *CRC Crit. Rev. Toxicol.,* 9, 21, 1981.

115. **Cooke, R. and Cock, J.,** Cassava crops up again, *New Scientist,* 17, 63, 1989.

116. **Jayaraman, K. S.,** Neurolathyrism remains a threat in India, *Nature,* 339, 495, 1989.

117. **Ames, B. N., Magaw, R., and Gold, L. S.,** Response to letter: risk assessment, *Science,* 237, 235, 1987.

118. **Pitot, H. C., Goldsworthy, T. L., Moran, S., Kennan, W., Glauert, H. P., Maronpot, R. R., and Campbell, H. A.,** A method to quantitate the relative initiating and promoting potencies of hepatocarcinogenic agents in their dose-response relationships to altered hepatic foci, *Carcinogenesis,* 8, 1491, 1987.

119. **Farber, E.,** Possible etiologic mechanisms in chemical carcinogenesis, *Environ. Health Perspect.,* 75, 65, 1987.

120. **Butterworth, B. and Slaga, T.,** A perspective on cell proliferation in rodent carcinogenicity studies: relevance to human beings, in *Chemically Induced Cell Proliferation: Implications for Risk Assessment,* Alan R. Liss, New York, in press.

121. **Ames, B. N.,** Mutagenesis and carcinogenesis: endogenous and exogenous factors, *Environ. Mol. Mutagen.,* 14 (Suppl. 16), 66, 1989.

122. **Ames, B. N.,** Endogenous oxidative DNA damage, aging, and cancer, *Free Rad. Res. Commun.,* 7, 121—128, 1989.

123. **Schiestl, R. H., Gietz, R. D., Mehta, R. D., and Hastings, P. J.,** Carcinogens induce intrachromosomal recombination in yeast, *Carcinogenesis,* 10, 1445, 1989.

124. **Liskay, R. M. and Stachelek, J. L.,** Evidence for intrachromosomal gene conversion in cultured mouse cells., *Cell,* 35, 157, 1983.

125. **Fahrig, R.,** The effect of dose and time on the induction of genetic alterations in *Saccharomyces cerevisiae* by aminoacridines in the presence and absence of visible light irradiation in comparison with the dose-effect-curves of mutagens with other types of action, *Mol. Gen. Genet.,* 144, 131, 1976.

126. **Ramel, C.,** Short-term testing—are we looking at wrong endpoints?, *Mutat. Res.,* 205, 13, 1988.

127. **Sasaki, M., Okamoto, M., Sato, C., Sugio, K., Soejima, J., Iwama, T., Ikeuchi, T., Tonomura, A., Miyaki, M., and Sasazuki, T.,** Loss of constitutional heterozygosity in colorectal tumors from patients with familial polyposis coli and those with nonpolyposis colorectal carcinoma, *Cancer Res.,* 49, 4402, 1989.

128. **Erisman, M. D., Scott, J. K., and Astrin, S. M.,** Evidence that the familial adenomatous polyposis gene is involved in a subset of colon cancers with a complementary defect in *c-myc* regulation, *Proc. Natl. Acad. Sci. U.S.A.,* 86, 4264, 1989.

129. **Vogelstein, G., Fearon, E. R., Kern, S. E., Hamilton, S. R., Preisinger, A. C., Nakamura, Y., and White, R.,** Allelotype of colorectal carcinomas, *Science,* 244, 207, 1989.

130. **Orr-Weaver, T. L. and Spradling, A. C.,** *Drosophila* chorion gene amplification requires an upstream region regulating *s18* transcription, *Mol. Cell. Biol.,* 6, 4624, 1986.

131. **Wilson, V. L., Smith, R. A., Ma, S., and Cutler, R. G.,** Genomic 5-methyldeoxycytidine decreases with age, *J. Biol. Chem.,* 262, 9948, 1987.

132. **Lu, L.-J. W., Liehr, J. G., Sirbasku, D. A., Randerath, E., and Randerath, K.,** Hypomethylation of DNA in estrogen-induced and -dependent hamster kidney tumors, *Carcinogenesis,* 9, 925, 1988.

133. **Trosko, J. E.,** A failed paradigm: carcinogenesis is more than mutagenesis, *Mutagenesis,* 4, 363, 1988.

134. **Trosko, J. E. and Chang, C. C.,** Non-genotoxic mechanisms in carcinogenesis: role of inhibited intercellular communication, in *Carcinogen Risk Assessment, Banbury Report 31,* Hart, R. W. and Hoerger, F. D., Eds., Cold Spring Harbor Laboratory, Cold Spring Harbor, NY 1988, 139.

135. **Yamasaki, H., Enomoto, K., Fitzgerald, D. J., Mesnil, M., Katoh, F., and Hollstein, M.,** Role of intercellular communication in the control of critical gene expression during multistage carcinogenesis, in *Cell Differentiation, Genes and Cancer,* IARC Scientific Publ. No. 92, Kakunaga, T., Sugimura, T., Tomatis, L., and Yamasaki, H., Eds., International Agency for Research on Cancer, Lyon, France, 1988, 57.

136. **Farber, E., Parker, S., and Gruenstein, M.,** The resistance of putative premalignant liver cell populations, hyperplastic nodules, to the acute cytotoxic effects of some hepatocarcinogens, *Cancer Res.,* 36, 3879, 1976.

137. **Farber, E.,** Cellular biochemistry of the stepwise development of cancer with chemicals: G. H. A. Clowes Memorial Lecture, *Cancer Res.,* 44, 5463, 1984.

138. **Joossens, J. V. and Geboers, J.,** Nutrition and gastric cancer, *Nutr. Cancer,* 2, 250, 1981.

139. **Furihata, C., Sato, Y., Hosaka, M., Matsushima, T., Furukawa, F., and Takahashi, M.,** NaCl induced ornithine decarboxylase and DNA synthesis in rat stomach mucosa, *Biochem. Biophsy. Res. Commun.,* 121, 1027, 1984.

140. **Tuyns, A. J.,** Salt and gastrointestinal cancer, *Nutr. Cancer,* 11, 229, 1988.

141. **Lu, J.-B. and Qin, Y.-M.,** Correlation between high salt intake and mortality rates for oesophageal and gastric cancers in Henan Province, China, *Int. J. Epidemiol.,* 16, 171, 1987.

142. **Furihata, C., Sudo, K., and Matsushima, T.,** Calcium chloride inhibits stimulation of replicative DNA synthesis by sodium choride in the pyloric mucosa of rat stomach, *Carcinogenesis,* 10, 2135—2137, 1990.

143. **Coggon, D., Barker, D. J. P., Cole, R. B., and Nelson, M.,** Stomach cancer and food storage, *J. Natl. Cancer Inst.,* 81, 1178, 1989.

144. **Charnley, G. and Tannenbaum, S. R.,** Flow cytometric analysis of the effect of sodium chloride on gastric cancer risk in the rat, *Cancer Res.,* 45, 5608, 1985.

145. **Karube, T., Katayama, H., Takemoto, K., and Watanabe, S.,** Induction of squamous metaplasia, dysplasia and carcinoma *in situ* of the mouse tracheal mucosa by inhalation of sodium chloride mist following subcutaneous injection of 4-nitroquinoline 1-oxide, *Jpn. J. Cancer Res.,* 80, 698, 1989.

146. **Ward, J. F., Limoli, C. L., Calabro-Jones, P., and Evans, J. W.,** Radiation vs. chemical damage to DNA, in *Anticarcinogenesis and Radiation Protection,* Cerutti, P. A., Nygaard, O. F., and Simic, M. G., Eds., Plenum Press, New York, 1987.

147. **Crawford, D. and Cerutti, P.,** Expression of oxidant stress-related genes in tumor promotion of mouse epidermal cells JB6, in *Anticarcinogenesis and Radiation Protection,* Nygaard, O., Simic, M., and Cerutti, P., Eds., Plenum Press, New York, 1988, 183.

148. **Sieweke, M. H., Stoker, A. W., and Bissell, M. J.,** Evaluation of the cocarcinogenic effect of wounding in Rous Sarcoma virus tumorigenesis, *Cancer Res.,* 49, 6419, 1989.

149. **Chan, T. M., Chen, E., Tatoyan, A., Shargill, N. S., Pleta, M., and Hochstein, P.,** Stimulation of tyrosine-specific protein phosphorylation in the rat liver plasma membrane by oxygen radicals, *Biochem. Biophys. Res. Commun.,* 139, 439, 1986.

150. **Crave, P. A., Pfanstiel, J., and DeRubertis, F. R.,** Role of activation of protein kinase C in the stimulation of colonic epithelial proliferation and reactive oxygen formation by bile acids, *J. Clin. Invest.,* 79, 532, 1987.

151. **Weitzman, S. A. and Gordon, L. J.,** Inflammation and cancer: role of phagocyte-generated oxidants in carcinogenesis, *Blood,* 76, 655—663, 1990.

152. **Demopoulos, H. B., Pietronigro, D. D., Flamm, E. S., and Seligman, M. L.,** The possible role of free radical reactions in carcinogenesis, *J. Environ. Pathol. Toxicol.,* 3, 273, 1980.

153. **Templeton, A.,** Pre-existing, non-malignant disorders associated with increased cancer risk., *J. Environ. Pathol. Toxicol.,* 3, 387, 1980.

154. **Lewis, J. G. and Adams, D. O.,** Inflammation, oxidative DNA damage, and carcinogenesis, *Environ. Health Perspect.,* 76, 19, 1987.

155. **Petruska, J., Marsh, J. P., Kagan, E., and Mossman, B. T.,** Release of superoxide by cells obtained from bronchoalveolar lavage after exposure of rats to either crocidolite or chrysotile asbestos, *Am. Rev. Respir. Dis.,* 137, 403, 1988.

156. **Yeh, F.-S., Mo, C.-C., Luo, S., Henderson, B. E., Tong, M. J., and Yu, M. C.,** A seriological case-control study of primary hepatocellular carcinoma in Guangxi, China, *Cancer Res.,* 45, 872, 1985.

157. **Wu, T. C., Tong, M. J., Hwang, B., Lee, S.-D., and Hu, M. M.,** Primary hepatocellular carcinoma and hepatitis B infection during childhood, *Hepatology,* 7, 46, 1987.

158. **Peto, R. and zur Hausen, H., Eds.,** *Banbury Report 21. Viral Etiology of Cervical Cancer,* Cold Spring Harbor Laboratory, Cold Spring Harbor, NY, 1986.

159. **Henderson, B. E., Ross, R., and Bernstein, L.,** Estrogens as a cause of human cancer: The Richard and Hinda Rosenthal Foundation Award Lecture, *Cancer Res.,* 48, 246, 1988.

160. **Iversen, O. H., Ed.,** *Theories of Carcinogenesis,* Hemisphere, Washington, D.C., 1988.

161. **Ledda-Columbano, G. M., Columbano, A., Curto, M., Ennas, M. G., Coni, P., Sarma, D. S. R., and Pani, P.,** Further evidence that mitogen-induced cell proliferation does not support the formation of enzyme-altered islands in rat lever by carcinogens, *Carcinogenesis,* 10, 847, 1989.

162. **Mirsalis, J. C. and Steinmetz, K. L.,** The role of hyperplasia in liver carcinogenesis, in *Mouse Liver Carcinogenesis: Mechanisms and Species Comparisons,* Stevenso, D., McClain, M., Popp, J., Slaga, T., Ward, J., and Pitot, H., Eds., Wiley-Liss, NY, 1990.

163. **Mirsalis, J. C., Tyson, C. K., Steinmetz, K. L., Loh, E. K., Hamilton, C. M., Bakke, J. P., and Spalding, J. W.,** Measurement of unscheduled DNA synthesis and S-phase synthesis in rodent hepatocytes following in vivo treatment: testing 24 compounds, *Environ. Mol. Mutagen.,* 14, 155, 1989.

164. **Zeiger, E.,** Carcinogenicity of mutagens: predictive capability of the *Salmonella* mutagenesis assay for rodent carcinogenicity, *Cancer Res.,* 27, 1287, 1987.

165. **Ashby, J. and Tennant, R. W.,** Chemical structure, *Salmonella* mutagenicity and extent of carcinogenicity as indicators of genotoxic carcinogenesis among 222 chemicals tested in rodents by the U.S. NCI/NTP, *Mutat. Res.,* 204, 17, 1988.

166. **Hoel, D. G. and Portier, C. J.,** Nonlinearity of dose-response functions for carcinogenicity, submitted for publication, 1990.

167. **Ames, B. N., Gold, L. S., and Magaw, R.,** Response to letter: risk assessment, *Science,* 237, 1399, 1987d.

168. **Swenberg, J. A., Richardson, F. C., Boucheron, J. A., Deal, F. H., Belinsky, S. A., Charbonneau, M., and Short, B. G.,** High- to low-dose extrapolation: critical determinants involved in the dose response of carcinogenic substances, *Environ. Health Perspect,* 76, 57, 1987.

169. **Jones, T. D.,** A unifying concept for carcinogenic risk assessments: comparison with radiation-induced leukemia in mice and men, *Health Phys.,* 4, 533, 1984.

170. **Little, J. B., Kennedy, A. R., and McGandy, R. B.,** Effect of the dose rate on the induction of experimental lung cancer in hamsters by α radiation, *Radiat. Res.,* 103, 293, 1985.

171. **Haseman, J. K.,** Issue in carcinogenicity testing: dose selection, *Fundam. Appl. Toxicol.,* 5, 66, 1985.

172. **Bernstein, L., Gold, L. S., Ames, B. N., Pike, M. C., and Hoel, D. G.,** Some tautologous aspects of the comparison of carcinogenic potency in rats and mice, *Fundam. Appl. Toxicol.,* 5, 79, 1985.

173. **Lipkin, M.,** Biomarkers of increased susceptibility to gastrointestinal cancer: new application to studies of cancer prevention in human subjects, *Cancer Res.,* 48, 235, 1988.

174. **Yang, C. S. and Newmark, H. L.,** The role of micronutrient deficiency in carcinogenesis, *CRC Crit. Rev. Oncol. Hematol.,* 7, 267, 1987.

175. **Pence, B. C. and Buddingh, F.,** Inhibition of dietary fat-promoted colon carcinogenesis in rats by supplemental calcium or vitamin D_3, *Carcinogenesis,* 9, 187, 1988.

176. **Reddy, B. S. and Cohen, L. A., Eds.,** *Diet, Nutrition, and Cancer: A Critical Evaluation,* Vols. I and II, CRC Press, Boca Raton, FL 1986.

177. **Joossens, J. V., Hill, M. J., and Geboers, J., Eds.,** *Diet and Human Carcinogenesis,* Elsevier Science, Amsterdam, 1986.

178. **Ames, B. N.,** Review of Evidence for Alcohol-Related Carcinogenesis, Report for Proposition 65 Meeting, Sacramento, CA, December 11, 1987.

179. **National Research Council, Diet and Health,** *Implications for Reducing Chronic Disease Risk,* National Academy Press, Washington, D.C., 1989.

180. **Ames, B. N., Profet, M., and Gold, L. S.,** II. Dietary pesticides (99.99% all natural), *Proc. Natl. Acad. Sci. U.S.A.,* 87, 1137, in press.

Chapter 4

MUTAGENICITY OF CHEMICALS ADDED TO FOODS

Errol Zeiger and Virginia C. Dunkel

Among the chemically defined substances that can be found in food are microbial toxins, natural toxicants, environmental contaminants, food additives, color additives, and pesticide residues. In general, microbial toxins (such as the various bacterial enterotoxins and botulin) or fungal toxins (such as aflatoxin) can be present in food because of spoilage from faulty procedures used in food preparation and storage. Natural toxicants on the other hand are those substances normally synthesized by plants and include such compounds as 5- and 8-methoxypsoralen, allyl isothiocyanate, p-hydrazinobenzoate, and many other so-called "natural pesticides". Other contaminants are found in food as one of the end results of environmental pollution and can include such substances as methyl mercury, arsenic, and lead. These microbial toxins, natural toxicants, and environmental contaminants find their way into the food chain either naturally or accidentally, while food additives, color additives, and pesticides are either specifically added to, or migrate into, food at some stage during the progression from raw agricultural commodity to the finished consumable product.

The substances considered to be food additives vary among countries, depending on the particular chemical usage definitions they adopt. For example, the Codex Alimentarius Commission[1] has concluded that a "Food additive means any substance not normally consumed as a food by itself and not normally used as a typical ingredient of the food, whether or not it has nutritive value, the intentional addition of which to food for technological (including organoleptic) purpose in the manufacture, processing, preparation, treatment, packing, packaging, transport or holding of such food results, or may be reasonably expected to result (directly or indirectly) in it or its by-products becoming a component of or otherwise affecting the characteristics of such foods. The term does not include 'contaminants' or substances added to food for maintaining or improving nutritional qualities."

In the U.S., the definition of a food additive is given in the Federal Food, Drug, and Cosmetic Act.[2] The Act [Section 201(s)] states that "The term 'food additive' means any substance the intended use of which results or may reasonably be expected to result, directly or indirectly, in its becoming a component or otherwise affecting the characteristics of any food (including any substance intended for use in producing, manufacturing, packing, processing, preparing, treating, packaging, transporting, or holding food; and including any source of radiation intended for any such use), if such substance is not generally recognized among experts qualified by scientific training and experience to evaluate its safety, as having been adequately shown through scientific procedures (or, in the case of a substance used in food prior to January 1, 1958, through either scientific procedures or experience based on common use in food) to be safe under the conditions of its intended use . . . ". The Act also specifies that this term does not include: (1) pesticides used on raw food or food in its natural state; (2) pesticides used during the production, storage or transportation of raw food or food in its natural state; (3) color additives; (4) substances granted 'prior sanction'; this "means an explicit approval granted with respect to use of a substance in food prior to September 6, 1958, by the Food and Drug Administration or by the United States Department of Agriculture pursuant to the Federal Food, Drug, and Cosmetic Act, the Poultry Products Inspection Act,[3] or the Meat Inspection Act.[4]"; and (5) new drugs used on food-producing animals.

The substances classified as food additives by the U.S. Food and Drug Administration (FDA) in the Code of Federal Regulations[5] fall into a number of categories (Table 1), and the total number of substances regulated is in the thousands. Many of these chemicals are

TABLE 1
Categories of U.S. Food Additives[a]

Category	21CFR§§
Color additives	73;74
Direct food additives	172
Secondary direct food additives	173
Indirect food additives	174—178
Irradiation in the production, processing, and handling of food	179
Food additives permitted in food on an interim basis . . .	180
Prior-sanctioned food ingredients	181
Substances generally recognized as safe (GRAS)	182; 184; 186

    [a] As listed in 21CFR.

listed in more than one category of use, and each use may be bound by a different concentration level. Some of the chemicals, especially the indirect additives of the type used in the manufacture of paper, cardboard, plastics and rubber products, as well as adhesives, lubricants, and biocides, are highly reactive industrial chemicals that are identical with, or structurally similar to, well-studied mutagens and toxins. However, for the majority of these indirect additives, extremely small amounts are allowed in articles that may contact food and in all probability many may not migrate into food.

Color additives, although added directly to food, are not included within the definition of "food additive". Under section 201(t) of the Food, Drug, and Cosmetic Act,[2] "The term 'color additive' means a material which is a dye, pigment, or other substance made by a process of synthesis or similar artifice, or extracted, isolated or otherwise derived, with or without intermediate or final change of identity, from a vegetable, animal, mineral or other source, and when added or applied to a food, drug, or cosmetic, or to the human body or any part thereof, is capable (alone or through reaction with other substance) of imparting color thereto . . . ".

In this review, only those chemicals added directly to food will be considered. This will include direct food additives, substances "generally recognized as safe" (GRAS), and approved color additives. The list of chemicals reviewed was derived from the U.S. FDA lists of substances in these categories contained in the Code of Federal Regulations (CFR).[5] Natural flavoring substances that are extracts of plants, such as spices, natural seasonings, essential oils, oleoresins, and other plant extracts (21CFR 172.510, 182.10, 182.20, 182.40, and 182.50),[5] have been excluded from this master list because they are not well-defined chemically, and there is little, if any, published mutagenicity test data on these substances. In general, the natural products extracted from plants and added to foods are those that are generally known to be safe for human consumption. Such plant-derived substances include agar-agar, clove, dill, and licorice derivatives, as well as extracts from a wide variety of roots, leaves, and flowers. Other food plants contain substances classified as glycosides, hydrazides, quinones, and phenols; chemicals in these classes contain many bacterial mutagens, mammalian cell clastogens, and rodent carcinogens.[6-10] Additionally, although "flavoring agents and adjuvants" are listed by the U.S. FDA under both direct food additives (21CFR 172.515) and GRAS substances (21CFR 182.60), they have been separated into a group by themselves for the purposes of this review because they constitute such a large number of compounds.

The primary emphasis in this review is on the U.S. food additives, because more substances are defined as food additives in the U.S. than under the Codex Alimentarius Commission or the regulations of most other countries.[11] There are approximately 3000 chemicals listed in 21CFR for direct food additives, color additives, and GRAS substances alone. This includes different salts of the same chemical and mixtures of substances. Because of the extensive number of publications from Japanese laboratories on the mutagenicity of Japanese food additives, these data were also considered. Although the U.S. FDA has studied the mutagenicity of a number of food additives and can request mutagenicity data on proposed new food additives, there is no published compilation of these mutagenicity results.

Before a chemical is approved as a new direct food or color additive, it must be demonstrated by appropriate methods to be safe for its proposed use. Among the test methods that have been initially used in the development of a potential new additive are the Salmonella mutagenicity (Ames) and mammalian cell chromosome aberration tests. These test systems are available in many laboratories, are convenient to use, and are relatively inexpensive. The Salmonella mutagenicity assay measures reversion from histidine dependence to histidine independence in a series of tester strains of *Salmonella typhimurium*. The changes are induced by agents that cause base-pair substitutions or frameshifts in genes of the histidine operon.[12] In tests for chromosome aberrations, mammalian cells in culture are challenged with the test chemical in the presence and absence of a metabolic activation system. A positive response in this test is indicated by an increased number of structural chromosomal aberrations observable in cells in metaphase.[13,14] The results from these two assays together with data from other mutagenicity tests are generally considered as indicative of the potential of the chemical to induce adverse effects such as germ cell mutagenicity or carcinogenicity and signal the need for additional chronic animal studies. Often the cost of such additional testing, plus the availability of other nonmutagenic chemicals that can be used as alternative food or color additives, results in the mutagen being removed from consideration as a commercial chemical. The majority of chemicals that are mutagenic in Salmonella and clastogenic in mammalian cells have been shown to be rodent carcinogens.[15] Because the Delaney Amendment to the Federal Food Drug and Cosmetic Act $(409(c)(3)(A))^2$ prohibits the use of a known carcinogen as a food or color additive, it is less likely that a chemical that is mutagenic or clastogenic *in vitro* will be developed or proposed for these purposes.

In general, it can be inferred that because chemicals are allowed to be directly added to foods as preservatives, flavors, colors, sequestrants, etc. after review of their toxicological properties by the U.S. FDA, they are more likely to be innocuous than chemicals in the other categories (e.g., indirect additives). For those chemicals that are categorized as GRAS, safety is generally assumed through experience based on common use in foods as well as upon published and generally available corroborative, unpublished data and information. Overall, chemicals in the GRAS category have not been subjected to the quantity or quality of testing procedures required for approval of a chemical classified as a food additive.

The results of Salmonella mutagenicity tests and those for chromosomal aberrations in mammalian cells in culture on direct food additives, GRAS substances, and approved color additives are summarized in Table 2. Of the 237 chemicals for which there were Salmonella test data, only 19 (8.0%) were reported to be mutagenic. The chemicals inducing positive responses include sodium nitrite and 2′,4′,5′-trihydroxybutyrophenone (THBP) (food preservatives): L-cysteine and L-cysteine monohydrochloride (special dietary and nutritional additives); allyl isothiocyanate, cadinene, 2-hexenal, lepidine, maltol, o-methoxycinnamaldehyde, and pyruvaldehyde (synthetic flavoring substances and adjuvants); potassium bromate (specific usage additives); azodicarbonamide (multipurpose additives); caramel (multipurpose GRAS food substances); zinc chloride (dietary supplements); hydrogen peroxide (affirmed GRAS substances); and carmine, FD&C Green No. 3, and Citrus Red No.2 (color additives).

In contrast to the Salmonella test results, 44 of the 166 chemicals (26.5%) tested for their ability to induce chromosome aberrations *in vitro* were reported as positive. This discrepancy is not unexpected. Previous comparisons of results from these two test systems, for the purpose of distinguishing between carcinogens and noncarcinogens, have shown that there is a higher proportion of chemicals positive for induction of chromosomal aberrations

TABLE 2

Summary of Salmonella (SAL) Mutagenicity and *In Vitro* Chromosome Aberration (ABS) Results on Food Additives

Food additive category	21CFR (§§)	No.[a]	SAL		ABS	
			No. tested	% + (no.)	No. tested	% + (no.)
Color additives	73;74	36	18	22.2(4)	13	46.2(6)
Direct additives	172	127	24	16.7(4)	20	25.0(5)
Synthetic flavorings	172.515; 182.60	750	99	7.1(7)	53	26.4(14)
GRAS	182; 182.1	135	44	2.3(1)	40	17.5(7)
GRAS direct additives	184	157	52	5.8(3)	40	30.0(12)
Summation		(1205)	237	8.0(19)	166	26.5(44)
Ishidate et al.[15]			200	7.0(14)	242	22.3(54)

[a] Total number of chemicals or mixtures listed in the designated sections. The same chemical may be listed in more than one section; however, none of the positive chemicals appears in more than one of these sections.

in mammalian cells than mutations in Salmonella.[15,16] However, a high proportion of the chemicals detected as positive in *in vitro* chromosomal aberration tests are often noncarcinogens.[15] The *in vitro* test for chromosomal aberrations is sensitive to changes in pH of the culture medium and high osmotic pressure, which can lead to increases in the frequency of chromosomal aberrations.[17] The results of tests in which there are increases in aberrations under these conditions should be considered as possible "artifactual positives".

In a survey of the mutagenicity of Japanese food additives, Ishidate et al.[18,19] reported similar proportions of positive responses; 14 of 200 (7.0%) chemicals tested in Salmonella and 54 of 242 (22.3%) chemicals tested for induction of chromosomal aberrations *in vitro* were positive. In a subsequent study, Hayashi et al.[20] evaluated the ability of a 40-chemical subset of the 242 food additives tested by Ishidate et al. to induce micronuclei in mouse bone marrow. Among the chemicals selected for this subset, 67.5% were positive for induction of chromosomal aberrations *in vitro* as compared to 22.3% in the original sample of 242. Although a high proportion of chemicals in the subset induced chromosomal damage *in vitro,* only 5 (12.5%) were positive for micronuclei induction *in vivo.*

It has been estimated that nearly 3,000 substances are intentionally added to processed foods in the U.S. and an additional 12,000 may enter foods indirectly through their use in food processing and packaging.[6] Relatively few of the chemicals added directly to food have been reported in the literature as having been tested for mutagenicity, although the number actually tested may be quite high because researchers are often reluctant to report negative results. In contrast, mutagenicity test information is available for a greater proportion of indirect additives. Such chemicals are more likely to have been tested because many of them are reactive industrial intermediates or known biocides.

Of the chemicals tested, only 8.0% of those classified as direct food additives, approved color additives, or as GRAS chemicals were reported to be mutagenic in Salmonella. Based on the types of chemicals in these groups, such as natural substances, caramel, aldehydes, inorganic salts, and color additives, it is not anticipated that the proportions of chemicals giving positive responses will increase as additional chemicals are tested. Indeed, it is probable that the proportions of mutagens will decrease as more chemicals are tested for mutagenicity prior to their regulatory approval because a number of the mutagens already identified are those chemicals whose structures led investigators to believe that they would be positive if tested. Many of the chemicals that apparently have not been tested are the plant extracts, physiological chemicals, and inorganic salts. The indirect additives and pesticides generally comprise entirely different classes of chemicals. They are often chemically reactive, in the case of such indirect additives as plasticizers, paper manufacture components, adhesives, etc., or biologically reactive, such as pesticides and slimicides. It is therefore to be anticipated that these classes of chemicals would contain a higher proportion of mutagens and clastogens.

ACKNOWLEDGMENT

Thanks to Elizabeth von Halle of EMIC for providing a computer search of chemicals tested in *Salmonella* and *in vitro* cytogenetics, and to Drs. Sam Shibko and Charles Kokoski of the U.S. FDA Center for Food Safety and Applied Nutrition for their helpful comments.

REFERENCES

1. Codex Alimentarious Commission, Procedural Manual, 4th ed. 1975.
2. Federal Food, Drug, and Cosmetic Act, as amended, 1980, 21 USC 301, et seq.
3. Poultry and Poultry Products Inspection Act, Amended 1968, 21 USC 451, et seq.
4. Meat Inspection Act of March 4, 1907, as amended and extended, 1978, 21 USC 603, et seq.
5. Code of Federal Regulations (CFR) Title 21, Office of the Federal Register, National Archives and Records Service, General Services Administration, Washington, D.C., 1989.
6. *Diet, Nutrition, and Cancer,* Committee on Diet, Nutrition, and Cancer, Assembly of Life Sciences, National Research Council, National Academy Press, Washington, D.C., 1982.
7. **Ames, B. N.,** Dietary carcinogens and anticarcinogens, *Science,* 221, 1256, 1983.
8. **Prival, M. J.,** Carcinogens and mutagens present as natural components of food or induced by cooking, *Nutr. Cancer,* 6, 236, 1985.
9. **Yamanaka, H., Nagao, M., Sugimura, T., Furuya, T., Shirai, A., and Matsushima, T.,** Mutagenicity of pyrrolizidine alkaloids in the Salmonella/mammalian-microsome test, *Mutat. Res.,* 68, 211, 1979.
10. **Creasey, W. A.,** *Diet and Cancer,* Lea & Febiger, Philadelphia, 1985, 175.
11. Proposed System of Food Safety Assessment, Final Report of the Scientific Committee of the Food Safety Council, Washington, D.C., 1980.
12. **Maron, D. M. and Ames, B. N.,** Revised methods for the Salmonella mutagenicity test, *Mutat. Res.,* 113, 173, 1983.
13. **Preston, R. J., Au, W., Bender, M. A., Brewen, J. G., Carrano, A. V., Heddle, J. A., McFee, A. F., Wolff, S., and Wassom, J. S.,** Mammalian in vivo and in vitro cytogenetic assays: a report of the U.S. EPA's Gene-Tox Program, *Mutat. Res.,* 87, 143, 1981.
14. **Galloway, S. M., Bloom, A. D., Resnick, M., Margolin, B. H., Nakamura, F., Archer, P., and Zeiger, E.,** Development of a standard protocol for in vitro cytogenetic testing with Chinese hamster ovary cells: comparison of results for 22 compounds in two laboratories, *Environ. Mutagen.,* 7, 1, 1985.
15. **Tennant, R. W., Margolin, B. H., Shelby, M. D., Zeiger, E., Haseman, J. K., Spalding, J., Caspary, W., Resnick, M., Stasiewicz, S., Anderson, B., and Minor, R.,** Prediction of chemical carcinogenicity in rodents from in vitro genetic toxicity assays, *Science,* 236, 933, 1987.
16. **Auletta, A. and Ashby, J.,** Workshop on the Relationship Between Short-Term Test Information and Carcinogenicity; Williamsburg, Virginia, January 20—23, 1987, *Environ. Mol. Mutagen.,* 11, 135, 1988.
17. **Brusick, D.,** Genotoxic effects in cultured mammalian cells produced by low pH treatment conditions and increased ion concentrations, *Environ. Mutagen.,* 8, 879, 1986.
18. **Ishidate, M., Jr., Sofuni, T., Yoshikawa, K., Hayashi, M., Nohmi, T., Sawada, M., and Matsuoka, A.,** Primary mutagenicity screening of food additives currently used in Japan, *Food Chem. Toxicol.,* 22, 623, 1984.
19. **Ishidate, M., Jr., Harnois, M. C., and Sofuni, T.,** A comparative analysis of data on the clastogenicity of 951 chemical substances tested in mammalian cell cultures, *Mutat. Res.,* 195, 151, 1988.
20. **Hayashi, M., Kishi, M., Sofuni, T., and Ishidate, M., Jr.,** Micronucleus tests in mice on 39 food additives and eight miscellaneous chemicals, *Food Chem. Toxicol.,* 26, 487, 1988.

Chapter 5

MUTAGEN PRECURSORS IN FOOD

Chapter 5.1

MUTAGEN FORMATION IN MUSCLE MEATS AND MODEL HEATING SYSTEMS

James S. Felton and Mark G. Knize

TABLE OF CONTENTS

I. INTRODUCTION

The discovery of the formation of potent mutagenic activity in cooked foods by Sugimura et al.[1] and then by Commoner et al.[2] led a world-wide effort to determine the parameters responsible for mutagen formation.

Many groups have found that cooking temperature and time are important determinants for the mutagenic response. The commonly eaten meat types — beef, pork, chicken, and fish — all show some level of mutagen formation following cooking, although the food preparation methods can have an important influence.

Jägerstad et al.[3] suggested in 1983 that creatine, amino acids, and sugars derived from muscle are important precursors in the production of the amino-imidazo mutagens found in cooked meats. The involvement of these precursors has been confirmed in many laboratories. As its structural analogy suggests, creatine has proven to be the essential precursor for mutagen formation and produces a variety of mutagens when added to meats or heated with natural meat components.

Understanding the precursors and reaction conditions for mutagen formation during cooking is important, as this information is necessary for devising strategies to reduce or prevent mutagen formation.

II. MUTAGEN FORMATION IN MEATS WITH OR WITHOUT ADDITIVES

A. MEATS

When high levels of mutagenic activity were first discovered in cooked meats, experiments were undertaken to determine the parameters for the mutagen-forming reactions. The heating temperature was shown to be important because an increase in temperature has a marked positive effect on total mutagen production in foods.[2,4-8]

Figure 1 shows the effect of cooking temperature and time on mutagen formation. Initially, the cooking temperature has a large influence on the rate of mutagen formation, but after the first 4 to 10 min it has much less of an effect. Interestingly, although mutagenic activity increases greatly, the same set of mutagenic compounds is produced in similar relative amounts.[8]

B. MEATS WITH ADDED CREAT(IN)INE, MILK, OR AMINO ACIDS

Work by Nes[9] and Becher et al.[10] showed a dramatic increase in mutagenic activity by adding the natural meat component creatine to a recipe combining pork, beef, veal, starch, and milk.

An investigation in our laboratory of mutagen production in beef is shown in Table 1. Mutagen formation in beef is greatly increased with added milk and creatine, but the starch has little effect. Creatine or milk added to beef alone are not as stimulatory as both added together. These recent results, together with those cited above, suggest that creatine is rate limiting. Free amino acids and sugars in the milk may also contribute to mutagen formation.

Överik et al.[11] showed that the individual addition of 15 different amino acids to ground pork before frying increased mutagenic activity from 1.5 to 43-fold. This sharply contrasts with the report of Ashoor et al.[12] in which 17 amino acids showed no effect when added to ground beef before frying. Only the addition of proline caused an increase (eight-fold) in mutagenic activity.

There are a number of differences in experimental procedure that may explain these contradictory results. Ashoor et al.[12] used ground beef containing 19% fat and cooked the patties for 2 min per side at 191°C, adding 0.5 or 1.0 mmol of amino acid dissolved in 2 ml water, to a 35-g patty (approximately 0.4 g per 100 g meat). Övervik at al.[11] used ground

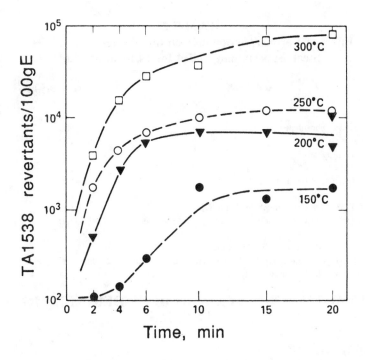

FIGURE 1. The effect of cooking temperature and time on mutagen production in fried beef patties, 100 g equivalent weight before frying (gE).

TABLE 1
Mutagenic Activity of Fried Beef with Added Creatine, Milk, and Starch

	TA1538 rev/g
Beef	1,861
Beef + milk + creatine + starch	12,100
Beef + milk + creatine	12,700
Beef + milk	1,080
Beef + creatine	5,018

pork containing 4% fat cooked for 6 min on the first and 4 min on the second side at 200°C. Amino acids were added neat, at 1 g per 100 g of meat.

The discrepancy in the effect of added amino acids can easily be explained by the cooking-time and water-content differences. For example, Figure 1 shows, particularly at the lowest temperature, that mutagen formation increases dramatically after an initial lag of 2 to 4 min. The cooking times in the linear range of mutagen formation used by Övervik et al.[11] would be more likely to give positive effects with additives. Taylor et al.[13,14] showed that water is an inhibitor of mutagen formation. It appears that the addition of water with amino acids and the very short cooking time used by Ashoor et al.[12] affected the mutagen-forming reactions, possibly negating the effects of the added amino acids.

The work by Övervik et al.[11] supports the idea that free amino acids are involved in the production of mutagens in meat and, like creatine, they can be rate limiting for mutagen formation.

TABLE 2
Effect of Various Compounds on the Mutagenicity of Beef
Steak Supernatant Boiled for 14 hr at pH 4.0[a]

Additions[b]	TA1538 rev/g dry beef
None	90
Creatine phosphate	600
Creatine	107
L-Tryptophan	1400
Proline	250
D-Glucose + creatine phosphate	540
FeSO$_4$	135
L-Trp + creatine phosphate + FeSO$_4$	5500
L-Trp + creatine phosphate + FeSO$_4$ + D-glucose	3400

[a]   Data adapted from Taylor et al., 1986.
[b]   L-Tryptophan (L-Trp) = 10 mM, creatine phosphate = 2.5 mM, FeSO$_4$ = 0.5
    mM, and D-glucose = 10 mM concentrations.

III. MUTAGEN FORMATION IN BEEF-DERIVED MODELING SYSTEMS

A. AQUEOUS BOILING

Taylor et al.,[13] in order to identify the precursors and elucidate the reaction conditions that yield heterocyclic amine mutagens in cooked meat products and fish, used a supernatant system derived from water-homogenized beef steak. When the supernatant was boiled for 30 hr at a pH of 4.0, 600 revertants (rev)/g of dry beef were formed (TA1538). When creatine phosphate is added to the supernatant, 1400 rev/g of dry beef were formed. The addition of a mixture of tryptophan, creatine phosphate, and FeSO$_4$ to the supernatant was the most stimulating; 24,000 rev/g dry weight were produced. Similar results can be seen in Table 2 where the boiling takes place for only 14 hr; only the absolute values are lower with the shorter time. It is worth noting that the addition of D-glucose had little effect on the mutagenic response. In fact, it lowers the mutagenicity somewhat. In both boiling experiments mutagenic activity is primarily due to IQ, Trp-P-2, and MeIQ (trace) as determined by mass spectral analysis following purification by HPLC and electrophoresis. Trp-P-1 was also produced, but only in the mixture containing all the constituents. Mutagen formation is dependent on a fraction containing precursors of less than 500 molecular weight in the supernatant. Creatine phosphate alone or a number of amino acids alone (proline, histidine, glycine) gave very slight increases compared to L-tryptophan when added to the supernatant.

This same group also pressure heated the supernatant for 2 hr at pH 4.0 and analyzed the mutagenic material that was formed. This method produced 42,000 rev/g dry beef supernatant which contained mutagens that co-eluted with IQ/MeIQx, MeIQ, and Trp-P-2 on reverse-phase HPLC. It is important to note that not only is water content important in these supernatant boiling studies, but also mutagenicity is affected dramatically (up to ninefold) with different initial water content in the beef patty itself.[4]

B. DRY HEATING

The high mutagenic activity of the pan residue after cooking meat seen by Övervik et al.[11,15] and Knize et al.[16] could arise from the water-soluble precursors that drain from the meat upon heating. They might evaporate and then be essentially dry heated on the pan surface. It is also possible that grinding the meat helps to release precursors and increase mutagenic activity over the activity found with meat steaks, but the comparative study has not been made.

TABLE 3
Mutagens Made from Creatin(in)e and Amino Acids

Mutagen	Amino acid	Sugar	Heating	Ref.
IQ	Glycine	Glucose	Water/gly[a]	18
	Proline	—	Dry[b]	19
	Phenylalanine	—	Dry	20
	Serine	—	Dry	21
	Phenylalanine	Glucose	Dry	20
MeIQ	Alanine	Fructose	Water/gly	22
MeIQx	Glycine	Glucose	Water/gly	23
	Glycine	Glucose	Water/gly	22
	Tyrosine	—	Dry	11
	Alanine	—	Dry	11
	Serine	—	Dry	11
4,8-DiMeIQx	Alanine	Fructose	Water/gly	18
IQx	Serine	—	Dry	21
7,8-DiMeIQx	Glycine	Glucose	Water/gly	24
PhIP	Phenylalanine	Glucose	Water/gly	25
	Phenylalanine	—	Dry	20
	Phenylalanine	Glucose	Dry	20
	Phenylalanine	—	Dry	11
	Leucine	—	Dry	11

[a] Water/ethylene glycol mixture at 128°C.
[b] Dry heated at 200°C.

Dry heating of the supernatant fraction of a 1:1 aqueous homogenate of beef steak was performed to see if the mutagenic products were similar to those produced when beef is fried well-done.[17] The supernatant (48 g) was freeze-dried and heated for 2 hr at 200°C. Mutagenic activity in TA1538 was calculated for the entire 48 g to be 45×10^6 revertants. The material was extracted and purified using HPLC, paper chromatography, and paper electrophoresis and six mutagens were identified from the 45 million revertants. They are IQ, MeIQ, DiMeIQx, PhIP, and oxygen-containing mutagens with molecular weights of 202 and 216. As in the fried beef, PhIP was the predominant product, 9700 ng/g dry weight. This is more than 100 times the concentration of the PhIP per gram dry weight in a beef patty after 300°C cooking for 10 min a side. Clearly, the production of the identical mutagens in both fried ground beef and a dry-heated beef-supernatant system makes this a valid model for study of 2-amino-imidazole mutagens.

IV. MUTAGEN FORMATION IN ARTIFICIAL MODEL SYSTEMS

A. LIQUID ETHYLENE GLYCOL REFLUX SYSTEMS

Achieving the temperatures necessary for mutagen production with aqueous systems requires the use of a high-temperature boiling solvent and many laboratories have used diethylene glycol and water mixtures. Table 3 lists mutagens made from heating mixtures of creatine or creatinine (collectively abbreviated as creatin(in)e), an amino acid, and a sugar. The structures of the mutagens are shown in Figure 2. All of the seven mutagens have been found in cooked meat, and the formation of common products suggests that the mutagen-forming reactions in the model systems are similar to those in cooked meat.

B. DRY HEATING

Table 3 also lists the mutagens found in cooked foods that have been modeled in dry heating reactions with or without sugars and with a variety of amino acids. Early reports[4]

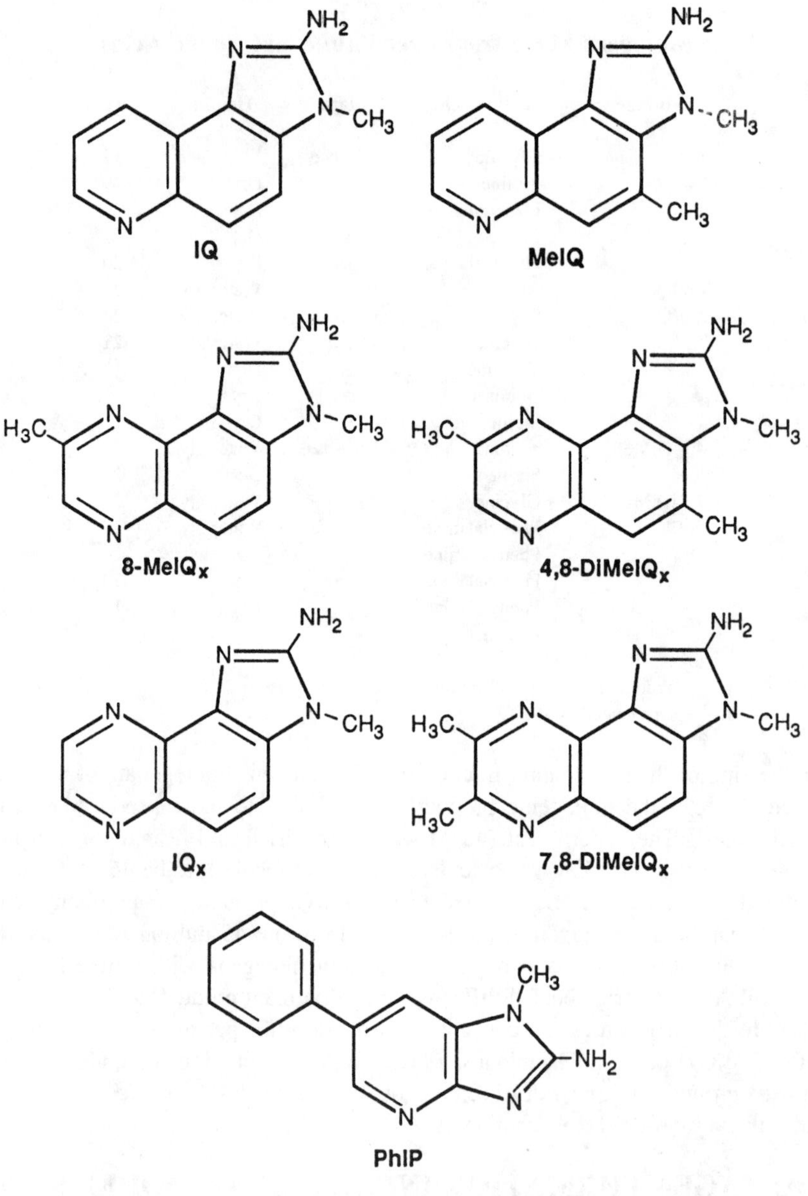

FIGURE 2. Structures of mutagenic compounds identified from creatine model heating systems. IQ = 2-amino-3-methylimidazo[4,5-*f*]quinoline; MeIQ = 2-amino-3,4-dimethylimidazo[4,5-*f*]quinoline; MeIQx = 2-amino-3,8-dimethylimidazo[4,5-*f*]quinoxaline; 4,8-DiMeIQx = 2-amino-3,4,8-trimethylimidazo[4,5-*f*]quinoxaline; IQx = 2-amino-3-methylimidazo[4,5-*f*]quinoxaline; 7,8-DiMeIQx = 2-amino-3,7,8-trimethylimidazo[4,5-*f*]quinoxaline; PhIP = 2-amino-1-methyl-6-phenylimidazo[4,5-*b*]pyridine.

suggested that water was essential for mutagen production, although food mutagens have been made both with and without water present.

It is clear from Table 3 that IQ can be made from any of four amino acids and that glucose or water is not required. There may be many routes of formation for IQ and for the other mutagens. All of these model reactions are low yielding and complex, since heating serine or phenylalanine with creatin(in)e gives at least two mutagenic products.

In quantitative terms, dry heating the amount of phenylalanine (5 mg) and creatine (440 mg) that is found in 100 g of raw beef yields 18 ppb of PhIP.[26] This is similar to the 15

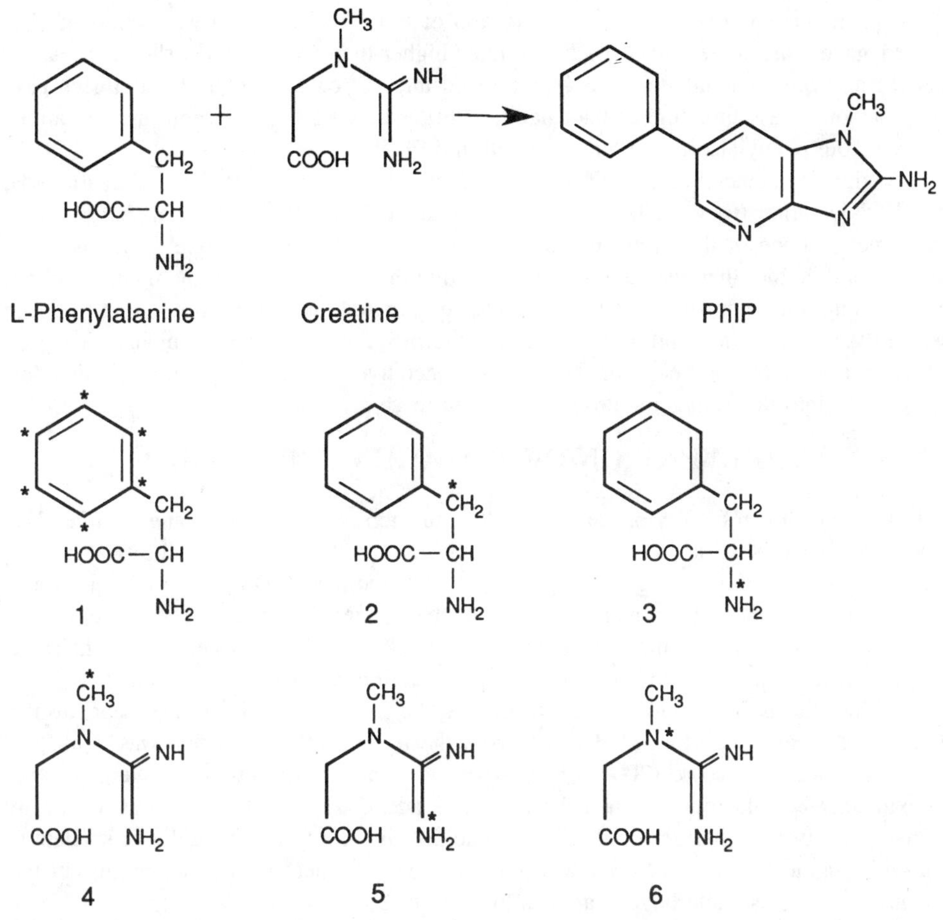

FIGURE 3. Structures of L-phenylalanine, creatine, and PhIP and heavy-isotope-labeled L-phenylalanine and creatine. The position of isotopically labeled atoms is shown by an asterisk (*).

ppb of PhIP determined to be produced from 100 g of fried beef[27] and shows that simple dry heating has comparable yields to the cooking process.

The role of sugars in mutagen formation in meats is unclear. Laser-Reuterswürd et al.[28] examined the content of creatine, creatinine, monosaccharides, and free amino acids in bovine tissues before and after cooking. The amounts of creatine and creatinine correlated best with the mutagen production. In mutagen modeling systems glucose and fructose have been shown to be necessary for mutagen production in aqueous-ethylene glycol refluxing experiments. However, as Table 3 shows, many of the mutagens found in foods have also been formed from dry heating reactions without sugars. These data suggest some involvement of sugars or their breakdown products although they are not necessary for mutagen formation. Determining if atoms from sugar molecules are incorporated into the mutagenic molecules would be an important step in clarifying their role in mutagen formation.

C. ISOTOPE LABELING OF MUTAGENIC PRODUCTS

The relatively efficient formation of PhIP in dry heating reactions and the availability of heavy-isotope-labeled phenylalanine and creatine made it possible for Taylor et al.[26] to show incorporation of specific atoms into the PhIP molecule, shown in Figure 3.

Separate batches of PhIP were generated by heating creatine (for 2 hr at 200°C) with L-[ring U-^{13}C]phenylalanine (Figure 3, compound 1), DL-[-3-^{13}C]phenylalanine (2), or L-

[^{15}NH$_2$]phenylalanine (3). Upon the purification of PhIP, mass spectra were obtained. Reaction 1 gave a molecule with mass 5 to 6 units higher than natural PhIP, showing that the phenyl ring from phenylalanine was incorporated intact. Reactions with 2 and 3 each gave a product one mass unit higher than natural PhIP, showing the 3-carbon and the amino nitrogen from phenylalanine are incorporated into PhIP.

In a similar manner, [methyl-^{13}C]creatine (Figure 3, compound 4), [^{15}NH$_2$]creatine (5), and [1-^{15}N]creatine (6) were heated with phenylalanine, purified, and analyzed by mass spectrometry. Each of the products from the isotopically labeled creatine reactions had a mass one unit higher than the natural PhIP, showing that the 1-nitrogen, the methyl-carbon, and the amino-nitrogen from creatine are each incorporated into PhIP. Although it has been assumed that creatin(in)e and amino acids are the precursors for amino-imidazo mutagens in food, these are the first experiments to prove unequivocally that the source of the atoms incorporated into the mutagenic product are from these precursors.

V. PREVENTION OF MUTAGEN FORMATION

It was also shown by Taylor et al.[14] that the mutagen precursors are water soluble. The movement of water-soluble precursors (creatine, free amino acids, sugars) to the meat surface followed by dry heating during cooking supports the findings of Dolara et al.[5] who found that the outside surface of the meat had most of the mutagenic activity. The movement of the precursors to the surface may also explain why thick and thin meat patties have the same mutagenic activity per gram of meat and not per square centimeter surface area.[8]

This hypothesis led Taylor et al.[14] to microwave meat 1 to 1.5 min and separate the juice appearing at the bottom of the dish from the meat itself. When this was done, the meat was cooked as before (200°C, 6 min per side) and the resulting mutagenicity was approximately 50-fold lower. When the juice was added back to the meat and recooked, the mutagenicity was then restored. This simple cooking control methodology, if adapted by a wide segment of the society that normally eats their meat well-done, might cut the exposure risk to these heterocyclic amines to near zero.

Clearly, not overcooking beef, chicken, and fish is another preventive measure. Alternate cooking methods such as boiling and baking are also good methods of prevention, as the temperatures do not normally reach those needed for mutagen formation.[29] (see Chapter 10.2 for further discussion).

VI. CONCLUSIONS

It is now apparent why cooked nonmuscle foods such as tofu, beans, and cheese have little or no mutagenic activity. They lack the creatine that is present at about 0.5% by weight in muscle meats. Removing the creatine and water-soluble precursors before cooking is a way to sharply reduce the mutagenic activity produced during cooking.

The reactions that produce mutagens are not merely a process for the random coalescence of small fragments, but specific condensation reactions. Among the possible quinoline and quinoxaline mutagen structures known, only mutagens with methyl groups at the 3-, 4-, or 8-position have been found. Isomers with methyl groups at the 1-, 5-, and 7-position are potent mutagens,[30-32] but are yet to be seen as products of natural reactions. Thus, there appears to be specificity inherent in the precursors that directs the formation to a limited set of mutagenic products.

ACKNOWLEDGMENTS

Work performed under the auspices of the U.S. Department of Energy by the Lawrence Livermore National Laboratory under Contract W-7405-ENG-48 and supported by IAG NIEHS 222Y01-ES-10063 and NCI grant RO1-CA40811.

REFERENCES

1. **Sugimura, T., Nagao, M., Kawachi, T., Honda, M., Yahagi, T., Seino, Y., Sato, S., Matsukura, N., Matsushima, T., Shirai, A., Sawamura, M., and Matsumoto, H.**, Mutagen-carcinogens in foods with special reference to highly mutagenic pyrolytic products in broiled foods, in *Origins of Human Cancer*, Hiatt, H. H., Watson, J. D., and Winsten, J. A., Eds., Cold Spring Harbor Laboratory, Cold Spring Harbor, NY, 1977, 1561.

2. **Commoner, B., Vithayathil, A. J., Dolara, P., Nair, S., Madyastha, P., and Cuca, G. C.**, Formation of mutagens in beef and beef extract during cooking, *Science*, 201, 913, 1978.

3. **Jägerstad, M., Laser-Reuterswärd, A., Oste, R., and Dahlqvist, A.**, Creatinine and Malliard reaction products as precursors of mutagenic compounds formed in fried beef, in *The Maillard Reaction in Foods and Nutrition, ACS Symp. Ser. No. 215*, Waller, G. R. and Feather, M. S., Eds., American Chemical Society, Washington, D. C., 1983, 507.

4. **Bjeldanes, L. F., Morris, M. M., Timourian, H., and Hatch, F. T.**, Effects of meat composition and cooking conditions on mutagenicity of fried ground beef, *Agric. Food Chem.*, 31, 18, 1983.

5. **Dolara, P., Commoner, B., Vithayathil, A. J., Cuca, G. C., Tuley, E., Madyastha, P., Nair, S., and Kriebel, D.**, The effect of temperature on the formation of mutagens in heated beef stock and cooked ground beef, *Mutat. Res.*, 60, 231, 1979.

6. **Spingarn, N. E. and Weisburger, J. H.**, Formation of mutagens in cooked food. I. Beef, *Cancer Lett.*, 7, 259, 1979.

7. **Pariza, M. W., Ashoor, S. H., Chu, F. S., and Lund, D. B.**, Effects of temperature and time on mutagen formation in pan-fried hamburger, *Cancer Lett.*, 7, 63, 1979.

8. **Knize, M. G., Andresen, B. D., Healy, S. K., Shen, N. H., Lewis, P. R., Bjeldanes, L. F., Hatch, F. T., and Felton, J. S.**, Effect of temperature, patty thickness and fat content on the production of mutagens in fried ground beef, *Food Chem. Toxicol.*, 23, 1035, 1985.

9. **Nes, I. F.**, Mutagen formation in fried meat emulsion containing various amounts of creatine, *Mutat. Res.*, 175, 145, 1986.

10. **Becher, G., Knize, M. G., Nes, I. F., and Felton, J. S.**, Isolation and identification of mutagens from a fried Norwegian meat product, *Carcinogenesis*, 9, 247, 1988.

11. **Övervik, E., Kleman, M., Berg, I., and Gustafsson, J.-Å.**, Influence of creatine, amino acids and water on the formation of the mutagenic heterocyclic amines found in cooked meat, *Carcinogenesis*, 10, 2293, 1989.

12. **Ashoor, S. H., Dietrich, R. A., Chu, F. S., and Pariza, M. W.**, Proline enhances mutagen formation in ground beef during frying, *Life Sci.*, 26, 1801, 1980.

13. **Taylor, R. T., Fultz, E., and Knize, M. G.**, Mutagen formation in a model beef supernatant fraction. IV. Properties of the system, *Environ. Health Perspect.*, 67, 59, 1986.

14. **Taylor, R. T., Fultz, E., and Knize, M. G.**, Mutagen formation in a model beef supernatant fraction. Elucidation of the role of water in fried ground beef mutagenicity, *Environ. Mutagen.*, 8 (Suppl. 6), 65, 1986.

15. **Övervik, E., Nilsson, L., Fredholm, L., Levin, O., Nord, C.-E., and Gustafsson, J.-Å.**, Mutagenicity of pan residues and gravy from fried meat, *Mutat. Res.*, 187, 47, 1987.

16. **Knize, M. G., Shen, N. H., and Felton, J. S.**, A comparison of mutagen production in fried ground chicken and beef: effect of supplemental creatine, *Mutagenesis*, 3, 503, 1988.

17. **Taylor, R. T., Fultz, E., and Knize, M. G.**, Mutagen formation in a model beef supernatant fraction: purification and identification of the major mutagenic activities produced by dry heating, in Proc. Mutagens and Carcinogens in the Diet, Madison, WI, July 1989, 26.

18. **Grivas, S., Nyhammar, T., Olsson, K., and Jägerstad, M.**, Isolation and identification of the food mutagens IQ and MeIQx from a heated model system of creatinine, glycine and fructose, *Food Chem. Toxicol.*, 20, 127, 1986.

19. **Yoshida, D., Saito, Y., and Mizusaki, S.**, Isolation of 2-amino-3-methyl imidazo [4,5-f]quinoline as mutagen from the heated product of a mixture of creatine and proline, *Agric. Biol. Chem.*, 48, 241, 1984.

20. **Taylor, R. T., Fultz, E., Knize, M. G., and Felton, J. S.**, Formation of the fried ground beef mutagens 2-amino-3-methylimidazo[4,5-f]quinoline (IQ) and 2-amino-1-methyl-6-phenylimidazo[4,5-b]pyridine (PhIP) from L-phenylalanine (Phe) + creatinine (Cre) (or creatine), *Environ. Mutagen.*, 9 (Suppl. 8), 106, 1987.

21. **Knize, M. G., Shen, N. H., and Felton, J. S.**, The production of mutagens in foods, *Proc. Air Pollut. Control Assoc.*, 88-130.3, 1, 1988.

22. **Grivas, S., Nyhammar, T., Olsson, K., and Jägerstad, M.**, Formation of a new mutagenic DiMeIQx compound in a model system by heating creatinine, alanine and fructose, *Mutat. Res.*, 151, 177, 1985.

23. **Jägerstad, M., Olsson, K., Grivas, S., Negishi, C., Wakabayashi, K., Tsuda, M., Sato, S., and Sugimura, T.**, Formation of 2-amino-3,8-dimethlimidazo[4,5-f]quinoxaline in a model system by heating creatinine, glycine and glucose, *Mutat. Res.*, 126, 239, 1984.

24. **Negishi, C., Wakabayashi, K., Tsuda, M., Sato, S., Sugimura, T., Saito, H., Maeda, M., and Jägerstad, M.,** Formation of 2-amino-3,7,8-trimethylimidazo[4,5-f]quinoxaline, a new mutagen, by heating a mixture of creatinine, glucose and glycine, *Mutat. Res.,* 140, 55, 1984.
25. **Shioya, M., Wakabayashi, K., Sato, S., Nagao, M., and Sugimura, T.,** Formation of a mutagen 2-amino-1-methyl-6-phenylimidazo[4,5-b]-pyridine (PhIP) in cooked beef, by heating a mixture containing creatinine, phenylalanine and glucose, *Mutat. Res.,* 191, 133, 1987.
26. **Taylor, R. T., Fultz, E., Morris, C., Knize, M. G., and Felton, J. S.,** Model system phenylalanine (Phe) and Creatine (Cr) heavy-isotope-labeling of the fried ground beef mutagen 2-amino-1-methyl-6-phenylimidazo[4,5-b]pyridine (PhIP), *Environ. Mutagen.,* 11 (Suppl. 11), 104, 1988.
27. **Felton, J. S., Knize, M. G., Shen, N. H., Andresen, B. D., Bjeldanes, L. F., and Hatch, F. T.,** Identification of the mutagens in cooked beef, *Environ. Health Perspect.,* 67,17,1986.
28. **Laser-Reuterswärd, A., Skog, K., and Jägerstad, M.,** Mutagenicity of pan-fried bovine tissue in relation to their content of creatine, creatinine, monosaccharides and free amino acids, *Food Chem. Toxicol.,* 25, 755, 1987.
29. **Bjeldanes, L. F., Morris, M. M., Felton, J. S., Healy, S., Stuermer, D., Berry, P., Timourian, H., and Hatch, F. T.,** Mutagens from the cooking of food: survey by Ames/Salmonella test of mutagen formation in the major protein-rich foods of the American diet, *Food Chem. Toxicol.,* 20, 357, 1982.
30. **Nagao, M., Wakabayashi, K., Kasai, H., Nishimura, S., and Sugimura, T.,** Effect of methyl substitution on mutagenicities of 2-amino-3-methylimidazo[4,5-f]quinoline, isolated from broiled sardine, *Carcinogenesis,* 2, 1147, 1981.
31. **Kaiser, G., Harnasch, D., King, M.-T., and Wild, D.,** Chemical structure and mutagenic activity of aminoimidazoquinolines and aminonaphthimidazoles related to 2-amino-3-methylimidazo[4,5-f]quinoline, *Chem. Biol. Interact.,* 57, 97, 1986.
32. **Knize, M. G., Happe, J., Healy, S. K., and Felton, J. S.,** Identification of the mutagenic quinoxaline isomers from fried ground beef, *Mutat. Res.,* 178, 25, 1987.

Chapter 5.2

NITROSATABLE PRECURSORS OF MUTAGENS IN FOODS

Kiyomi Kikugawa and Minako Nagao

TABLE OF CONTENTS

I. INTRODUCTION

A good correlation between nitrate intake and gastric cancer mortality in various countries has been documented.[1] Nitrate taken from diet and secreted into saliva is converted into nitrite by bacteria in the oral cavity,[2,3] which may serve as an agent to produce carcinogenic compounds by reaction with their precursors under gastric conditions. Recently, nitrite and nitrate have been found to be produced from arginine in mammalian cells, and immuno-stimulants are known to enhance the synthesis of nitrite and nitrate.[4] Dialkylamines and trialkylamines are well-known precursors for the *N*-nitrosamines in foods. These amines are generally found in fish meats and they can undergo nitrosation by reaction with nitrite under gastric conditions to form carcinogenic dialkylnitrosamines.[5,6] Direct-acting genotoxic *N*-nitrosamidines and nitrosamides such as *N*-alkyl-*N'*-nitro-*N*-nitrosoguanidine and *N*-alkyl-*N*-nitrosourea have been known to induce cancer in the glandular stomach of experimental animals.[7-10] These types of direct-acting genotoxic compounds can be produced by reaction of their precursors if present in foods with nitrite under acidic conditions similar to those of gastric juice. Recent hypotheses for the development of gastric cancer suggest that exposure in the stomach to direct-acting genotoxic *N*-nitroso compounds, formed endogenously, may be involved.[11]

Several research groups looked for the presence of nitrosatable mutagen precursors in foods, to produce direct-acting mutagens by reaction with nitrite. Fava beans commonly eaten in Colombia[12] where stomach cancer incidence is high showed direct-acting muta-genicity towards Salmonella strains after nitrite treatment. Japan also has a high incidence of stomach cancer, and Japanese fish, soy sauce, bean paste, fish sauce, and Chinese cabbage which are favorites of the Japanese showed direct-acting mutagenicity on *S. typhimurium* TA100 after treatment with nitrite.[13-15] Since then several nitrosatable mutagen precursors have been isolated and their structures have been determined. All the nitrosatable precursors found so far are aromatic or heterocyclic compounds rather than alkylnitrosamides.

In this chapter, the occurrence of these newly identified nitrosatable precursors and their identification is briefly summarized. In addition, the biological properties of nitrosated products are also discussed.

II. PRESENCE OF POTENT PRECURSORS OF DIRECT-ACTING MUTAGENS

In order to find nitrosatable precursors in foods taken by people in high risk areas for developing gastric cancer, several research groups tested the mutagenicity of various foods after nitrite treatment.

Weisburger and his associate have described a working concept on the mechanism whereby gastric cancer in man may arise[16] and suggested that endogenous formation of an alkylnitrosamide type may be important.[7-10] They found that ether extracts from Japanese raw fish (Sanma hiraki) homogenate treated with 5000 ppm nitrite showed direct-acting mutagenicity to *S. typhimurium*.[13] This extract was also reported to induce tumors including adenocarcinomas of the glandular stomach in rats,[17] although neither adenoma nor adeno-carcinomas were induced by extract from fish not treated with nitrite. The alkylating activity on 4-(*p*-nitrobenzyl)pyridine of extracts from several processed fish with or without nitrite in simulated gastric juice was investigated.[18] Some of the extracts had strong alkylating potency after nitrite treatment, although structures of active compounds produced were not determined.

A nutrition survey carried out in a high risk area for gastric cancer in Colombia revealed a positive correlation between the consumption of fava beans and gastric cancer.[19] Incubation of fava beans in nitrite-containing acidic solutions resulted in the formation of direct-acting mutagens as detected with the Salmonella/microsome assay.[12,20]

TABLE 1
Mutagenicities of Various Kinds of Pickled
Vegetables Treated with Nitrite[22]

Type of pickle	rev/g original material
Vegetable pickled in bean paste	
Cucumber	18,000
Vegetable pickled in rice bran	
Radish root	3,100
Cabbage	5,900
Cucumber	2,900
Vegetable pickled in lees	
Radish root	1,900
Oriental melon	12,000
Vegetable pickled with salt	
Turnip leaves	6,100
Chinese cabbage	
A	2,500
B	4,400
C	3,200
Sliced vegetables pickled in soy sauce	11,000
Kimchi	
A	5,500
B	4,600
C	2,300
D	13,000
E	9,500
F	4,800

Note: All pickled vegetables other than kimchi were produced in Japan. All kimchis were produced in Korea. Mutagenicity was tested on TA100 without S9 mix.

The incidence of stomach cancer and the intake of nitrate of Japanese people are both high. Foodstuffs in Japan were classified into 13 groups according to market basket method, and it was found that total daily intake of net mutagenicity on *S. typhimurium* with and without metabolic activation from nitrite-treated foodstuffs was 2.5 and 1.7 times those from nitrite-untreated foodstuffs, respectively.[21] Wakabayashi et al. found that soy sauce, bean paste, fish sauce, and Chinese cabbage in Japan showed direct-acting mutagenicity on *S. typhimurium* TA100 after treatment with nitrite.[14,15] Chinese cabbage is a popular vegetable in Japan and its average consumption is 45 g/day per capita. Water extracts of fresh Chinese cabbage and Chinese cabbage pickled with salt were not mutagenic to *S. typhimurium* TA100. However, after treatment with nitrite in acidic conditions, they became mutagenic to *S. typhimurium* TA100 without S9 mix. Wakabayashi et al. have made extensive studies on the appearance of direct-acting mutagenicity of various foodstuffs produced in Japan and Southeast Asia on nitrite treatment[22] (Table 1 and 2). After nitrite treatment, various kinds of pickled vegetables and sun-dried fishes produced in Japan showed direct-acting mutagenicity on *S. typhimurium* TA100, inducing 1,900 to 18,000 revertants/g. Kimchis, sun-dried fishes, sun-dried squid, soy sauces, fish sauces, bean pastes, and shrimp paste produced in Korea, the Philippines, and Thailand also showed direct-acting mutagenicity after nitrite treatment. They also found that various broiled meats and fish contained nitrosatable mutagen precursors.[23] Broiled chicken, pork, mutton, beef, and sun-dried sardine were found to yield direct-acting mutagenicity after nitrite treatment (Table 3). In contrast, raw meat showed no mutagenicity after nitrite treatment. Treatment of broiled chicken with 0.5 to 3 mM

TABLE 2
Mutagenicities of Various Sun-Dried Sea
Foods Treated with Nitrite[22]

Food	Country of production	rev/g original materials
Herring	Japan	5500
Sardine	Japan	2400
Siganid	Philippines	3500
Surgeon fish	Philippines	1400
Fimbriated herring	Philippines	4200
Squid	Thailand	2400

Note: Mutagenicity was assayed on TA100 without S9 mix.

TABLE 3
Mutagenicity of Cooked Foods after
Nitrite Treatment[23]

Food	rev/g original material	
	TA100	TA98
Meat		
Chicken	12,800	33,300
Beef	7,400	22,600
Mutton	5,700	43,600
Pork	3,800	15,000
Fish		
Sun-dried sardine	17,900	20,200

Note: Meat and fish were broiled over a gas flame. The samples were treated with 50 mM sodium nitrite at pH 3.0 for 1 hr at 37°C, and their mutagenicities were assayed on TA100 and TA98 without S9 mix. The mutagenicity was calculated from the linear portion of the dose-response curve.

nitrite, which is a physiologically feasible concentration in the human stomach under some conditions, induced direct-acting mutagenicity.[23]

Hayatsu and Hayatsu treated an aqueous homogenate of boiled rice with nitrous acid at pH 3 and found that mutagens were formed. The presence of the mutagens was demonstrated by isolating the mutagenic fractions through blue-rayon adsorption, a method used to extract polycyclic compounds, and subsequent high-performance liquid chromatography. The mutagens were active in *S. typhimurium* TA100 and TA98 without metabolic activation.[24]

Soy sauce is widely used as a seasoning in Southeast Asia. In Japan, the daily consumption of soy sauce is about 30 ml per capita. Studies have been done on the mutagenicity of soy sauce after nitrite treatment. After treatment with 50 mM nitrite, eight kinds of Japanese soy sauce were strongly mutagenic to *S. typhimurium* TA100 without metabolic activation.[14] Since soybeans themselves, and a soy sauce which is produced by acidic hydrolysis of soybeans, showed no mutagenicity with nitrite treatment, the nitrosatable precursors in soy sauce must be formed during the fermentation process. In contrast, most soy sauce in the U.S. had only a little mutagenicity after treatment with nitrite. Lin et al.[25] investigated the mutagenicity of soy sauce and found that in the Salmonella microsome test soy sauce treated with nitrite in the range 1,000 to 10,000 ppm was mutagenic in direct relation to nitrite concentration. However, soy sauce was not mutagenic in the test when it was treated with 100, 500, or 1000 ppm nitrite.[26] There is a report indicating that soy sauce is unlikely to be significantly mutagenic since it does not become mutagenic at a level of 50 ppm nitrite present in saliva and gastric juice.[27] Soy sauce treated with nitrite was found to be more mutagenic to *Escherichia coli* WP2 *uvrA*/pKM101 than *S. typhimurium* TA100 without S9 mix.[28] The mutagenicity of soy sauce treated with nitrite is affected by the concentration of soy sauce in the nitrosation mixture, and a concentration of 5% resulted in the highest activity. By incubating soy sauce at a concentration of 5% in a solution of 1 m*M* nitrite at pH 3 for 1 hr at 37°C, the equivalent of 1 ml of soy sauce induced 2790 revertants of *E. coli* WP2 *uvrA*/pKM101 without S9 mix. The effects of feeding male Fischer 344 rats with soy sauce and nitrite were studied.[29] This experiment showed that soy sauce alone and soy sauce plus nitrite affected the stomach mucosa of rats. There is a report showing that soy sauce alone induced mucus loss and altered the nuclear chromatin pattern of cells of the surface of the epithelium and gastric pit of rats.[30]

Namiki et al.[31] found that sorbic acid, a commonly used food additive, can react with nitrite to yield mutagens. They also tested the mutagenicity of spices after treatment with nitrite.[32,33] Among several spices treated with nitrite, pepper exhibited the strongest mutagenic activity by the Ames method, and nutmeg, chili pepper, and laurel strong activities. No mutagenicity was observed for spices alone. Mutagen production was observed between pH 2 and 6, with the maximum between pH 3 and 3.5, and the reaction was very fast at 40°C or above, even at the low levels of nitrite permitted by legal regulations. The mutagenicities of spice-nitrite reaction products were completely inactivated by S9 mix, but the activity of pepper-nitrite products toward TA100 remained unchanged.

Ohshima et al.[34] have screened European food items to identify those that show direct-acting genotoxicity after nitrosation *in vitro*. They found that nitrosation of smoked foods, the frequent consumption of which has been associated with an increased risk of stomach cancer,[35] leads to such direct-acting genotoxicity.

The interaction of orally administered drugs with nitrite under mildly acidic conditions has been considered from a safety point of view of the drugs. Common drugs including aminopyrine (tertiary amines) react with nitrite to form dimethylnitrosamine (or dialkylnitrosamines).[36,37] Several amine drugs have been tested as to whether they can produce nitrosamines, *N*-nitroso, and C-nitroso compounds.[38-49] Phenolic drugs including bamethan, acetaminophen, and etilefrin also became mutagenic on treatment with nitrite under mildly acidic conditions.[50-52]

Tobacco snuff, one form of smokeless tobacco, is commonly used as a substitute for smoking by several occupational groups, e.g., coal miners. In the general population, snuff dipping is becoming increasingly popular among young male athletes and students in high school and college.[53] Polar solvent extracts of tobacco snuff under acidic conditions were mutagenic in *S. typhimurium*.[54] After acid treatment, nitroso compounds in the amount corresponding to the nitrite concentration were detected. The mutagenic potency of the acid-treated extracts was consistent with the content of nitroso compounds generated. The results indicate that a nitrosation process was involved in snuff extracts during acid treatment. Studies related to the source of nitrite in tobacco snuff demonstrated that snuff contained bacteria which were able to reduce nitrate to nitrite and that the amount of nitrite in snuff extracts could be further increased by incubation of the extracts with the bacteria. Since snuff contains a considerable amount of nitrate, it seems that reduction of nitrate in snuff to nitrite by bacteria and nitrosation of certain constituents in snuff by nitrite under acidic conditions to form mutagenic nitroso compounds are possible mechanisms responsible for the acid-mediated mutagenicity of snuff extracts.

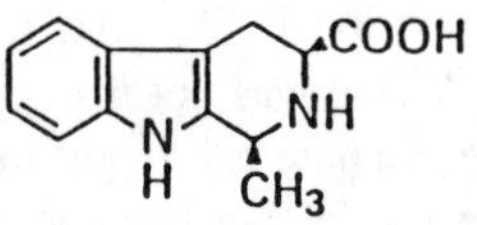

FIGURE 1.   Nitrosatable 1-methyl-1,2,3,4-tetrahydro-β-carboline-3-carboxylic acid in soy sauce.

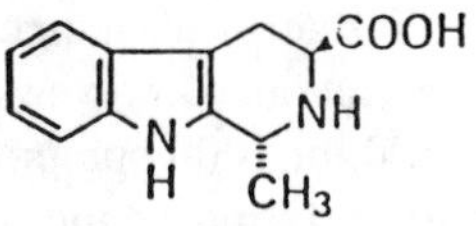

III. INDOLES AND RELATED COMPOUNDS

After treatment with nitrite, Japanese soy sauce was strongly mutagenic to *S. typhimurium* TA100 without S9 mix.[14] Two precursors of the mutagen were isolated from Japanese soy sauce, and these were identified as (−)-(1*S*,3*S*)-1-methyl-1,2,3,4-tetrahydro-β-carboline-3-carboxylic acid [(−)-1*S*,3*S*)-MTCA] (I) and its stereoisomer (−)-(1*R*,3*S*)-MTCA (II) (Figure 1). After treatment with 50 m*M* nitrite, 1-mg samples of these compounds induced 17,400 and 13,000 revertants of TA100 without S9 mix. Quantitative analysis of various kinds of soy sauces produced in Japan showed the presence of 82 to 678 μg of MTCA per milliliter. The mutagenicities of these compounds with nitrite accounted for 16 to 61% of the total mutagenicity of soy sauce with nitrite. The amounts of these precursors in most soy sauces produced in the U.S., of which the mutagenicities after nitrite treatment were weaker, were far less or below the detection limit. A major reaction product of (−)-(1*S*,3*S*)-MTCA and nitrite was a compound having a nitroso substitution at position N-2, but this compound was not mutagenic. Thus, the mutagen(s) formed from (−)-(1*S*,3*S*)-MTCA and nitrite was a minor product(s), and its specific mutagenic activity must be very high. The MTCAs are condensation products of L-tryptophan and acetaldehyde, and they are probably formed during fermentation and brewing of soy sauce. MTCA was also detected in Japanese sake.[55] The results of mutagen testing of reaction mixtures of nitrite and representative structures of tetrahydro-β-carbolines, tetrahydroisoquinolines, and tryptophols using *S. typhimurium* strains TA100, TA98, and TA97A were presented.[56] The highest mutagenic activity was observed with strain TA98 with the exception of the reaction mixture of 5-hydroxytryptophol which was more active with TA100.

Fava beans *(Vicia faba)*, upon treatment with nitrite under simulated gastric conditions, form a direct-acting bacterial mutagen, comparable in specific activity to the most potent known mutagens for several strains of *S. typhimurium*.[57] The precursor of the mutagen was isolated and identified as 4-chloro-6-methoxyindole (III). The major nitrosation product of III may be a stable α-hydroxy *N*-nitroso compound (IX) (Figures 2 and 3). Studies on nitrosation kinetics indicated that the nitrosation of indoles is a relatively fast reaction. Both the structural and rate studies give strong support to the hypothesis that intragastric nitrosation of fava beans yield this putative gastric carcinogen in the high-risk area in Colombia. Under identical conditions 4-chloroindole was found to yield two products, one of which showed similar mutagenicity. 5-Chloroindole, on the other hand, gave only nonmutagenic products.[57] On treatment with dilute mineral acid at slightly elevated temperatures these mutagenic nitrosamines are irreversibly converted to the corresponding 4-formylindazoles, and it is suggested that the nitrosation of indoles to yield indazoles generally proceeds via 2-hydroxy-*N'*-nitrosoindolin-3-one oximes.[58]

After treatment with nitrite, Chinese cabbage showed direct-acting mutagenicity on *S. typhimurium* TA100 inducing 3100 revertants per g.[13] Three precursors that became mutagenic after nitrite treatment were isolated and identified as indole-3-acetonitrile (IV), 4-methoxyindole-3-acetonitrile (V), and 4-methoxyindole-3-aldehyde (VI).[15,59] After treatment with nitrite, 1 mg of IV, V, and VI induced 17,400, 31,800 and 156,900 revertants of TA100 without S9 mix, respectively. IV was treated with 50 m*M* nitrite at pH 3.0 for 1 hr

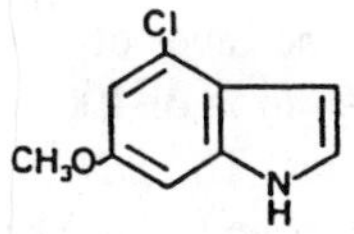

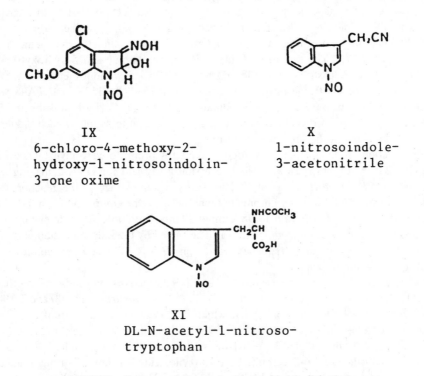

FIGURE 2. Naturally occurring nitrosatable indole derivatives.

III
4-chloro-6-methoxyindole
(fava beans)

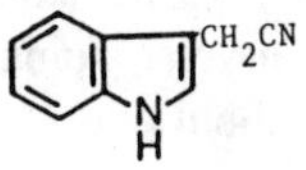

IV
indole-3-acetonitrile
(Chinese cabbage)

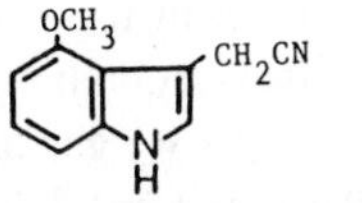

V
4-methoxyindole-3-
acetonitrile
(Chinese cabbage)

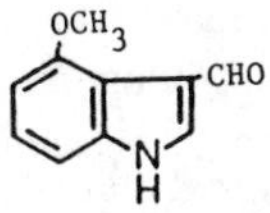

VI
4-methoxyindole-3-
acetaldehyde
(Chinese cabbage)

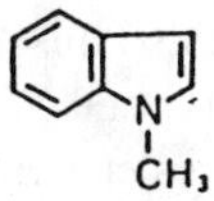

VII
1-methylindole
(cigarette smoke
condensate)

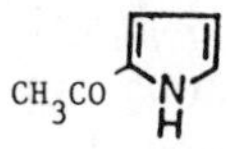

VIII
2-acetylpyrrole

IX
6-chloro-4-methoxy-2-
hydroxy-1-nitrosoindolin-
3-one oxime

X
1-nitrosoindole-
3-acetonitrile

XI
DL-N-acetyl-1-nitroso-
tryptophan

FIGURE 3. Structures of nitrosated products of indole derivatives.

and the resultant 1-nitrosoindole-3-acetonitrile (X) was isolated and identified.[60] It induced 45 and 30 revertants in the Salmonella mutation test without S9 mix using strains TA100 and TA98, respectively. The mutagenicity was reduced to one tenth by 100 μl S9.

Administration of X at doses of 40 to 300 mg/kg body weight by gastric intubation to male F344 rats induced up to a 100-fold increase in ornithine decarboxylase activity with a maximum after 24 hr and up to a 10-fold increase in DNA synthesis with a maximum after 16 hr in the pyloric mucosa of the stomach.[61] These results suggest that it has a potential tumor-promoting activity in carcinogenesis in the glandular stomach. Modification of DNA by X was investigated. ^{32}P-Postlabeling analysis clearly demonstrated the formation of DNA adducts in the stomach of rats after intragastric administration of X.[62] The level of DNA adducts in both the forestomach and glandular stomach 2 hr after administration of 100 mg/ kg body weight of the compound was about one adduct per 10^7 nucleotides. The DNAs of the forestomach and glandular stomach gave six common spots on two-dimensional chromatography, three of which were also produced by *in vitro* reaction of this compound with DNA. These results suggest that X has *in vivo* tumor-initiating activity in the stomach.

DL-*N*-Acetyl-*N'*-nitrosotryptophan (XI) and its methyl ester were readily formed under mild conditions by the reaction of nitrite with *N*-acetyltryptophan or its methyl ester.[63] Both compounds were assayed for mutagenicity in a series of *E. coli* WP2 strains and in several strains of *S. typhimurium,* in the presence and absence of S9 mix. XI was mutagenic to *E. coli* strains WP2, WP2*uvrA,* WP2pKM101, and WP2*uvrA*pKM101 and to *S. typhimurium* TA1535, TA98, and TA100. The methyl ester was consistently less mutagenic than XI to the *E. coli* strains, inactive in *S. typhimurium* TA98 and TA100, but more active than XI in TA1535. Addition of S9 did not enhance the mutagenicity of either compound and in some cases reduced the mutagenic effect.

Tryptophan treated with nitrite under acidic conditions was found to be mutagenic to *S. typhimurium* AT100 and TA98 both in the presence and absence of S9 mix.[64] Tryptamine, glycyltryptophan, and indole were also mutagenic when treated with nitrite, suggesting that the appearance of mutagenic activity from tryptophan was attributable to the reaction of nitrite with the indole ring. Three naturally occurring indoles (tryptophan, tryptamine, and 5-hyroxytryptamine) were nitrosated to products which were directly mutagenic to *S. typhimurium* TA1537.[65] The products of nitrosation of tryptamine and 5-hydroxytryptamine were also mutagenic to strains TA1538, TA98, and TA1535 without metabolic activation. The sensitivities of the frameshift-detecting strains TA1537, TA1538, and TA98 were of particular interest, since the nitroso compounds are characteristically base-substitution mutagens. The mutagenic effects of the products formed after nitrosation of each indole at pH 3.6 were eliminated in the presence of S9 mix.

The reaction products from L-tryptophan treated with nitrite were investigated for mutagenic activity with the *S. typhimurium* reversion assay and for DNA-damaging activity using the *Bacillus subtilis* rec-assay.[66] The diethyl ether extract of the reaction mixture showed eight spots on thin-layer chromatography (TLC). One compound from the TLC had a high mutagenic activity to strain TA98 without S9 mix, with little DNA-damaging activity. The mutagenic activity of the compound was determined by the induced mutation frequency method; the induced mutation frequency was about 19.2×10^{-5} at a concentration of 800 μg/ml for a 30-min treatment.

Indole and its 7 derivatives, L-tryptophan and its 9 derivatives, and carboline (norharman) and its 11 derivatives were tested for mutagenicity to *S. thypimurium* TA100 and TA98 after nitrite treatment.[67] 1-Methylindole (VII), which is present in cigarette smoke condensate, was the most mutagenic to TA100 without S9 mix after nitrite treatment, inducing 615,000 revertants/mg. 2-Methylindole, 1-methyl-DL-tryptophan, harmaline, and $(-)$-(1*S*,3*S*)-1,2-dimethyl-1,2,3,4-tetrahydro-β-carboline-3-carboxylic acid also showed strong mutagenicity after nitrite treatment, inducing 129,000, 184,000, 103,000 and 197,000 revertants/mg, respectively. These mutagenic potencies were comparable with those of benzo(*a*)pyrene, 3-methylcholanthrene, and 2-amino-9*H*-pyrido[2,3-*b*]indole (AαC). On 31 compounds tested, 22 were mutagenic after nitrite treatment. Since various indole compounds are ubiquitous in our environment, especially in plants, the presence of their mutagenicities after nitrite treatment warrants further studies, including those on their *in vivo* carcinogenicities.

As shown by Coughlin et al.,[68] Maillard reaction product 1-(*N*-L-tryptophan)-1-deoxy-D-fructose (Trp-Fru) can undergo nitrosation by nitrite under mildly acidic conditions to produce NO-Trp-Fru which displays considerable mutagenic activity in *S. typhimurium* strains TA98 and TA100. HeLa S3 cells in suspension were incubated at 37°C with various concentrations of Trp-Fru, its nitrosated analogue NO-Trp-Fru, and sodium nitrite, for varying periods of time, and were assayed for viability and for intracellular DNA, RNA, and protein synthesis.[69] None of the compounds tested had any effect on cell viability or on RNA and protein synthesis, apart perhaps from a slight inhibitory action. While Trp-Fru also remained ineffective as far as intracellular DNA synthesis was concerned, both NO-Trp-Fru and nitrite had a major effect on DNA synthesis. With nitrite, stimulation of DNA synthesis occurred at concentrations above 1 mM in the growth medium, but with NO-Trp-Fru synthesis increased at concentrations below 1 μM. The excess DNA synthesis (i.e., synthesis above control activity) observed with NO-Trp-Fru and also with nitrite was due to DNA repair. This was verified by keeping the cells under conditions that prevented normal semiconservative replication but permitted DNA repair ("unscheduled DNA synthesis").

Three 2-substituted pyrroles (2-acetylpyrrole (VIII), pyrrole-2-carboxaldehyde, and pyrrole-2-carboxylic acid), which are products of the Maillard browning reaction, were reacted with nitrite in buffer solution (pH 3) at 50°C for 24 hr.[70] The reaction mixtures were extracted with methylene chloride and the extracts were tested for mutagenicity using S. typhimurium strains TA97, TA98, TA100, TA102, and TA104, with and without S9 mix. The methylene chloride extract of the VIII-nitrite reaction mixture showed strong mutagenicity to all the tester strains, both in the presence and absence of S9 mix. The reaction product of pyrrole-2-carboxaldehyde with nitrite only gave a weak mutagenic response with strain TA100, while the pyrrole-2-carboxylic acid-nitrite reaction product did not produce a mutagenic response in any of the tester strains. Two mutagenically active fractions, separated by TLC, were found in the reaction of VIII with nitrite. The formation of mutagenic products in the latter reaction was found to vary with reaction pH, time and temperature, nitrite level, and VIII concentration.

It is known that nitrosation of indoles occurs preferentially at the C-3 position, but if this position is substituted, nitrosation occurs at the N-1 position.[71] In accordance with this preference, the indole derivatives mentioned above reacted with nitrous acid to yield mutagenic nitrosoindoles. In addition, the nitrosation of indoles proceeded very rapidly at physiologically feasible concentrations of nitrite found in saliva or gastric juice.

IV. PHENOLICS

Tyramine was newly identified as a mutagen precursor in Japanese soy sauce.[72] The mutagenic compound was determined to be 4-(2-aminoethyl)-6-diazo-2,4-cyclohexadienone(3-diazotyramine) (XII) (Figure 4), and its specific mutagenic activity was 112 revertants/μg towards S. typhimurium TA100 without S9 mix. The yield of XII from 5 mM tyramine and 50 mM nitrite at pH 1 for 1 hr at 37°C was around 25%. The XII concentration decreased by 40% when kept as an aqueous solution at 22°C in the dark for 4 days. The nonmutagenic product was identified as 3-nitrotyramine. The effect of nitrite concentration on the formation of XII from tyramine and nitrite was also studied. When 5 mM tyramine was reacted with 10, 25, and 50 mM nitrite, XII formation was increased dose dependently. However, XII was not produced by incubation of 5 mM tyramine with an equivalent concentration of nitrite. Nonmutagenic 3-nitrotyramine was produced under all of the conditions tested. The tyramine levels in eight kinds of Japanese soy sauce ranged from 10 to 2250 μg/mℓ.[22,72] Tyramine is formed by decarboxylation of tyrosine. On the basis of the annual consumption of soy sauce in Japan, the intake by Japanese is estimated to be around 27 mℓ per person per day (Food Balance Sheet in Japan, 1981). Therefore, the daily ingestion of tyramine from soy sauce is calculated to be at most about 60 mg. Besides soy sauce, tyramine is also found in several foods, such as cheese, soybean paste, meat extract, and beer.[73-75] Some kinds of cheese contain large amounts of tyramine, their levels being comparable to those in Japanese soy sauce.

Soy sauces and fish sauces in Korea, the Philippines, and Thailand also showed direct-acting mutagenicity after nitrite treatment.[22] They contained as much tyramine as 17 to 1020

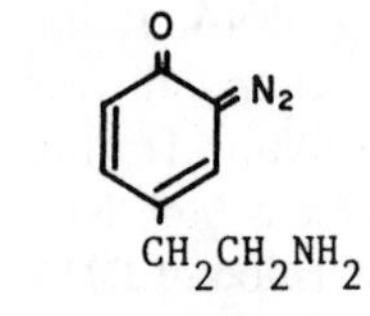

XII
3-diazotyramine

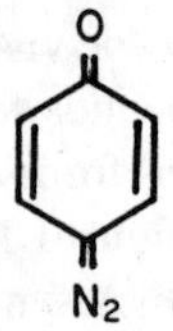

XIII
p-diazoquinone

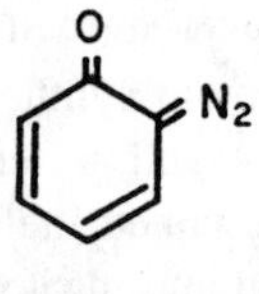

XIV
o-diazoquinone

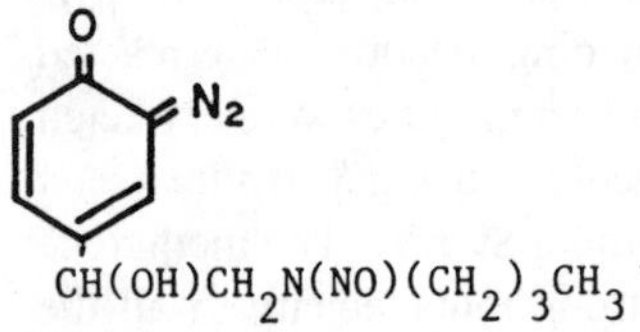

XV
3-diazo-N-nitroso-
bamethan

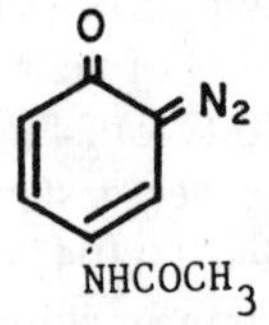

XVI
3-diazoacetaminophen

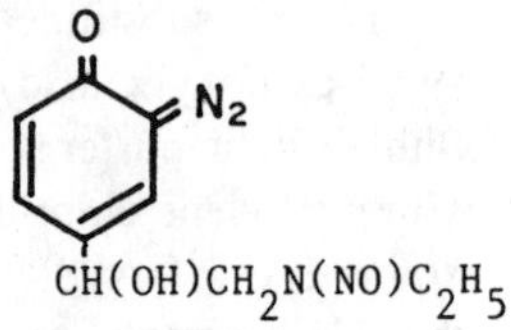

XVII
3-diazo-N-nitroso-
etilefrin

FIGURE 4.   Structures of nitrosated products of phenolics.

TABLE 4
Mutagenicities after Treatment with Nitrite and Tyramine
Contents in Soy and Fish Sauces[22]

	Country of production	Rev/ml original materials	Tyramine (μg/ml original materials)
Soy sauce			
A	Korea	5,900	17
B	Korea	5,500	25
C	Korea	21,300	1,020
D	Korea	7,500	570
E	Korea	2,700	24
F	Korea	14,800	590
G	Philippines	3,000	34
H	Philippines	3,400	61
I	Philippines	3,500	25
Fish sauce			
A	Philippines	2,700	30
B	Philippines	1,800	66

Note: Mutagenicity was tested on TA100 without S9 mix.

μg/ml (Table 4), and the mutagenicity expected by tyramine was 1 to 30% of the total mutagenicity of the soy and fish sauces after nitrite treatment. MTCA (I and II) was detected only in one soy sauce (B), in which its concentration was only 10 μg/ml.

Tahira et al.[28] suggested that some unknown precursor(s) may be present in soy sauce because MTCA and tyramine could not account for the mutagenicity of soy sauce treated with a low concentration of nitrite. It has been reported that soy sauce contains tyramine, MTCA, 1-methyl-1,2,3,4-tetrahydro-β-carboline, and a factor that augments the mutagenicity of XII ninefold and that tyramine and the augment factor account for almost all mutagenicity of soy sauce treated with 50 mM nitrite, of 25 brands produced in Japan.[76]

Carcinogenicity of XII was tested in male Fischer 344 rats. The animals, 6 weeks old at the start of the experiment, were given *ad libitum* 0.1% XII dissolved in deionized water in a light-proof container throughout the experiment. The experiment was terminated after 116 weeks. Squamous cell carcinomas of the oral cavity developed in 19 of the 28 treated rats.[77] These carcinomas originated in the epithelium of the floor of the oral cavity, close to the root of the tongue. No tumors were observed in any other organs. Thus, XII specifically induced oral cavity cancer under the above conditions.

When phenol was treated with four equivalents of nitrite in aqueous solution, three compounds were produced. They were nonmutagenic *p*-nitrosophenol and mutagenic *p*-diazoquinone (XIII) and *o*-diazoquinone (XIV).[78] XIII showed direct-acting mutagenicity in *S. typhimurium* TA100 and TA98 and also genotoxicity on the SOS Chromotest.[34,78] XIV exhibited direct-acting genotoxicity in the SOS Chromotest but no appreciable mutagenicity in Salmonella strains. When treated with an equimolar amount of nitrite little mutagenicity was observed and nonmutagenic *p*-nitrosophenol was the preferred product. *p*-Nitrosophenol was furthermore shown to react with additional nitrite to produce XIII.[78] *o*-Nitrosophenol was not detected under any of the above conditions. However, *o*-nitrosophenol is also presumed to be formed with phenol and nitrite. This nitroso compound would be very rapidly converted to XIV in the presence of nitrite (see Chapter 10.2).

After treatment with 50 m*M* nitrite, aqueous extracts of various smoked fish and meat products, including smoked salmon, herring, halibut, and ham, showed direct-acting genotoxicity in the SOS Chromotest.[34] Similar genotoxic properties were also observed in nitrosated aqueous extracts of various kinds of wood-smoke condensates. A variety of phenolic compounds, including phenol, catechol, and vanillin, were identified as being the nitrosatable precursors of genotoxic substances in the commercial hickory-smoke condensate[34] (See Chapter 5.3 for further discussion).

Drugs with phenol moiety were demonstrated to show nitrosatable activity. Bamethan [1-(4-hydroxyphenol)-1-hydroxy-2-butylaminoethane], which is used to treat cardiovascular diseases, reacted with nitrite and showed direct-acting mutagenicity, inducing 5,800 revertants in TA98 and 53,000 revertants in TA100 per micromole bamethan.[50] This drug is administered orally for long periods. A typical bamethan dose is about 75 mg/day.

A mixture of 2 mmol bamethan and 8 mmol sodium nitrite was incubated in 50 ml aqueous solution at pH 3 at 37°C for 4 hr, after which the products were separated by HPLC. Two compounds, a mutagen and a nonmutagen, were produced as the major and minor products, respectively. The structure of the mutagen was determined to be 1-(3-diazo-4-oxo-1,5-cyclohexadienyl)-1-hydroxy-2-(*N*-nitrosoamino)ethane (3-diazo-*N*-nitrosobamethan) (XV), and the nonmutagen was 1-(4-hydroxyphenyl)-1-hydroxy-2-(*N*-nitrosoamino)ethane (*N*-nitrosobamethan).[50] Direct-acting mutagen XV was easily decomposed to non or very weak mutagenic compound(s) by exposure to air and light. If an equimolar (2 mmol) mixture of bamethan and nitrite was incubated, the nonmutagenic mononitroso-substituted compound, *N*-nitrosobamethan, was mainly produced.

To test the tumor-initiating and -promoting activities of XV in the glandular stomach of rats, the induction of unscheduled DNA synthesis, ornithine decarboxylase, and replicative DNA synthesis were measured after administration by gastric intubation.[79] When doses of 75, 120, and 225 mg/kg body weight were given, unscheduled DNA synthesis was induced dose dependently in the pyloric mucosa of the stomach 2 hr after administration. XV also clearly induced ornithine decarboxylase activity and replicative DNA synthesis in the pyloric mucosa of the rat stomach, reaching a maximum 16 hr after administration at doses of 75 to 225 mg/kg body weight. Thus, this suggested that this diazo compound has tumor-initiating and -promoting activities in the glandular stomach of the rat.

Acetaminophen (paracetamol) showed mutagenicity to *S. typhimurium* TA100 and TA98 either with or without S9 mix when treated with excess nitrite in acidic solution. The mutagen

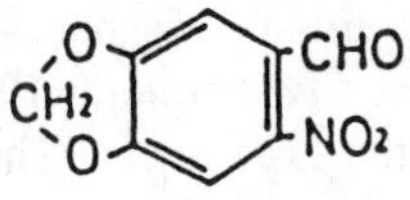

XXI
6-nitropiperonal

FIGURE 5.   Structure of nitrosated product of piperine.

was isolated and identified as 4-acetylamino-6-diazo-2,4-cyclohexadienone (3-diazo-acet-aminophen) (XVI).[51]

Reaction of an antihypotensive drug, etilefrin [α-{(ethylamino)methyl}-*m*-hydroxybenzyl alcohol], with nitrite under mildly acidic conditions produced α-{(N-nitrosoethylamino)methyl}-*m*-hydroxybenzyl alcohol (*N*-nitrosoetilefrin) and 1-(4-diazo-3-oxo-1,5-cyclohexadienyl-2-(*N*-nitrosoethylamino)ethanol (3-diazo-*N*-nitrosoetilefrin) (XVII).[52] The latter showed mutagenicity to *S. typhimurium* TA98 and TA100 strains without S9 mix. Specific mutagenic activity of the compound was 300 revertants for both TA98 and TA100 strains with a dose of 1.0 μmol. Addition of a S9 mix affected the activity little.

Butylated hydroxyanisole, a food additive, was treated with two equivalents of nitrite in aqueous ethanol at pH 2 for 1 hr at ambient temperature. Two direct-acting mutagens towards TA100 and TA98, namely 2-*tert*-butyl-*p*-quinone and its dimer [3,3′-di-*tert*-butyl-biphenylquinone-(2,5,2′,5′)] were produced.[80] This is an example of nitrite acting as an oxidizing agent rather than as a nitrosating agent. Production of DNA-breaking substance on treatment of monophenols with nitrite and then with dimethyl sulfoxide has been demonstrated.[81]

The mutagenic nitrosated products of phenol derivatives including tyramine, bamethan, and phenol were demonstrated to be diazo compounds. The reaction sequence proceeds via *o*- and/or *p*-nitrosophenol compounds. Then these nitrosophenol compounds are converted to diazo derivatives by further reaction with nitrite. Thus, relatively high amounts of nitrite may be required for the formation of the diazo compounds *in vitro* and presumably also *in vivo* in the stomach. Tyramine, bamethan, and etilefrin have a common partial structure: a phenolic structure with an aminoethyl group at the *p*- or *o*-position. All three compounds have been shown to be diazotized by reaction with nitrite into *ortho* diazoquinone derivatives regardless of whether the amino group is nitrosated, and the diazoquinone derivatives with aminoethyl groups were highly mutagenic to *S. typhimurium* strains.

V. SORBIC ACID AND PIPERINE

Sorbic acid, a food preservative, reacts rapidly with nitrite in acidic media.[31,82-84] The reaction afforded at least ten compounds, among which ethylnitrolic acid (XVIII), 1,4-dinitro-2-methylpyrrole (XIX), and a furoxan derivative (XX) and its precursors were obtained. It should be noted that XVIII and XIX are mutagenic.[31]

Osawa et al. investigated the pepper-nitrite system[32] and found the formation of highly active mutagens other than *N*-nitrosopiperidine (Figure 5). Strong mutagenic activity was detected, especially in a nitrite-treated piperine solution, by the rec-assay and the Ames test.[85] Because the same mutagenicity profile was observed in a piperic acid-nitrite reaction mixture, a large-scale reaction of nitrite with piperic acid was carried out and four mutagenic reaction products were isolated. One of the main mutagens was identified as 6-nitropiperonal (XXI). Mutagen formation by nitrite-spice reaction has been extensively investigated.[33]

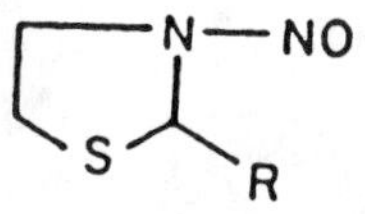

XXII
N-nitrosothiazolidines

FIGURE 6. *N*-Nitrosothiazolidines.

VI. THIAZOLIDINES AND THIOPROLINE

Sakaguchi and Shibamoto[86] observed the formation of thiazolidine and its alkyl derivatives by heating an aqueous cysteamine and acetaldehyde mixture. Thiazolidines have a strong roasted flavor and can be readily nitrosated with nitrite. The resultant nitroso derivatives showed some mutagenicity on *S. typhimurium*.[87] *N*-Nitrosothiazolidine was detected in fried bacon.[88] The presence in bacon appears to be associated with smokehouse processing.[89-91]

The effect of the alkyl side-chain substituents in *N*-nitrosothiazolidines (XXII) on their mutagenicities was investigated (Figure 6).[92,93] Some showed positive mutagenic responses toward *S. typhimurium* TA100, and the order of mutagenic potency relative to their 2-alkyl substituents was as follows: unsubstituted > isopropyl > propyl > ethyl > butyl > isobutyl > methyl. The metabolic activation system was not required to detect mutagenicity. In fact, the addition of S9 mix strongly suppressed the mutagenic activity of XXII.

Fiddler et al.[94] have reported that *N*-nitrosothiazolidine is not mutagenic to *S. typhimurium* TA100 over a 2-log dose range.

Ohshima et al.[95] and Tsuda et al.[96,97] found that *N*-nitrosothioproline (XXIII) and *N*-nitroso-2-methylthioproline are the main *N*-nitroso compounds usually present in human urine. These sulfur-containing *N*-nitrosamino acids can be formed by *in vivo* nitrosation of the corresponding amino acids and are excreted without further metabolism.[98] Tsuda et al. found that cigarette smoking increased the urinary levels of these *N*-nitrosamino acids.[99,100] XXIII is nonmutagenic to *S. typhimurium* strains[101] and probably noncarcinogenic, like *N*-nitrosoproline[102,103] which is also an *N*-nitroso compound usually found in human urine.[98-100] The *in vivo* nitrosation of thioproline (XXIV) is 1000-fold faster than that of proline.[101] XXIV is thus an effective nitrite-trapping agent in the human body.

When five volunteers were given food containing cod and vegetables (a traditional Japanese food, called *tara-chiri*), their urinary excretion of XXIII increased from 7.9 ± 4.2 (S.D.) μg/day to 110 ± 64.5 μg/day.[104] This increase was accounted for by *in vivo* nitrosation of XXIV by nitrite formed from nitrate in the vegetables. This finding was confirmed by results on a volunteer who ate boiled cod and Japanese radish (daikon) (a simple version of the food containing cod and vegetables). Boiled cod was found to contain 300 to 500 μg/100 g of XXIV and the level nearly doubled when the cod was boiled with Japanese radish. This increase occurred during the cooking of cod with Japanese radish by the reaction of formaldehyde in the cod with cysteine in the radish. The nitrosation of XXIV was estimated to be 1000-fold that of proline in the human body. Thus XXIV is a very sensitive probe of *in vivo* nitrosation. XXIV formation either *in vivo* or *in vitro* may have the following two roles in reducing tumorigenesis in humans: (1) detoxication of formaldehyde, which is genotoxic and (2) blocking the formation of carcinogenic *N*-nitroso compounds by trapping nitrite and then being excreted in the urine (Figure 7).

VII. HETEROCYCLIC AMINES

Aminoimidazoquinoline, aminoimidazoquinoxaline, and aminoimidazopyridine com-

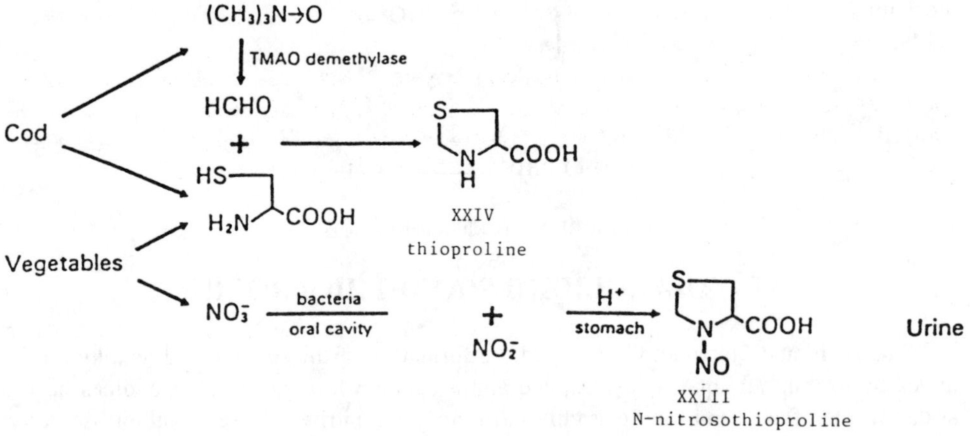

FIGURE 7.   Possible pathways of formation of thioproline and *N*-nitrosothioproline.

pounds, which are found in cooked foods and are mutagens to bacteria with S9 mix before nitrite treatment, became mutagenic to TA98, TA100, and *E. coli* WP2 *uvrA* without S9 mix following treatment with 50 m*M* nitrite.[105] Formation of a nitro derivative by nitrite treatment was observed with mutagenic and carcinogenic 2-amino-3-methylimidazo[4,5-*f*]-quinoline (IQ). IQ was converted to 3-methyl-2-nitroimidazo[4,5-*f*]quinoline, showing mutagenicity towards Salmonella strains without S9 mix, after 50 m*M* nitrite treatment at pH 3. In contrast, treatment of IQ with a much lower amount of nitrite (2 m*M*) produced no effect.[106] The other aminoimidazoquinoline, aminoimidazoquinoxaline, and aminoimidazopyridine compounds must react with nitrite in a similar manner to IQ.

VIII. CONCLUSION

In this chapter, the occurrence of nitrosatable compounds such as indole and phenol derivatives in foods and drugs and the structures and biological properties of their nitrosated products are discussed.

The indoles are among the most rapidly nitrosated compounds found in foods, so there is a real possibility that nitrosoindoles are formed in the human stomach under normal conditions. Direct-acting nitrosoindole compounds forms DNA adducts and also induce ornithine decarboxylase activity in the rat stomach. Therefore, it is very important to elucidate the carcinogenicity of nitrosoindoles, which so far is not known.

Among diazo compounds showing direct-acting mutagenicity, 3-diazotyramine was shown to be carcinogenic in rats. It was also suggested that 3-diazo-*N*-nitrosobamethan has tumor-initiating and -promoting activities in rat glandular stomach mucosa. Thus, diazo compounds might induce damage in DNA and other biomolecules in the stomach and other organs if the compounds are produced endogenously and also exogenously. However, relatively high amounts of nitrite are required to form diazo compounds from phenolic compounds under acidic conditions. Therefore, the formation of diazo compounds in the human stomach could be limited to special cases.

Thioproline is of special interest. The nitrosation of the compound was very rapid, and *N*-nitrosothioproline is nonmutagenic. It may act by blocking the formation of carcinogenic or mutagenic *N*-nitroso or diazo compounds by trapping nitrite and then being excreted in the urine.

Nitrite is reported to inactivate the mutagenicity of heterocyclic amines other than IQ type by converting the amino-group to a hydroxy-group.[106] Thus, the reaction of nitrite under mildly acidic conditions may have two aspects: one is the production of mutagens and the other is the destruction of mutagens.

Since stomach cancer remains such a common neoplastic disease in many parts of the world, investigation must be continued to elucidate whether nitrosatable compounds are involved in the development of human cancer, particularly of the stomach.

REFERENCES

1. **Fine, D. H., Challis, B. C., Hartman, P., and Van Ryzin, J.**, Endogenous syntheses of volatile nitrosamines: model calculations and risk assessment, in *N-Nitroso Compounds: Occurrence and Biological Effects*, Bartsch, H., O'Neill, I. K., Castegnaro, M., Okada, M., and Davis, W., Eds., IARC Scientific Publ. No. 41, International Agency for Research on Cancer, Lyon, 1982.
2. **Spiegelhalder, B., Eisenbrand, G., and Preussmann, R.**, Influence of dietary nitrate on nitrite content on human saliva: possible relevence to in vivo formation of *N*-nitroso compounds, *Food Cosmet. Toxicol.*, 14, 545, 1976.
3. **Tannenbaum, S. R., Weismann, M., and Fett, D.**, The effect of nitrate intake on nitrite formation in human saliva, *Food Cosmet. Toxicol.*, 14, 549, 1976.
4. **Marletta, M. A.**, Nitric oxide: biosynthesis and biological significance, *TIBS*, 14, 488, 1989.
5. **Sander, J. and Burkle, G.**, Induktion maligner Tumoren bei Ratten durch gleichzeitige Verfutterung von Nitrit und sekundaren Aminen, *Z. Krebsforsch.*, 73, 54, 1969.
6. **Mirvish, S. S.**, Formation of *N*-nitroso compounds: chemistry, kinetics and in vivo occurrence, *Toxicol. Appl. Pharmacol.*, 31, 325, 1975.
7. **Sugimura, T. and Fujimura, S.**, Tumour production in glandular stomach of rats by *N*-methyl-*N'*-nitro-*N*-nitrosoguanidine, *Nature (London)*, 216, 943, 1967.
8. **Sugimura, T. and Kawachi, T.**, Experimental stomach cancer, in *Methods in Cancer Research*, Vol. VII, Busch, H., Ed., Academic Press, New York, 1967.
9. **Drückrey, H., Ivankovic, S., and Preussmann, R.**, Selective Erzeugung von Carcinomen des Drusen-magens bei Ratten durch orale Gabei von N-methyl-N-nitroso-N'-acetylharnstoff (AcMNH), *Z. Krebs-forsch.*, 74, 23, 1970.
10. **Hirota, N., Aunuma, T., Yamada, S., Kawai, T., Saito, K., and Yokoyama, T.**, Selective induction of glandular stomach carcinoma in F344 rats by *N*-methyl-*N*-nitrosourea, *Jpn J. Cancer Res. (Gann)*, 78, 634, 1987.
11. **Mirvish, S. S.**, The etiology of gastric cancer. Intragastric nitrosamide formation and other theories, *J. Natl Cancer Inst.*, 71, 631, 1983.
12. **Piacek-Llanes, B. G. and Tannenbaum, S. R.**, Formation of an activated *N*-nitroso compound in nitrite-treated fava beans *(Vicia faba)*, *Carcinogenesis*, 3, 1379, 1982.
13. **Marquardt, H., Rufino, F., and Weisburger, J. H.**, Mutagenic activity of nitrite-treated foods: human stomach cancer may be related to dietary factors, *Science*, 196, 1000, 1977.
14. **Wakabayashi, K., Ochiai, M., Saito, H., Tsuda, M., Suwa, Y., Nagao, M., and Sugimura, T.**, Presence of 1-methyl-1,2,3,4-tetrahydro-β-carboline-3-carboxylic acid, a precursor of a mutagenic nitroso compound, in soy sauce, *Proc. Natl. Acad. Sci. U.S.A.*, 80, 2912, 1983.
15. **Wakabayashi, K., Nagao, M., Ochiai, M., Tahira, T., Yamaizumi, Z., and Sugimura, T.**, A mutagen precursor in Chinese cabbage, indole-3-acetonitrile, which becomes mutagenic on nitrite treatment, *Mutat. Res.*, 143, 17, 1985.
16. **Weisburger, J. H. and Raineri, R.**, Assessment of human exposure and response to N-nitroso compounds. New view on the etiology of digestive tract cancers, *Toxicol. Appl. Pharmacol.*, 31, 369, 1975.
17. **Weisburger, J. H., Marquardt, H., Hirota, H., Mori, H., and Williams, G. M.**, Induction of cancer of the glandular stomach in rats by an extract of nitrite-treated fish, *J. Natl. Cancer Inst.*, 64, 163, 1980.
18. **Yano, K.**, Alkylating activity of processed fish products treated with sodium nitrite in simulated gastric juice, *Jpn. J. Cancer Res. (Gann)*, 72, 451, 1981.
19. **Correa, P., Cuello, C., Fajardo, L. F., Haenszel, W., Bolanos, O., and de Ramirez, B.**, Diet and gastric cancer: nutrition survey in a high-risk area, *J. Natl. Cancer Inst.*, 70, 673, 1983.
20. **Van der Hoeven, J. C. M., Lagerweij, W. J., Van Gastel, A., Huitink, J., De Dreu, R., and Van Broekhoven, L. W.**, Intercultivar difference with respect to mutagenicity of fava beans *(Vicia faba L)* after incubation with nitrite, *Mutat. Res.*, 130, 391, 1984.

21. **Hashizume, T., Yokoyama, T., Kamiki, T., Nakamura, Y., Kinae, N., and Tomita, I.,** Mutagenicity of daily foodstuffs and the effect of nitrite treatment, *J. Jpn. Food Hyg. Soc.,* 24, 369, 1983.

22. **Wakabayashi, K., Nagao, M., Chung, T. H., Yin, M., Karai, I., Ochiai, M., Tahira, T., and Sugimura, T.,** Appearance of direct-acting mutagenicity of various foodstuffs produced in Japan and Southeast Asia in nitrite treatment, *Mutat. Res.,* 158, 119, 1985.

23. **Yano, M., Wakabayashi, K., Tahira, T., Arakawa, N., Nagao, M., and Sugimura, T.,** Presence of nitrosable mutagen precursors in cooked meat and fish, *Mutat. Res.,* 202, 119, 1988.

24. **Hayatsu, H. and Hayatsu, T.,** Mutagenicity arising from boiled rice on treatment with nitrous acid, *Jpn. J. Cancer Res. (Gann),* 80, 1021, 1989.

25. **Lin, J. Y., Wang, H.-I., and Yen, Y.-C.,** The mutagenicity of soy sauce, *Food Cosmet. Toxicol.,* 17, 329, 1979.

26. **Shibamoto, T.,** Possible mutagenic constituents in nitrite-treated soy sauce, *Food Chem. Toxicol.,* 21, 745, 1983.

27. **Nagahara, A., Ohshita, K., and Nasuno, S.,** Relation of nitrite concentration to mutagen formation in soy sauce, *Food Chem. Toxicol.,* 24, 13, 1986.

28. **Tahira, T., Fujita, Y., Ochiai, M., Wakabayashi, K., Nagao, M., and Sugimura, T.,** Mutagenicity of soy sauce treated with a physiologically feasible concentration of nitrite, *Mutat. Res.,* 174, 255, 1986.

29. **Nagao, M., Wakabayashi, K., Fujita, Y., Tahira, T., Ochiai, M., Takayama, S., and Sugimura, T.,** Nitrosatable precursors of mutagens in vegetables and soy sauce, in *Diet Nutrition and Cancer.,* Hayashi, Y., Nagao, M., Sugimura, T., Takayama, S., Tomatis, L., Wattenberg, L. W., and Wogan, G. N., Eds., Japan Scientific Societies Press, Tokyo, VNU Science Press, Utrecht, 1986.

30. **MacDonald, W. C. and Dueck, J. W.,** Long-term effect of shoyu (Japanese soy sauce) of the gastric mucosa on the rat, *J. Natl. Cancer Inst.,* 56, 1143, 1976.

31. **Namiki, M., Osawa, T., Ishibashi, H., Namiki, K., and Tsuji, K.,** Chemical aspects of mutagen formation by sorbic acid-sodium nitrite reaction, *J. Agric. Food Chem.,* 29, 407, 1981.

32. **Osawa, T., Ishibashi, H., Namiki, M., Yamanaka, M., and Namiki, K.,** Formation of mutagens by pepper-nitrite reaction, *Mutat. Res.,* 91, 291, 1981.

33. **Namiki, K., Yamanaka, M., Osawa, T., and Namiki, M.,** Mutagen formation by nitrite-spice reaction, *J. Agric. Food Chem.,* 32, 948, 1984.

34. **Ohshima, H., Friesen, M., Malaveille, C., Brouet, I., Hautefeuille, A., and Bartsch, H.,** Formation of direct-acting genotoxic substances in nitrosated smoked fish and meat products: identification of simple phenolic precursors and phenyldiazonium ions as reactive products, *Food Chem. Toxicol.,* 27, 193, 1989.

35. **Howson, C. P., Hiyama, T., and Wynder, E. L.,** The decline in gastric cancer: epidemiology of unplanned triumph, *Epidemiol. Rev.,* 8, 1, 1986.

36. **Lijinsky, W., Conrad, E., and Van de Bogart, R.,** Carcinogenic nitrosamines formed by drug/nitrite interactions, *Nature (London),* 39, 165, 1972.

37. **Lijinsky, W.,** Reaction of drugs with nitrous acid as a source of carcinogenic nitrosamines, *Cancer Res.,* 34, 255, 1974.

38. **Rao, G. S. and Krishna, G.,** Drug-nitrite interactions: formation of N-nitroso, C-nitroso, and nitro compounds from sodium nitrite and various drugs under physiological conditions, *J. Pharm. Sci.,* 64, 1579, 1975.

39. **Gold, B. and Mirvish, S. S.,** N-Nitroso derivatives of hydrochlorothiazine, niridazole, and tolbutamide, *Toxicol. Appl. Pharmacol.,* 40, 131, 1977.

40. **Arisawa, M., Fujiu, M., Suhara, Y., and Maruyama, H. B.,** Differential mutagenicity of reaction products of various pyrazolones with nitrite, *Mutat. Res.,* 57, 287, 1978.

41. **Andrews, A. W., Fornwald, J. A., and Lijinsky, W.,** Nitrosation and mutagenicity of some amine drugs, *Toxicol. Appl. Pharmacol.,* 52, 237, 1980.

42. **Lijinsky, W., Reuber, M. D., and Blackwell, B.-N.,** Liver tumors induced in rats by oral administration of the antihistaminic methapyrilene hydrochloride, *Science,* 209, 817, 1980.

43. **Takeda, Y. and Kanaya, H.,** Formation of nitroso compounds and mutagens from tranquilizers by drug/ nitrite interaction, *Cancer Lett.,* 12, 81, 1981.

44. **Takeda, Y. and Kanaya, H.,** Formation of nitroso compounds and mutagens from cinnarizine, ethambutol, piromidic acid, pyridinol carbamate and tiaramide by drug/nitrite interaction, *Cancer Lett.,* 15, 53, 1982.

45. **Kanaya, H. and Takeda, Y.,** N-Mononitrosopyridinol carbamate: the principal mutagen formed by nitrosation of pyridinol carbamate, *Cancer Lett.,* 18, 143, 1983.

46. **Andrews, A. W., Lijinsky, W., and Snyder, S. W.,** Mutagenicity of amine drugs and their products of nitrosation, *Mutat. Res.,* 135, 105, 1984.

47. **Gillatt, P. N., Hart, R. J., and Walters, C. L.,** Susceptibilities of drugs to nitrosation under standardized chemical conditions, *Food Chem. Toxicol.,* 22, 269, 1984.

48. **Gillatt, P. N., Palmer, R. C., Smith, P. L. R., and Walters, C. L.,** Susceptibilities of drugs to nitrosation under simulated gastric conditions, *Food Chem. Toxicol.,* 23, 849, 1985.

83

49. **Kammerer, R. C., Froines, J. R., and Price, T.,** Mutagenicity studies of selected antihistamines, their metabolites and products of nitrosation, *Food Chem. Toxicol.,* 24, 981, 1986.
50. **Kikugawa, K., Kato, T., and Takeda, Y.,** Formation of a highly mutagenic diazo compound from the bamethan-nitrite reaction, *Mutat. Res.,* 177, 35, 1987.
51. **Ohta, T., Oribe, H., Kameyama, T., Goto, Y., and Takitani, S.,** Formation of a diazoquinone-type mutagen from acetaminophen treated with nitrite under acidic conditions, *Mutat. Res.,* 209, 95, 1988.
52. **Kikugawa, K., Kato, T., and Takeda, Y.,** Formation of a direct mutagen, diazo-N-nitrosoetilefrin, by interaction of etilefrin with nitrite, *Chem. Pharm. Bull.,* 37, 1600, 1989.
53. **Christen, A. G.,** The case against smokeless tobacco: five facts for the health professional to consider, *J. Am. Dent. Assoc.,* 101, 464, 1980.
54. **Whong, W.-Z., Stewart, J. D., Wang, Y.-K., and Ong, T.,** Acid-mediated mutagenicity of tobacco snuff: its possible mechanism, *Mutat. Res.,* 117, 241, 1987.
55. **Sato, S., Tadenuma, M., Takahashi, K., and Nakamura, N.,** Configuration of sake taste. VI. Rough taste components. 3. Tetrahydroharman-3-carboxylic acid, *Nippon Jozo Kyokai Zasshi,* 70, 821, 1975.
56. **Valin, N., Haybron, D., Groves, L., and Mower, H. F.,** The nitrosation of alcohol-induced metabolites produces mutagenic substances, *Mutat. Res.,* 158, 159, 1985.
57. **Yang, D., Steven, R., Tannenbaum, S. R., Büchi, G., and Lee, G. C. M.,** 4-Chloro-6-methoxyindole is the precursor of a potent mutagen (4-chloro-6-methoxy-2-hydroxy-1-nitroso-indole-3-one oxime) that forms during nitrosation of the fava bean *(Vicia faba), Carcinogenesis,* 5, 1219, 1984.
58. **Büchi, G., Lee, G. C. M., Yang, D., and Tannenbaum, S. R.,** Direct acting, highly mutagenic, α-hydroxy N-nitrosamines from 4-chloroindoles, *J. Am. Chem. Soc.,* 108, 4115, 1986.
59. **Wakabayashi, K., Nagao, M., Tahira, T., Yamaizumi, Z., Katayama, M., Marumo, S., and Sugimura, T.,** 4-Methoxy indole derivatives as nitrosable precursors of mutagens in Chinese cabbage, *Mutagenesis,* 1, 423, 1986.
60. **Wakabayashi, K., Nagao, M., Tahira, T., Saito, H., Katayama, M., Marumo, S., and Sugimura, T.,** 1-Nitrosoindole-3-acetonitrile, a mutagen produced by nitrite treatment of indole-3-acetonitrile, *Proc. Jpn. Acad. Ser. B,* 61, 190, 1985.
61. **Furihata, C., Sato, Y., Yamakoshi, A., Takimoto, M., and Matsushima, T.,** Induction of ornithine decarboxylase and DNA synthesis in rat stomach mucosa by 1-nitrosoindole-3-acetonitrile, *Jpn. J. Cancer Res. (Gann),* 78, 432, 1987.
62. **Yamashita, K., Wakabayashi, K., Kitagawa, Y., Nagao, M., and Sugimura, T.,** ^{32}P-Postlabeling analysis of DNA adducts in rat stomach with 1-nitrosoindole-3-acetonitrile, a direct-acting mutagenic indole compound formed by nitrosation, *Carcinogenesis,* 9, 1905, 1988.
63. **Venitt, S., Crofton-Sleigh, C., Ooi, S. L., and Bonnett, R.,** Mutagenicity of nitrosated α-amino acid derivatives N-acetyl-N'-nitroso-tryptophan and its methyl ester in bacteria, *Carcinogenesis,* 1, 523, 1980.
64. **Ohta, T., Isa, M., Suzuki, Y., Yamahata, N., Suzuki, S., and Kurechi, T.,** Formation of mutagens from tryptophan by the reaction with nitrite, *Biochem. Biophys. Res. Commun.,* 100, 52, 1981.
65. **Gatehouse, D. and Wedd, D.,** The bacterial mutagenicity of three naturally occurring indoles after reaction with nitrous acid, *Mutat. Res.,* 124, 35, 1983.
66. **Ohara, A., Mizuno, M., Danno, G., Kanazawa, K., Yoshioka, T., and Natake, M.,** Mutagen formation from tryptophan treated with sodium nitrite in acidic solution, *Mutat. Res.,* 206, 65, 1988.
67. **Ochiai, M., Wakabayashi, K., Sugimura, T., and Nagao, M.,** Mutagenicities of indole and 30 derivatives after nitrite treatment, *Mutat. Res.,* 172, 189, 1986.
68. **Coughlin, J. R., Wei, C. I., Hsieh, D. P. H., and Russell, G. F.,** Synthesis, Mutagenicity and Human Health Implications of N-Nitroso Amadori Compounds from Maillard Browning Reactions in the Presence of Nitrite, presented at the American Chemical Society/Chemical Society of Japan Chemical Congress, Honolulu, 1979.
69. **Lynch, S. C., Gruenwedel, D. W., and Russell, G. F.,** Mutagenic activity of a nitrosated early Maillard product: DNA synthesis (DNA repair) induced in HeLa S3 carcinoma cells by nitrosated 1-(N-L-tryptophan)-1-deoxy-D-fructose, *Food Chem. Toxicol.,* 21, 551, 1983.
70. **Yen, G.-C. and Lee, T.-C.,** Mutagen formation in the reaction of Maillard browning products, 2-acethylpyrrole and its analogues, with nitrite, *Food Chem. Toxicol.,* 24, 1303, 1986.
71. **Bonnett, R. and Nicolaidou, P.,** Nitrosation and nitrosylation of haemoproteins and related compounds. II. The reaction of nitrous acid with the side chains of α-acylamino-acid esters, *J. Chem. Soc. Perkin 1,* 1969, 1979.
72. **Ochiai, M., Wakabayashi, K., Nagao, M., and Sugimura, T.,** Tyramine is a major mutagen precursor in soy sauce, being convertible to a mutagen by nitrite, *Jpn. J. Cancer Res. (Gann),* 75, 1, 1984.
73. **Blackwell, B. and Mabbitt, L. A.,** Tyramine in cheese related to hypertensive crises after monoamine-oxidase inhibition, *Lancet,* i, 938, 1965.
74. **Yamamoto, S., Wakabayashi, S., and Makita, M.,** Gas-liquid chromatographic determination of tyramine in fermented food products, *J. Agric. Food Chem.,* 28, 790, 1980.
75. **Smith, T. A.,** Amines in food, *Food Chem.,* 6, 169, 1981.

76. **Higashimoto, M., Matano, K., and Ohnishi, Y.,** Augmenting effect of a nonmutagenic fraction in soy sauce on mutagenicity of 3-diazotyramine produced in the nitrite-treated sauce, *Jpn. J. Cancer Res. (Gann),* 79, 1284, 1988.

77. **Fujita, Y., Wakabayashi, K., Takayama, S., Nagao, M., and Sugimura, T.,** Induction of oral cavity cancer by 3-diazotyramine, a nitrosated product of tyramine present in foods, *Carcinogenesis,* 8, 527, 1987.

78. **Kikugawa, K. and Kato, T.,** Formation of a mutagenic diazoquinone by interaction of phenol with nitrite, *Food Chem. Toxicol.,* 26, 209, 1988.

79. **Furihata, C., Yamakoshi, A., Matsushima, T., Kato, T., and Kikugawa, K.,** Possible tumor-initiating and promoting activities of 3-diazo-N-nitrosobamethan in rat stomach mucosa, *Mutagenesis,* 3, 299, 1988.

80. **Mizuno, M., Ohara, A., Danno, G., Kanazawa, K., and Natake, M.,** Mutagens formed from butylated hydroxyanisole treated with nitrite under acidic conditions, *Mutat. Res.,* 176, 179, 1987.

81. **Yamada, K., Murakami, H., Yasumura, K., Shirahata, S., Shinohara, K., and Omura, H.,** Production of DNA-breaking substance after treatment of monophenols with sodium nitrite and then with dimethyl sulfoxide, *Agric. Biol. Chem.,* 51, 247, 1987.

82. **Namiki, M. and Kada, T.,** Formation of ethylnitrolic acid by the reaction of sorbic acid with sodium nitrite, *Agric. Biol. Chem.,* 39, 1335, 1975.

83. **Kito, Y. and Namiki, M.,** A new N-nitrosopyrrole: 1,4-Dinitro-2-methylpyrrole, formed by the reaction of sorbic acid with sodium nitrite, *Tetrahedron,* 34, 505, 1978.

84. **Osawa, T., Kito, Y., and Namiki, M.,** A new furoxan derivative and its precursors formed by the reaction of sorbic acid with sodium nitrite, *Tetrahedron Lett.,* 1979, 4399.

85. **Osawa, T., Namiki, M., and Namiki, K.,** Mutagen formation in the nitrite-piperic acid reaction, *Agric. Biol. Chem.,* 46, 3105, 1982.

86. **Sakaguchi, M. and Shibamoto, T.,** Formation of heterocyclic compounds from the reaction of cysteamine and D-glucose, acetaldehyde, or glyoxal, *J. Agric. Food Chem.,* 26, 1179, 1978.

87. **Mihara, S. and Shibamoto, T.,** Mutagenicity of products obtained from cysteamine-glucose browning model systems, *J. Agric. Food Chem.,* 28, 63, 1980.

88. **Kimoto, W. I., Pensabene, J. W., and Fiddler, W.,** Isolation and identification of N-nitrosothiazolidine in fried bacon, *J. Agric. Food Chem.,* 30, 757, 1982.

89. **Pensabene, J. W. and Fiddler, W.,** Factors affecting the N-nitrosothiazolidine content of bacon, *J. Food Sci.,* 48, 1452, 1983.

90. **Pensabene, J. W. and Fiddler, W.,** N-Nitrosothiazolidine in cured meat products, *J. Food Sci.,* 48, 1870, 1983.

91. **Mandagere, A. K., Gray, J. I., Skrypec, D. J., Booren, A. M., and Pearson, A. M.,** Role of woodsmoke in N-nitrosothiazolidine formation in bacon, *J. Food Sci.,* 49, 658, 1984.

92. **Sekizawa, J. and Shibamoto, T.,** Mutagenicity of 2-alkyl-N-nitrosothiazolidines, *J. Agric. Food Chem.,* 28, 781, 1980.

93. **Umano, K., Shibamoto, T., Fernando, S. Y., and Wei, C.-I.,** Mutagenicity of 2-hydroxyalkyl-N-nitrosothiazolidines, *Food Chem. Toxicol.,* 22, 253, 1984.

94. **Fiddler, W., Miller, A. J., Pensabene, J. W., and Doerr, R. C.,** Investigation on the mutagenicity of N-nitrosothiazolidine using the Ames Salmonella test, paper presented at the 8th Int. Meet. N-Nitroso Compounds — Occurrence and Biological Effects, Banff, Canada, 1983.

95. **Ohshima, H., Friesen, M., O'Neill, I. K., and Bartsch, H.,** Presence in human urine of a new N-nitroso compound, N-nitrosothiazolidine 4-carboxylic acid, *Cancer Lett.,* 20, 183, 1983.

96. **Tsuda, M., Hirayama, T., and Sugimura, T.,** Presence of N-nitroso-L-thioproline and N-nitroso-L-methylthioproline in human urine as major N-nitroso compounds, *Jpn. J. Cancer Res. (Gann),* 74, 331, 1983.

97. **Tsuda, M., Kakizoe, T., Hirayama, T., and Sugimura, T.,** New type of N-nitrosamino acids, N-nitroso-L-thioproline and N-nitroso-L-methylthioproline, found in human urine as major N-nitroso compounds, *IARC Sci. Publ.,* 57, 87, 1984.

98. **Ohshima, H., O'Neill, I. K., Friesen, M., Bereziat, J.-C., and Bartsch, H.,** Occurrence in human urine of new sulfur-containing N-nitrosamino acids, N-nitrosothiazolidine 4-carboxylic acid and its 2-methyl derivative, and their formation, *J. Cancer Res. Clin. Oncol.,* 108, 121, 1984.

99. **Tsuda, M., Niitsuma, J., Sato, S., Hirayama, T., Kakizoe, T., and Sugimura, T.,** Increase in the levels of N-nitrosothioproline and N-nitroso-2-methylthioproline in human urine by cigarette smoking, *Cancer Lett.,* 30, 117, 1986.

100. **Tsuda, M., Nagai, A., Suzuki, H., Hayashi, T., Ikeda, M., Kuratsune, M., Sato, S., and Sugimura, T.,** Effect of cigarette smoking and dietary factors on the amounts of N-nitrosothiazolidine 4-carboxylic acid and N-nitroso-2-methylthiazolidine 4-carboxylic acid in human urine, *IARC Sci. Publ.,* 84, 446, 1987.

101. **Tahira, T., Tsuda, M., Wakabayashi, K., Nagao, M., and Sugimura, T.,** Kinetics of nitrosation of thioproline, the precursor of a nitro compound in human urine, and its role as a nitrite scavenger, *Jpn. J. Cancer Res. (Gann),* 75, 889, 1984.

102. **Mirvish, S. S., Bulay, O., Runge, R. G., and Patil, K.,** Study of the carcinogenicity of large doses of dimethylnitramine, N-nitroso-L-proline, and sodium nitrite administered in drinking water to rats, *J. Natl. Cancer Inst.,* 64, 1435, 1973.

103. **Lijinsky, W. and Reuber, M. D.,** Transnitrosation by nitrosamines *in vivo, IARC Sci. Publ.,* 57, 625, 1984.

104. **Tsuda, M., Frank, N., Sato, S., and Sugimura, T.,** Marked increase in the urinary level of N-nitrosothioproline after ingestion of cod with vegetables, *Cancer Res.,* 48, 4049, 1988.

105. **Sasagawa, C., Muramatsu,M., and Matsushima, T.,** Formation of direct mutagens from amino-imidazoazaarenes by nitrite treatment, (Selected Abstracts of the 16th Annu. Meet. Environmental Mutagen Society of Japan, 27—28 October 1987, Kyoto, Japan), *Mutat. Res.,* 203, 386, 1988.

106. **Tsuda, M., Negishi, C., Makino, R., Sato, S., Yamaizumi, Z., Hirayama, T., and Sugimura, T.,** Use of nitrite and hypochlorite treatments in determination of the contributions of IQ type and non-IQ type heterocyclic amines to the mutagenicities in crude pyrolyzed materials, *Mutat. Res.,* 147, 335, 1985.

Chapter 5.3

NITRITE-REACTIVE PHENOLS PRESENT IN SMOKED FOODS AND AMINO-SUGARS FORMED BY THE MAILLARD REACTION AS PRECURSORS OF GENOTOXIC ARENEDIAZONIUM IONS OR NITROSO COMPOUNDS

Helmut Bartsch, Hiroshi Ohshima, Brigitte Pignatelli, Christian Malaveille, and Marlin Friesen

TABLE OF CONTENTS

I. INTRODUCTION

Stomach cancer, while declining in most populations in recent years, still remains the most common cancer in both sexes on a world-wide basis.[1] Data from several types of epidemiological studies, such as case-control, ecologic, migration, and family studies, have revealed various etiological factors for stomach cancer, including a diet rich in salted or smoked food and low in vitamin C.[2] Exposure to nitrate and nitrite, precursors of carcinogenic *N*-nitroso compounds, has also been correlated with stomach cancer mortality.[3] *N*-Nitroso compounds are among the most potent chemical carcinogens in laboratory animals, and certain *N*-nitrosamides have been shown to induce adenocarcinomas of the glandular stomach resembling human gastric cancer.[4,5] The *N*-nitrosamides formed in the human stomach by interaction between nitrite and suitable amine precursors have therefore been suspected of playing an important role in gastric carcinogenesis.[6] Several amines which show direct-acting genotoxicity, after nitrosation *in vitro,* have been identified in foods frequently consumed in high-risk areas for stomach cancer. They include 4-chloro-6-methoxyindole in fava beans consumed in Colombia,[7] several indole derivatives such as indole-3-acetonitrile in pickled Chinese cabbage,[8] and tyramine and β-carboline derivatives in Japanese soy sauce.[9] However, these food items are infrequently consumed in other high-risk areas for stomach cancer, especially northern Europe. For these reasons, we have screened European food items which show direct-acting genotoxicity after nitrosation *in vitro.* We found that nitrosation of smoked foods, frequent consumption of which has been associated with an increased risk of stomach cancer,[2,10] leads to such direct-acting genotoxicity. Various simple phenolic compounds have been identified as the precursors responsible for genotoxicity after nitrosation.[11]

In another series of investigations, we have examined the genotoxicity of nitrosated amines that are formed during the Maillard reaction.[12] In this reaction, reducing sugars react with amino groups of amines, amino acids, peptides, or proteins to give glycosylamines, glycoamino acids, peptides, or protein. These products can be further transformed into substances involved in "nonenzymatic browning reactions"[13] in complex reactions that occur during the heat processing, drying, storage, cooking, frying, and baking of various foods. Glycosylamines and Amadori compounds (*N*-substituted-1-amino-1-deoxy-2-ketoses) that are formed at an early stage of the nonenzymatic browning reaction can be nitrosated. Since most *N*-nitroso compounds are carcinogenic in experimental animals,[14] *N*-nitrosoglycosylamines and *N*-nitroso-Amadori compounds, which may be present in human diet or may be formed in the stomach from the corresponding amine precursors and nitrosating species, may constitute a hazard to human health. Of the synthetic *N*-nitrosofructose-amino acids, only derivatives of tryptophan, histidine, and threonine have been reported to be direct-acting mutagens in bacteria, and the mechanism has not been elucidated.[15-18]

II. NITRITE-REACTIVE PHENOLS PRESENT IN SMOKED FOOD

A. FORMATION OF DIRECT-ACTING GENOTOXIC SUBSTANCES IN NITROSATED SMOKED FOODS AND SMOKE CONDENSATES

Table 1 shows genotoxicity of aqueous extracts of various fish and meat products, which were incubated under acidic conditions with or without sodium nitrite. Although there were some variations between food products of the same type, in general much higher direct-acting genotoxicity was observed in the nitrosated smoked fish products than in the corresponding fresh samples. Three samples of smoked meat products and three samples of Japanese smoked and dried fish products "Katsuo-bushi" also exhibited potent direct-acting genotoxicity. Treatment with nitrite under acidic conditions was required to elicit such genotoxicity and only a few samples exhibited even weak activity without nitrosation. Thus,

TABLE 1
SOS-Inducing Potency (SOSIP) of Nitrosated Smoked Foods[a]

	Food	Sample no.	SOSIP[b] With NaNO$_2$	Without NaNO$_2$
Salmon	Smoked	1	440	65
		2	190	46
	Fresh		71	21
Herring	Smoked	1	300	0[c]
		2	240	0
		3	68	0
	Fresh		46	44
Mackerel	Smoked		64	39
	Sun-dried		45	0
	Fresh		47	30
Halibut	Smoked		200	34
	Fresh		53	0
Ham	Smoked		600	0
Bacon	Smoked		480	0
Lard	Smoked		130	0
Chicken	Smoked		270	0
Japanese smoked and dried fish				
"Katsuo-bushi"			430	0
Bonito		1	230	0
Mackerel		2	270	0

[a] From Ohshima, H., Friesen, M., Malaveille, C., Brouet, I., Hautefeuille, A., and Bartsch, H., *Food Chem. Toxicol.*, 27, 193, 1989. With permission. Experimental methods described therein.

[b] SOSIP was expressed as SOS-inducing factor per gram food and was calculated on the basis of two to four duplicate determinations of SOS-inducing factor with various concentrations of the sample. The maximal SOS-inducing factor for positive samples ranged from 1.7 to 7.0; the highest induction of β-galactosidase activity ranged from 1.1 to 5.6-fold that of the control value.

[c] Below detection limit. Under the present conditions, the lowest SOSIP that could be detected by the SOS Chromotest was 9.8 per gram food.

the formation of genotoxic substances in nitrosated smoked foods is clearly associated with the smoking process.

Similar direct-acting genotoxicity was also observed in the nitrosated samples of various wood-smoke condensates obtained either commercially or experimentally (Table 2). The SOS-inducing potency (SOSIP) varied from 9 for experimentally produced fir-smoke concentrate to 120 for bamboo-smoke concentrate. Two purified phenol fractions were among the most genotoxic. As with the smoked fish and meat products, nitrosation with nitrite was needed to elicit the direct-acting genotoxicity.

B. PURIFICATION AND ISOLATION OF PRECURSOR SUBSTANCES

Since the nitrosated samples of various wood-smoke condensates exhibited potent direct-acting genotoxicity similar to that observed in the nitrosated smoked food products, precursor substances in commercially available hickory-smoke condensate have been examined. The precursors in both hickory-smoke condensate and smoked herrings were extractable by ethyl acetate under neutral conditions. They were not adsorbed onto blue-cotton (Funakoshi Chemical Co., Tokyo, Japan), suggesting that they are neither polycyclic aromatic hydrocarbons such as benzo(*a*)pyrene nor mutagenic heterocyclic amines produced during cooking of foods.[19] The ethyl acetate extracts were fractionated by a semipreparative ODS-3 HPLC column. Many fractions were found to be genotoxic after nitrosation with nitrite, indicating

TABLE 2
SOS-Inducing Potency (SOSIP) of Nitrosated Smoke Condensates[a]

		SOSIP[b]	
Sample	No.	With NaNO$_2$	Without NaNO$_2$
Hickory	A 1	36	0.6
(commercial)	2	47	0[c]
	3	65	0
	4	67	0
	mixture	51	0
	B 1	30	0
	C 1	10	0
Hickory (exp.)[d]		68	0
Bamboo (exp.)		120	0
Beech (exp.)		62	0
Pine (exp.)		10	0
Fir (exp.)		9	0
White-beech (exp.)		15	0
Purified phenol fraction[d] (PAH-free)[e]		9,900	0
Purified monohydroxy phenol fraction[e]		27,000	0

[a] From Ohshima, H., Friesen, M., Malaveille, C., Brouet, I., Hautefeuille, A., and Bartsch, H., *Food Chem. Toxicol.*, 27, 193, 1989. With permission. Experimental methods described therein.

[b] SOSIP was expressed as SOS-inducing factor per microliter sample and was calculated on the basis of two to four duplicate determinations of SOS-inducing factor with various concentrations of the sample. The maximal SOS-inducing factor for positive samples ranged from 3.4 to 16; the highest induction of β-galactosidase activity ranged from 1.0- to 11-fold control value.

[c] Below detection limit. Under the present conditions, the lowest SOSIP which could be detected by the SOS Chromotest was 0.3 per microliter smoke condensate.

[d] Experimental smoke condensates produced in a laboratory smoke producer.

[e] Purified phenol fraction (PAH-free) and purified monohydroxyphenol fraction were prepared from white-beech wood as described previously[20] and kindly provided by K. Potthast, Institute of Chemistry and Physics, Federal Center for Meat Research, Kulmbach, FR Germany.

that there are several precursor compounds of the genotoxic substances in the hickory-smoke concentrate. GC-MS analysis showed that the positive fractions contained various simple phenolic compounds, including phenol and vanillin. Nitrosation of such phenolic compounds led to potent direct-acting genotoxicity in the SOS Chromotest. Therefore, the genotoxicity of various phenolic compounds before or after treatment with nitrite was studied.

C. GENOTOXICITY OF NITROSATED PHENOLIC COMPOUNDS

Various phenolic compounds reported to be present in smoke condensates[20] and other related compounds were nitrosated with nitrite and their genotoxicity was examined by the SOS Chromotest (Tables 3 and 4). Out of about 70 different phenols tested, 27 compounds showed genotoxicity in the absence of metabolic activation (S9 mix) system, with SOSIPs ranging from 0.006 to 14 (Table 3). Among the compounds known to occur in smoke condensates, 3-methoxycatechol, phenol, vanillin, and hydroquinone were especially strongly genotoxic after nitrosation, showing SOSIPs of 1.3, 0.80, 0.21, and 0.14, respectively. Under conditions similar to those of the present study, *N*-methyl-*N'*-nitro-*N*-nitrosoguanidine and methyl methanesulfonate exhibited SOSIPs of 0.9 and 0.02, respectively.[21]

TABLE 3

SOS-Inducing Potency (SOSIP) of Phenols Present in Smoke Condensates and Related Compounds after Nitrosation *In Vitro*[a]

Compound	SOSIP	Compound	SOSIP[b]
4-Aminophenol	14	2,4-Dimethylphenol[c,d]	0.046
2-Aminophenol	5.9	Guaiacol[c,d]	0.040
3-Methoxycatechol[c]	1.3	2-Allylphenol[c]	0.034
Phenol[a]	0.80	4-Hydroxy-3-methoxyben-zyl alcohol	0.029
Vanillin[c]	0.21	4-Ethylresorcinol[c]	0.025
Tyramine	0.16	Resorcinol	0.019
Hydroquinone[c]	0.14	*o*-Cresol[c]	0.019
p-Cresol[c]	0.13	4-Methylcatechol[c]	0.018
Aniline	0.12	Pyrogallol	0.016
Acetovanillone[c]	0.11	Caffeic acid[d]	0.013
4-Hydroxybenzaldehyde[c]	0.10	2,3,5-Trimethylphenol	0.012
m-Cresol[c]	0.059	2,3-Dimethylphenol[c,d]	0.008
3-Ethylphenol[c]	0.058	3-Methoxybenzaldehyde	0.006
Catechol[c]	0.056		

[a] From Ohshima, H., Friesen, M., Malaveille, C., Brouet, I., Hautefeuille, A., and Bartsch, H., *Food Chem. Toxicol.*, 27, 193, 1983. With permission. Experimental methods described therein.

[b] SOSIP was expressed as SOS-inducing factor per nanomole compound and was calculated on the basis of two to four duplicate determinations of SOS-inducing factor at various concentrations of the sample. The maximal SOS-inducing factor ranged from 1.5 to 12.4; the highest induction of β-galactosidase activity ranged from 1.3- to 9.8-fold control value.

[c] Present in wood smoke concentrates.[20]

D. IDENTIFICATION OF GENOTOXIC SUBSTANCES FORMED BY NITROSATION OF PHENOL

Since phenol was identified as one of the most important precursor substances contributing to the genotoxicity of the nitrosated smoke concentrates, we isolated and identified the genotoxic substances formed by nitrosation of phenol. Figure 1 shows a typical HPLC trace obtained for a nitrosated phenol sample. After 1 hr incubation at 37°C with sodium nitrite (pH 3.0), phenol was converted to several compounds, among which a major one (peak III) was identified as 4-nitrosophenol, formed with a yield of about 45%. However, the majority of the genotoxicity was observed in the fractions containing peaks I and II, which were eluted before 4-nitrosophenol, with retention times identical to those of nitrosation products of 4- and 2-aminophenol, respectively. These two fractions were subjected to azo-coupling reactions with *N*-ethyl-1-naphthylamine (NEN) and the reaction products were analyzed by TLC and MS according to Kolar and Schlesiger.[22] The NEN coupling products gave a violet spot at an R_F value of 0.23 and a bluish-pink spot at R_F 0.75. These were identical with regard to color and R_F value to the NEN coupling products formed with nitrosated 4- and 2-aminophenol, respectively. Furthermore, mass spectra of the NEN coupling products of peaks I and II showed ions at m/e 291 (parent ion) and 170 (stable NEN fragment) and were identical to those of the NEN coupling products formed with nitrosated 4- and 2-aminophenol. Kolar and Schlesiger[22] have reported that nitrosation of 4- and 2-aminophenol yielded corresponding hydroxyphenyl-diazonium ions which reacted with NEN to form hydroxyphenyl-azo-NEN derivatives. Consequently, the major genotoxic substances formed after nitrosation of phenol were identified as 4- and 2-hydroxyphenyldiazonium ions (Figure 2). Kikugawa and Kato have also recently shown that reaction of phenol with nitrite

TABLE 4
Compounds Which Were Not Genotoxic after Nitrosation[a]

Acetophenone	2-Hydroxy-3-methoxybenzaldehyde
4-Allylsyringol[b]	3-Hydroxy-4-methoxybenzoic acid
3-Aminobenzoic acid	
3-Aminophenol	4-(4-Hydroxy-3-methoxyphenyl)-3-buten-2-one[b]
Butylated hydroxyanisole	3-Hydroxy-2-methyl-4-pyrone[b]
Catechin	Isoeugenol[b]
Chlorogenic acid	3-Methyl-1,2-cyclopentanedione[b]
2,4-Dihydroxyacetophenone	Salicylaldehyde
1,4-Dimethoxybenzene[b]	Syringaldehyde[b]
2,4-Dimethylbenzyl alcohol	Syringic acid
2,5-Dimethylbenzyl alcohol	Syringol[b]
3,4-Dimethylbenzyl alcohol	3′,4′,5′-Trimethoxyacetophenone
3,5-Dimethylbenzyl alcohol	1,2,3-Trimethoxybenzene
2,5-Dimethylfuran[b]	1,2,4-Trimethoxybenzene
2,6-Dimethylphenol[b]	1,3,5-Trimethoxybenzene
Eugenol[b]	2,3,6-Trimethylphenol
Ferulic acid	2,4,6-Trimethylphenol
Homovanillic acid	3,4,5-Trimethylphenol
2′-Hydroxyacetophenone	L-Tyrosine
3-Hydroxybenzoic acid	Vanillic acid[b]

[a]   From Ohshima, H., Friesen, M., Malaveille, C., Brouet, I., Hautefeuille, A., and Bartsch, H., *Food Chem. Toxicol.*, 27, 193, 1989. With permission. Experimental details are given therein; the lowest SOSIP which could be detected by the SOS Chromotest was 0.006 per nanomole of the test compound.

[b]   Present in wood smoke concentrates.[20]

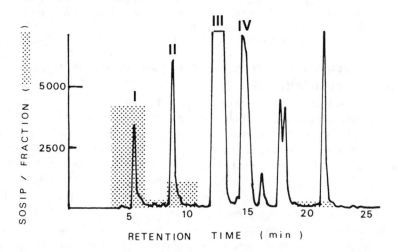

FIGURE 1.   A typical HPLC chromatogram obtained by the analysis of the nitrosated phenol and genotoxicity of fractions (shaded areas). Peaks I, II, III, and IV were identified as 4-hydroxyphenyldiazonium ion, 2-hydroxyphenyldiazonium ion, 4-nitrosophenol, and phenol, respectively. (From Ohshima, H., Friesen, M., Malaveille, C., Brouet, I., Hautefeuille, A., and Bartsch, H., *Food Chem. Toxicol.*, 27, 193, 1989. With permission.)

produces 4-nitrosophenol and 4- and 2-diazoquinone (hydroxyphenyldiazonium ion);[23] the latter compound was mutagenic in *Salmonella typhimurium* strains TA98 and TA100 without metabolic activation.

The NEN coupling reaction was also used to identify the diazo compounds generated in the nitrosated smoke concentrates. Each sample tested gave several colored spots on TLC,

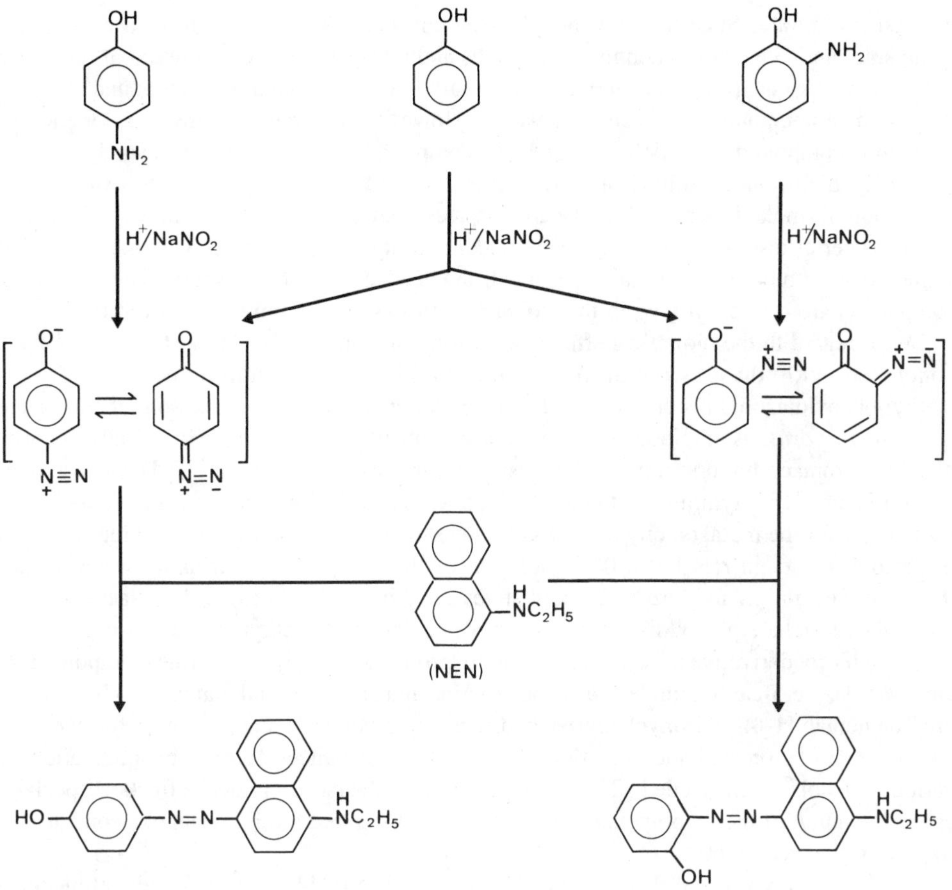

FIGURE 2. Scheme for the formation from phenolic precursors of hydroxyphenyldiazonium ions and their reaction with *N*-ethyl-1-naphthylamine (NEN) to yield corresponding stable diazo-coupling products. (From Ohshima, H., Friesen, M., Malaveille,C., Brouet, I., Hautefeuille, A., and Bartsch, H., *Food Chem. Toxicol.*, 27, 193, 1989. With permission.)

indicating that nitrosation of wood-smoke condensates formed diazonium compounds which reacted with NEN to yield colored azo-NEN coupling products. In particular, out of six samples tested, four smoke condensates (bamboo, beech, hickory, and monohydroxyphenol fraction) showed a violet spot at R_F 0.23, corresponding to the 4-(4-hydroxyphenylazo)-NEN coupling product. This indicates that nitrosation of the smoke condensates yields 4-hydroxyphenyldiazonium ion, via nitrosation of phenol, which is present in these samples at relatively high concentrations. The genotoxicity of nitrosated smoke condensates is, therefore, attributable in part to the formation of this diazonium compound.

E. TOXICOLOGICAL SIGNIFICANCE

We have demonstrated that nitrosation of smoked foods and smoke condensates produces relatively potent direct-acting genotoxic substances. Simple phenolic compounds present in smoke condensates at relatively high concentrations exhibit similar direct-acting genotoxicity after nitrosation. We have also shown that hydroxyphenyldiazonium ions formed after the nitrosation of phenol appear to be at least partly responsible for the genotoxicity of some of the nitrosated smoke condensates. The nitrosation of phenol to yield 4-nitrosophenol[24] and 4- and 2-hydroxyphenyldiazonium ions[25] has already been reported. The latter reaction was first described by Weselsky,[26] who treated phenol with nitrogen oxides and obtained diazophenol nitrate. Such hydroxyphenyldiazonium ions could well be formed during cigarette smoking, since both phenolic compounds and nitrogen oxides are present at relatively high concentrations in tobacco smoke.[27] Nitrosation of various phenolic compounds to form diazonium compounds has been reviewed.[25] However, the mechanisms and kinetics of diazonium compound formation from phenolic compounds have not been extensively studied. Relatively high concentrations of nitrite were required to produce the genotoxicity after nitrosation of smoked foods and smoke condensates, mainly because of the limited sensitivity of the bacterial test and the strong bactericidal activity of the smoke condensates, so that further kinetic studies and animal experiments are needed to establish whether the diazonium compounds are formed *in vivo* in the stomach with lower concentrations of nitrite.

With regard to the biological effects of diazonium compounds, a few studies have been conducted with those structurally similar to phenyldiazonium ions. 4-(Hydroxymethyl)phenyldiazonium salt, identified in the commonly cultivated mushroom of commerce *Agaricus bisporus,* is carcinogenic in mice after subcutaneous injection[28] and also induces glandular stomach tumors in mice after a single intragastric instillation.[29] 3-Diazotyramine, a reaction product of tyramine with nitrite which was recently identified as a major compound responsible for the mutagenicity of nitrosated soy sauce,[30] was also reported to induce tumors of the oral cavity in rats fed a 0.1% solution of the compound in drinking water.[31] Furthermore, several phenyldiazonium derivatives have been shown to be direct-acting mutagens. These include phenyldiazonium fluoroborate and its 4-methyl, 4-nitro, 4-chloro, and 2,4,6-trichloro derivatives,[12,32-33] diazotization products of a human urine component 4-aminoimidazole-5-carboxamide,[34] aminoantipyrine and aniline,[35] and that of a cardiovascular drug bamethan [1-(4-hydroxyphenyl)-1-hydroxy-2-butylaminoethane].[36] Thus, further studies on the isolation and identification of genotoxic substances formed by nitrosation of various phenols seem needed. The toxic and other adverse biological effects, especially carcinogenicity, of these compounds and of nitrosated smoked foods and smoke condensates require further research.

A recent study showed that the oral administration to F344 rats of commercial hickory-smoke condensate (HSC) alone caused DNA strand scission in rat pyloric mucosa[37] and also induced ornithine decarboxylase (ODC) (~500-fold at 16 hr after HSC dose) and replicative DNA synthesis (~50-fold). On the other hand, the co-administration of nitrite together with HSC induced markedly unscheduled DNA synthesis, but decreased the extent of DNA strand scission, the induction of ODC, and replicative DNA synthesis. These results indicate that HSC contains substances which have potent initiation and/or promotion activities for gastric carcinogenesis and some which are converted to initiators after the reaction with nitrite. On the basis of these results, investigation of the carcinogenicity of HSC in long-term animal experiments is planned. The active substances should also be isolated and identified.

III. MUTAGENICITY OF *N*-NITROSO DERIVATIVES OF GLYCOSYLAMINES AND AMADORI COMPOUNDS

As model compounds for the glycosylamines and Amadori compounds that may occur in foods, reaction products between amines (*p*-toluidine, *p*-nitroaniline, and tryptamine) and sugars (such as pentoses and glucose) and their *N*-nitrosated derivatives were synthesized and characterized to elucidate the structural parameters that determine mutagenic activity of *N*-nitroso glycosylamines and *N*-nitroso-Amadori compounds and to investigate the nature of the ultimate mutagen(s) derived from the parent compound.[12]

Several *N*-nitroso compounds (listed, with structures and abbreviations, in Table 5) were shown to be direct-acting mutagens. Their mutagenic activity in *Salmonella typhimurium* TA100 depends both on the structure of the amine and the sugar moiety; the presence of free hydroxyl groups in the sugar was required.

N-Nitrosated glycosyl-*p*-nitroaniline has been shown by Bognar and Puskas[38] to decom-

TABLE 5
N-Nitroso Derivatives of Glycosylamines and the Amadori Compound Tested for Mutagenic Activity

Structure	N-Nitroso derivative of glycosylamine or Amadori compound[a]	Abbreviation	Mutagenicity (no. rev/mg)[b]
	N-Nitroso-N-p-nitrophenyl-D-ribosylamine	N(NO)-NP-Rib	1,820[c]
	N-Nitroso-N-p-nitrophenyl-L-arabinosylamine	N(NO)-NP-Ara	4,550
	N-Nitroso-N-p-methylphenyl-D-arabinosylamine	N(NO)-MP-Ara	98
	N-Nitroso-N-p-nitrophenyl-D-xylosylamine	N(NO)-NP-Xyl	570
	N-Nitroso-N-p-methylphenyl-D-xylosylamine	N(NO)-MP-Xyl	<7
	N-Nitroso-N-p-nitrophenyl-D-lyxosylamine	N(NO)-NP-Lyx	2,770
	N-Nitroso-N-3-ethylindole-D-xylosylamine	N(NO)-T-Xyl[d]	21,900
	N-Nitroso-N-p-methylphenyl-1-deoxy-D-fructosylamine	N(NO)-MP-Fru[e]	76

TABLE 5 (continued)
N-Nitroso Derivatives of Glycosylamines and the Amadori Compound Tested for Mutagenic Activity

[a] For details on synthesis, see Reference 12.

[b] Specific mutagenicity was calculated from the (pseudo)-linear part of the dose-response curves. For details, see Reference 12.

[c] Unpublished results from Didier-Touron, M., *N*-Glycosylamines d'Origine Alimentaire: Synthèse, Nitrosation, Mutagénèse, Ph.D. thesis, Lyon, France, 1989.

[d] May contain trace amounts of a compound with a nitrosated indolyl-NH group and/or aromatic C-nitro compounds.

[e] *N*-Nitroso derivative of the Amadori compound.

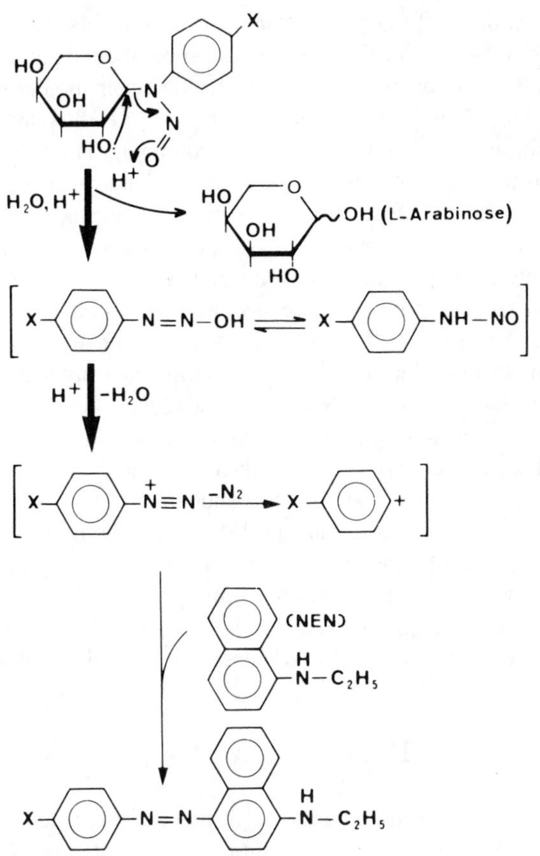

FIGURE 3. Scheme for the hydrolysis of *N*-nitrosoglycopyr-
anosylamines to yield arenediazonium cations and phenyl cations as ultimate mutagens. The coupling reaction with *N*-ethyl-
1-naphthylamine (NEN) was used to characterize the reactive
intermediate.

pose to arenediazonium cations through nonenzymatic hydrolysis (Figure 3). Experiments
were, therefore, carried out to determine whether the mutagenicity of *N*-nitrosoglycosylam-
ines or the *N*-nitroso-Amadori compound is related to the hydrolytic decomposition of the
compounds into arenediazonium cations. Both *p*-methylphenyldiazonium and *p*-nitrophen-
yldiazonium cations (tested as fluoroborate salts) which should be derived from *N*(NO)-MP-
Fru and *N*(NO)-NP-Ara, respectively, were found to be direct-acting mutagens in *S. typhimurium* TA100. When assayed under experimental conditions identical to those used to test
the two phenyldiazonium salts, *N*(NO)-MP-Fru was 170 times less mutagenic and *N*(NO)-
NP-Ara about 12 times less mutagenic than the corresponding phenyldiazonium salt.

To ascertain further the involvement of arenediazonium cations in the mutagenicity of
N-nitrosoglycosylamines or the *N*-nitroso-Amadori compound, we measured their formation
by azo-coupling to NEN up to pH 6.1. At pH 3.5, the yield of *p*-nitrophenyldiazonium
cation from *N*(NO)-NP-Ara or *N*(NO)-NP-Xyl was about 65 times higher than that of the
p-methylphenyldiazonium cation derived from *N*(NO)-MP-Fru. The acetylation of hydroxyl
groups in the sugar moiety of *N*(NO)-NP-Ara suppressed the formation of *p*-nitrophenyldi-
azonium cations. At pH 4.6, *N*(NO)-NP-Ara gave twice the yield of *p*-nitrophenyldiazonium
cations as did *N*(NO)-NP-Xyl.

Having shown that the specific mutagenicity of the arenediazonium cations was higher
than that of the corresponding *N*-nitrosoglycosylamines or *N*-nitroso-Amadori compound,

that the rate of formation (yield at pH 3.5 or 4.6) of arene(alkyl)diazonium cations from the N-nitrosoglycosylamines or the N-nitroso-Amadori compound parallelled their mutagenic activity, and that acetylation of hydroxyl groups in the sugar moiety suppressed the mutagenicity and formation of diazonium cations, we conclude that the mutagenicity of N-nitrosoglycosylamines and the N-nitroso-Amadori compound is attributable mainly to their hydrolytic decomposition into arenediazonium cations (and in the case of N(NO)-T-Xyl, into an alkyldiazonium cation), according to the mechanism proposed (Figure 3).

Recently, evidence for the formation of triazene adducts, resulting from the reaction between arenediazonium ion and the exocyclic amino groups of adenine and cytosine in nucleic acids, has been presented.[39] These decompose, probably into hypoxanthine and uridine, respectively. From these data and our own, we infer that the mutagenic action of N-nitrosoglycosylamines and N-nitroso-Amadori compounds involves the formation of arenediazonium cations that react with DNA bases; the adducts appear to be unstable and lead to deamination of bases, causing point mutations in DNA.

Humans are exposed to dietary N-nitrosoglycosylamines and N-nitroso-Amadori compounds or their precursors, which may undergo endogenous nitrosation, a reaction that has now been shown definitely to occur in humans.[40,41] It is thus important to gain more knowledge about the chemical and biological properties of this relatively unexplored class of N-nitroso compounds, particularly since the mutagenic potency of N(NO)-T-Xyl and N(NO)-NP-Ara (expressed as revertants in *S. typhimurium* TA100/mM concentration/min in liquid incubation) is close to that of N-nitroso-N-ethylurea,[42] a versatile carcinogen in many animal species.

IV. CONCLUSIONS

In conclusion, we have identified a number of precursor compounds that, after reaction with nitrite, yield DNA-reactive arenediazonium intermediates or N-nitroso compounds. Relatively little attention has been given to such precursors so far, to assess human exposure that could occur through ingestion of smoked or heated foods. As endogenous nitrosation reactions have now been unequivocally demonstrated to occur in humans,[40,41] it is possible that such reactive products also form intragastrically in man. The application of dosimetry methods that are currently available[43] or their development to assess macromolecular interactions with DNA and proteins in exposed humans is now warranted.

ACKNOWLEDGMENTS

We thank E. Bayle for secretarial help and J. Cheney for editing this manuscript. We also thank A. Hautefeuille and I. Brouet for technical assistance. This work was supported in part by an A.T.P. PIREN (Alimentation et Santé) from the Centre National de la Recherche Scientifique (CNRS) France by Contract No. ENV-654-F (SD) with the Commission of the European Communities. We are grateful to Dr. G. Kolar (Heidelberg, FR Germany) for his generous gift of the N-ethyl-1-naphthylamine and the phenyldiazonium fluoroborate salts and to Dr. K. Potthast (Kulmbach, FR Germany) for several wood-smoke condensates prepared experimentally in a laboratory smoke producer as well as purified phenol fractions. We acknowledge the contributions of Drs. C. Furihata and T. Matsushima, University of Tokyo, Japan; Professor G. Descotes, Laboratory of Organic Chemistry, University of Lyon I, France; and Dr. D. Piskorska, University of Gdansk, Poland, with whom we collaborated during various phases of the investigations presented herein.

REFERENCES

1. **Parkin, D. M. Läärä, E., and Muir, C. S.,** Estimates of the worldwide frequency of sixteen major cancers in 1980, *Int. J. Cancer,* 41, 184, 1988.
2. **Howson, C. P., Hiyama, T., and Wynder, E. L.,** The decline in gastric cancer: epidemiology of unplanned triumph, *Epidemiol. Rev.,* 8, 1, 1986.
3. **Mirvish, S. S.,** The etiology of gastric cancer: intragastric nitrosamide formation and other theories, *J. Natl Cancer Inst.,* 71, 629, 1983.
4. **Sugimura, T. and Fujimura, S.,** Tumour production in glandular stomach of rat by *N*-methyl-*N'*-nitro-*N*-nitrosoguanidine, *Nature (London),* 216, 943, 1967.
5. **Druckrey, H., Preussmann, R., Ivankovic, S., and Schmähl, D.,** Organotrope carcinogene Wirkungen bei 65 verschiedenen *N*-nitroso-Verbindungen an BD-Ratten, *Z. Krebsforsch.,* 69, 103, 1967.
6. **Shephard, S. E., Schlatter, C. H., and Lutz, W. K.,** Assessment of the risk of formation of carcinogenic *N*-nitroso compounds from dietary precursors in the stomach, *Food Chem. Toxicol.,* 25, 91, 1987.
7. **Yang, D., Tannenbaum, S. R., Büchi, G., and Lee, G. C. M.,** 4-Chloro-6-methoxyindole is the precursor of a potent mutagen (4-chloro-6-methoxy-2-hydroxy-1-nitroso-indolin-3-one oxime) that forms during nitrosation of the fava bean *(Vicia faba), Carcinogenesis,* 5, 1219, 1984.
8. **Wakabayashi, K., Nagao, M., Ochiai, M., Tahira, T., Yamaizumi, Z., and Sugimura, T.,** A mutagen precursor in Chinese cabbage, indole-3-acetonitrile, which becomes mutagenic on nitrite treatment, *Mutat. Res.,* 143, 17, 1985.
9. **Wakabayashi, K., Nagao, M., Ochiai, M., Fujita, Y., Tahira, T., Nakayasu, M., Ohgaki, H., Takayama, S., and Sugimura, T.,** Recently identified nitrite-reactive compounds in food: occurrence and biological properties of the nitrosated products, in *The Relevance of N-Nitroso Compounds to Human Cancer: Exposures and Mechanisms,* Bartsch, H., O'Neill, I. K., and Schulte-Hermann, R., Eds., IARC Scientific Publ. No. 84, International Agency for Research on Cancer, Lyon, France, 1987, 287.
10. **Hill, M. J.,** *Microbes and Human Carcinogenesis,* Edward Arnold, London, 1986, 36.
11. **Ohshima, H., Friesen, M., Malaveille, C., Brouet, I., Hautefeuille, A., and Bartsch, H.,** Formation of direct-acting genotoxic substances in nitrosated smoked fish and meat products: identification of simple phenolic precursors and phenyldiazonium ions as reactive products, *Food Chem. Toxicol.,* 27, 193, 1989.
12. **Pignatelli, B., Malaveille, C., Friesen, M., Hautefeuille, A., Bartsch, H., Piskorska, D., and Descotes, G.,** Synthesis, structure-activity relationships and a reaction mechanism for mutagenic *N*-nitroso derivatives of glycosylamines and Amadori compounds — model substances for *N*-nitrosated early Maillard reaction products, *Food Chem. Toxicol.,* 25, 669, 1987.
13. **Ericksson, C., Ed.,** *Maillard Reactions in Food: Chemical Physiological and Technological Aspects, Progress in Food and Nutritional Science,* Vol. 5, Pergamon Press, Oxford, 1982.
14. **Preussmann, R. and Stewart, B. W.,** *N*-Nitroso carcinogens, in *Chemical Carcinogens,* ACS Monogr. No. 182, American Chemical Society, Washington, D.C., 1984, 643.
15. **Coughlin, J. R.,** Formation of *N*-Nitrosamines from Maillard-Browning Reaction Products in the Presence of Nitrite, Thesis, University of California, 1979.
16. **Pool, B. L., Röper, H., Röper, S., and Romruen, K.,** Mutagenic studies on *N*-nitrosated products of the Maillard reaction: *N*-nitroso-fructose-amino acids, *Food Chem. Toxicol.,* 22, 797, 1984.
17. **Röper, H., Röper, S., Heyns, K., and Meyer, B.,** *N*-Nitroso sugar amino acids (N-NO-D-fructose-L-amino acids), in *N-Nitroso Compounds: Occurrence and Biological Effects,* Bartsch, H., O'Neill, I. K., Castegnaro, M., and Okada, M., Eds., IARC Scientific Publ. No. 41, International Agency for Research on Cancer, Lyon, France, 1982, 87.
18. **Röper, H., Röper, S., Heyns, K., and Meyer, B.,** Amadori and *N*-nitroso-Amadori compounds and their pyrolysis products. Chemical, analytical and biological aspects, in *N-Nitroso Compounds: Occurrence, Biological Effects and Relevance to Human Cancer,* O'Neill, I. K., von Borstel, R. C., Miller, C. T., Long, J., and Bartsch, H., Eds., IARC Scientific Publ. No. 57, International Agency for Research on Cancer, Lyon, France, 1984, 101.
19. **Hayatsu, H., Oka, T., Wakata, A., Ohara, Y., Hayatsu, T., Kobayashi, H., and Arimoto, S.,** Adsorption of mutagens to cotton bearing covalently bound trisulfocopper-phthalocyanine, *Mutat. Res.,* 119, 233, 1983.
20. **Toth, L. and Potthast, K.,** Chemical aspects of the smoking of meat and meat products, in *Advances in Food Research,* Vol. 29, Chichester, C. O., Mrak, E. M., and Schweigert, B. S., Eds., Academic Press, Orlando, FL, 1984, 87.
21. **Quillardet, P., Huisman, O., D'Ari, R., and Hofnung, M.,** SOS Chromotest, a direct assay of induction of an SOS function in *Escherichia coli* K-12 to measure genotoxicity, *Proc. Natl. Acad. Sci. U.S.A.,* 79, 5971, 1982.
22. **Kolar, G. F. and Schlesiger, J.,** Urinary metabolites of dimethyl-1-phenyltriazene, *Chem. Biol. Interact.,* 14, 301, 1976.

23. **Kikugawa, K. and Kato, T.,** Formation of a mutagenic diazoquinone by interaction of phenol with nitrite, *Food Chem. Toxicol.,* 26, 209, 1988.
24. **Challis, B. C.,** Rapid nitrosation of phenols and its implications for health hazards from dietary nitrites, *Nature (London),* 244, 466, 1973.
25. **Kazitsyna, L. A., Kikot, B. S., and Upadysheva, A. V.,** Quinone diazides and *p*-iminoquinone diazides, *Russ. Chem. Rev.,* 35, 388, 1966.
26. **Weselsky, P.,** Ueber die Einwirkung der salpetrigen Säure auf Phenol, *Ber.* 8, 98, 1875.
27. *IARC Monographs on the Evaluation of the Carcinogenic Risk of Chemicals to Humans,* Vol. 38, *Tobacco Smoking. Chemistry and Analysis of Tobacco Smoke,* International Agency for Research on Cancer, Lyon, France, 1986, 83.
28. **Toth, B., Patil, K., and Jae, H. S.,** Carcinogenesis of 4-(hydroxymethyl)-benzenediazonium ion (tetra-fluoroborate) of *Agaricus bisporus, Cancer Res.,* 41, 2444, 1981.
29. **Toth, B., Nagel, D., and Ross, A.,** Gastric tumourigenesis by a single dose of 4-(hydroxy-methyl)benzenediazonium ion of *Agaricus bisporus, Br. J. Cancer,* 46, 417, 1982.
30. **Ochiai, M., Wakabayashi, K., Nagao, M., and Sugimura, T.,** Tyramine is a major mutagen precursor in soy sauce, being convertible to a mutagen by nitrite, *Gann,* 75, 1, 1984.
31. **Nagao, M., Wakabayashi, K., Fujita, Y., Tahira, T., Ochiai, M., Takayama, S., and Sugimura, T.,** Nitrosatable precursors of mutagens in vegetables and soy sauce, in *Diet, Nutrition and Cancer,* Hayashi, Y., Nagao, M., Sugimura, T., Takayama, S., Tomatis, L., Wattenberg, L. W., and Wogan, G. N., Eds., Japan Scientific Societies Press, Tokyo, VNU Science Press, Utrecht, 1986, 77.
32. **Malaveille, C., Brun, G., Kolar, G., and Bartsch, H.,** Mutagenic and alkylating activities of 3-methyl-1-phenyltriazenes and their possible role as carcinogenic metabolites of the parent dimethyl compounds, *Cancer Res.,* 42, 1446, 1982.
33. **Gold, B. and Salmasi, S.,** Carcinogenicity tests of acetoxymethylphenyl-nitrosamine and benzenedi-azonium tetrafluoroborate in Syrian hamsters, *Cancer Lett.,* 15, 289, 1982.
34. **Lower, G. M., Lanpher, S. P., Jr., Johnson, B. M., and Bryan, G. T.,** Aryl and heterocyclic diazo compounds as potential environmental electrophiles, *J. Toxicol. Environ. Health,* 2, 1095, 1977.
35. **Boido, V., Bennicelli, C., Zanacchi, P., and De Flora, S.,** Formation of mutagenic derivatives from nitrite and two primary amines, *Toxicol. Lett.,* 6, 379, 1980.
36. **Kikugawa, K., Kato, T., and Takeda, Y.,** Formation of a highly mutagenic diazo compound from the bamethan-nitrite reaction, *Mutat. Res.,* 177, 35, 1987.
37. **Ohshima, H., Furihata, C., Matsushima, T., and Bartsch, H.,** Evidence of potential rumour-initiating and tumour-promoting activities of hickory smoke condensate when given alone or with nitrite to rats, *Food. Chem. Toxicol.,* 27, 511, 1989.
38. **Bognar, R. and Puskas, M. M.,** *N*-Glycosides. XVII. Preparation and structure of *N*-nitroso-*N*-arylgly-cosylamines, *Acta Chem. Hung.,* 76, 399, 1973.
39. **Koepke, S. R., Kroeger-Koepke, M. B., and Michejda, L. J.,** *N*-Nitroso-*N*-methylaniline: possible mode of DNA modification, in *The Relevance of N-Nitroso Compounds to Human Cancer: Exposures and Mechanisms,* Bartsch, H., O'Neill, I. K., and Schulte-Hermann, R., Eds., IARC Scientific Publ. No. 84, International Agency for Research on Cancer, Lyon, France, 1987, 68.
40. **Ohshima, H. and Bartsch, H.,** Quantitative estimation of endogenous nitrosation in humans by monitoring *N*-nitrosoproline excreted in the urine, *Cancer Res.,* 41, 3658, 1981.
41. **Bartsch, H., Ohshima, H., Pignatelli, B., and Calmels, S.,** Human exposure to endogenous *N*-nitroso compounds: quantitative estimates in subjects at high risk for cancer of the oral cavity, esophagus, stomach and urinary bladder, in *Nitrate, Nitrite and Nitroso Compounds in Human Cancer,* Forman, D. and Shuker, D., Eds., Cancer Surv., 8(2), 335, 1989.
42. **Bartsch, H., Terracini, B., Malaveille, C., Tomatis, L., Wahrendorf, J., Brun, G., and Dodet, B.,** Quantitative comparison of carcinogenicity, mutagenicity and electrophilicity of 10 direct-acting alkylating agents and of the initial O^6:7-alkylguanine ratio in DNA with carcinogenic potency in rodents, *Mutat. Res.,* 110, 181, 1983.
43. **Bartsch, H., Hemminki, K., and O'Neill, I. K., Eds.,** *Methods for Detecting DNA Damaging Agents in Humans: Applications in Cancer Epidemiology and Prevention,* IARC Scientific Publ. No. 89, International Agency for Research on Cancer, Lyon, France, 1988.

Chapter 6

METHODS FOR DETECTION OF MUTAGENS IN FOOD

Chapter 6.1

METHODS FOR SEPARATION AND DETECTION OF HETEROCYCLIC AMINES

Hikoya Hayatsu, Sakae Arimoto, and Keiji Wakabayashi

TABLE OF CONTENTS

I. OVERVIEW

The task of isolating and identifying food mutagens is, in general, laborious and time-consuming. The major reason for this is that these mutagens, notably the mutagenic heterocyclic amines, are present in food only in tiny amounts, at levels of parts per billion or below (see Section V). Their detection during the process of isolation is usually done by the use of bioassays, e.g., the Salmonella Ames test.[1] The Salmonella assay is very sensitive, which is a great advantage in this undertaking; on the other hand, since the assay utilizes the growth of bacteria on agar plates, it takes 2 days to obtain results and, therefore, one has to wait this period of time before determining the protocol of the next purification step.

The fact that a heterocyclic amine is present in food in a very small amount and yet is highly mutagenic in Salmonella gives rise to a situation in which the mutagenicity is easily detectable but characterization of the mutagen is difficult. For example, *katsuobushi,* the dried, heated fish meat popular among Japanese people, contains MeIQx at about 2 ng/g and the mutagenic activity of MeIQx is 145,000 revertants (rev)/μg in *Salmonella typhimurium* TA98 with metabolic activation. This setting makes it possible to detect the mutagenicity using less than 10 g of *katsuobushi* for preparing the test extract, but it also means that in order to obtain a quantity of purified MeIQx to be subjected to mass-spectroscopic characterization, a large amount of the material is necessary; in fact, 3.5 kg of *katsuobushi* was used for this purpose.[2]

As nitrogen-containing basic compounds, the heterocyclic amines are soluble in acid; therefore the first step in solubilizing them from food is usually extraction with acid, e.g., 1 N hydrochloric acid. In this regard, the mutagenic heterocyclic amines are similar to alkaloids, which are also nitrogenous basic compounds. It is interesting to note that the first isolation and structure determination of two heterocyclic amines, Trp-P-1 and Trp-P-2, was accomplished by Japanese workers who had long experience in alkaloid chemistry.[3]

Purification of the extracted heterocyclic amines is carried out with combinations of solvent-solvent partitions and various chromatographic techniques including high-performance liquid chromatography (HPLC). Use of selective adsorbents such as blue cotton is sometimes very effective to separate the mutagens from the bulk of food components (see Section IV). The Ames test is the most useful for monitoring the heterocyclic amines, especially because it can be used even in the early phase of the isolation, where a great variety of nonmutagenic materials are present in the sample at far greater amounts as compared with the amount of desired mutagens. However, it should be borne in mind that some of these nonmutagenic components could suppress the mutagenic activity of heterocyclic amines. For example, the mutagenicity of the basic fraction of heated meat extract is abolished when the acidic fraction is mixed with it, and this effect is ascribed to the presence of long-chain fatty acids in the acidic fraction (see Chapter 8 for further discussion on this point).[4]

Ultraviolet absorbance of the heterocyclic amines is generally intense, with ε values at the 10^4 level,[5] and therefore can be employed in the detection of these compounds. Detection by UV is especially useful in the last stages of the purification. UV absorption spectra are often used for a characterization purpose.

Fluorescence emission of heterocyclic amines is about one order of magnitude greater in sensitivity than UV absorption. Since Trp-P-1 (-2), Glu-P-1 (-2), AαC (MeAαC), and PhIP fluoresce when excited with light at their ultraviolet absorption wavelengths,[6] fluorescence is currently used in monitoring the HPLC profiles of these compounds.[7]

IQ and MeIQx derivatives, on the other hand, do not fluoresce. For these compounds, electrochemical activities have often been employed as a sensitive marker. This activity arises from the susceptibility of a compound to be oxidized or reduced electrochemically. IQ, MeIQ, MeIQx, and DiMeIQx are easily oxidizable by electrochemical treatment, and therefore this technique was used for monitoring these mutagens in HPLC purifications.[8]

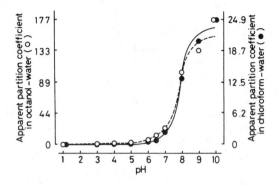

FIGURE 1.   Chloroform-water and *n*-octanol-water partitions of Trp-P-2.[12]

A powerful method for detecting heterocyclic amines in crude extracts of cooked food is an assay based on combined liquid chromatography-gas chromatography-mass spectrometry. In this method, a particular ion-peak characteristic of individual heterocyclic amines is selected and monitored. A pioneering study using this technique was done to quantify IQ and MeIQ in cooked salmon and beef.[9] More recently, MeIQx and 4,8-DiMeIQx were quantified in cooked beef by submitting to the analysis a solvent-extracted crude material prepared from only 2 g of the cooked patty.[10]

In several studies, specific sensitivity of a heterocyclic amine mutagenicity to the treatment with nitrous acid is used to determine whether the mutagens in question belong to the sensitive or the insensitive class.[11] Representatives of the sensitive class are Trp-P-2, Glu-P-1, and AαC, and those of the insensitive class are IQ and MeIQx.

II. ACID-BASE PARTITION

As discussed above, the mutagenic heterocyclic amines are in general soluble in aqueous acid but not in alkali. These aromatic amines in their free-base form are soluble in organic solvents such as dichloromethane and ethylacetate. A typical example is given in Figure 1. Trp-P-2 in its protonated form is insoluble in chloroform or *n*-octanol, but in its free form is readily soluble in these solvents.[12] Table 1 shows the pKa values of heterocyclic amines.

In almost all the work on the isolation and detection of heterocyclic amines, acid-base partitions have been incorporated as important steps in the process of purification.

III. COLUMN METHODS

A. KINDS OF COLUMNS USED FOR PURIFICATION OF HETEROCYCLIC AMINES

In earlier studies on the isolation and characterization of heterocyclic amines, chromatographic techniques with columns like silica gel and Sephadex were commonly used.[13-25] More recently, the XAD-2 resin adsorption-column technique was introduced. Furthermore, in current practice, HPLC has been very extensively used, particularly in the final stages of purification.

The usefulness of XAD-2 resin fractionation has been recently recognized. This cross-linked polysterene resin adsorbs lipophilic organic materials but not hydrophilic compounds or inorganic salts[26] and therefore is suited for concentrating heterocyclic amines.[27] Generally, a large volume of aqueous solution is passed through a small column packed with XAD-2, the resin is washed with water, and then the organic materials trapped in the resin are eluted with methanol or acetone. With this simple procedure, a high concentration of food-borne

TABLE 1
The pKa Values of Heterocyclic
Amines[a]

Compound	pKa
Trp-P-1	8.55
Trp-P-2	8.5
Glu-P-1	6.0
Glu-P-2	5.85
Phe-P-1	6.5
AαC	4.6
MeAαC	4.9
Lys-P-1	5.4
IQ	3.8, 6.6
	(3.5, 6.6 in the literature 80)
MeIQ	3.95, 6.4
MeIQx	<2, 6.25
	(6.4 in the literature 25)
4,8-DiMeIQx	<2, 6.25
7,8-DiMeIQx	<2, 6.45
PhIP	5.7

[a] The pKa values at room temperature (20°C) were determined spectroscopically. Buffers used for obtaining necessary absorption spectra were 0.04 *M* sodium phospate (pH 1 to 2, 5 to 7), sodium citrate (pH 3 to 4), sodium borate (pH 8 to 9), sodium carbonate (pH 10 to 11), and potassium phosphate (pH 12) with the presence of 0.8 *M* KCl in each buffer.

mutagens can be achieved. It is current practice, therefore, to use the XAD-2 resin concentration in the early phase of isolating heterocyclic amines from food.[28-40]

B. HPLC

Because of its high resolution power and time-saving effectiveness, HPLC has been used extensively in the study of heterocyclic amines.[41-69] HPLC has been commonly employed in the final purification steps. It has also been used extensively in analyzing metabolic and chemical changes of heterocyclic amines. Especially notable are the analyses of human excreta with HPLC for studying the fate of ingested mutagens.

A typical example of HPLC to show its resolution power is given in Figure 2.[55] Use of this procedure has allowed a simple quantification of IQ, MeIQ, MeIQx, 4,8-DiMeIQx, and PhIP.[55]

C. IMMUNOAFFINITY CHROMATOGRAPHY

The usefulness of immunoaffinity chromatography for the detection of food-borne mutagens was shown earlier in the studies of aflatoxin metabolites in human.[70] Also, immunoassays were developed for dioxin that contaminated some food.[71] Antibodies to individual mutagenic heterocyclic amines have been recently prepared. They can be used as such for immunoassays to detect and quantitate cognate heterocyclic amines or can be bound to chromatographic supports and used as affinity ligands. The features of these studies are summarized in Table 2. These methods have a great potential to minimize the laborious procedure that has been required for the detection of mutagens in food.

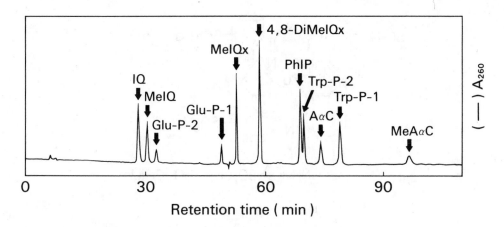

FIGURE 2.   Elution profile of heterocyclic amines by HPLC on an ODS column.[55]

TABLE 2
Immunochromatography and Immunoassays of Heterocyclic Amines

Mutagen	Process in which the antibodies are used	Results	Ref.
IQ	Affinity chromatography	Contents in beef extracts determined as 4.8 to 49 ng/g	81
MeIQx	Affinity chromatography	Contents in beef extracts determined as 30 to 69 ng/g	81
IQ	Titration (ELISA)[a]	IQ compounds in cooked beef detected	82
PhIP	Titration (ELISA)[a]	Contents in fried beef determined as 3 to 4200 ng/g	83

[a]   Enzyme-linked immunosorbent assay.

IV. BLUE-COTTON ADSORPTION

Blue cotton is cotton bearing covalently linked blue pigment, copper phthalocyanine trisulfonate.[44] The linkage connecting the pigment to cellulose is illustrated in Figure 3. Blue rayon is a recently developed, modified version of blue cotton. Both can be obtained from commercial sources.

The usefulness of blue cotton lies in its unique ability to adsorb aromatic compounds having three or greater fused rings. The adsorption capacity is strong, and the selectivity for adsorbing this class of compound is high.

Various mutagens dissolved in saline can be adsorbed to blue cotton and can be recovered from the cotton by elution with an organic solvent.[44] The adsorption is achieved simply by a batch process: shaking the mutagen solution with the blue cotton added. Elution with a mixture of methanol and a small amount of concentrated ammonia (usually used at a ratio of 50:1, but sometimes at a ratio of 1000:1) generally gives excellent recoveries of adsorbed mutagens. Table 3 shows results for several typical compounds. Trp-P-1, Trp-P-2, Glu-P-2, IQ, MeIQ, AαC, MeAαC, and PhIP can also be adsorbed to blue cotton with similar high efficiencies.[72] PhIP has a two-ring and a one-ring directly linked and conjugated and therefore would possess a planar size comparable to a three-ring system.

A large number of environmental mutagens and carcinogens have polycyclic aromatic

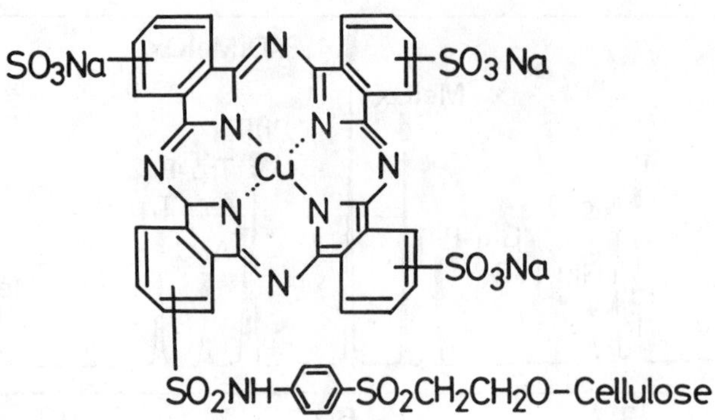

FIGURE 3.   Structure of blue cotton.[44]

TABLE 3
Adsorption and Recovery of Compounds by Blue Cotton[a,72]

	Adsorption (%)		Overall recovery by blue cotton (%)
Compound	Blue cotton	Plain cotton	
Glu-P-1	85	9	80
MeIQx	94	12	94
Daunorubicin	99	22	65
Aflatoxin B$_1$	84	26	63
4-Nitroquinoline-1-oxide	11	0	Not done
p-Nitrophenol·Na	5	0	Not done
[³H]Histidine	—	Not done	0.06
[¹⁴C]Nitrosodimethylamine	0	Not done	Not done

[a]   A 5-ml solution of a compound in 0.15 M NaCl (concentrations 0.1 to 100 μM) was treated with blue cotton (50 mg) by shaking at room temperature for 30 min. The cotton was taken up and squeezed to combine the cotton-held portion of the solution to the mother liquor. A fresh batch of blue cotton (50 mg) was added, and the treatment was repeated. The adsorption extent was determined by quantifying the compound that remained in the solution (either by spectrometry, fluorometry, or radioactivity measurement). The two 50-mg cotton samples were combined, wiped with a paper towel, washed twice with 0.15 M NaCl, freed of moisture by use of a paper towel, and then eluted with 5-ml methanol-concentrated ammonia (50:1) during gentle shaking for 30 min at room temperature. This elution was repeated once more, and the combined eluant was evaporated under reduced pressure. The residue was subjected to quantification of the compound. For the histidine adsorption, 1 mM [³H]histidine at 230,000 dpm was used, and the radioactivity recovered was 146 dpm, corresponding to 0.06% overall recovery.

structures; polycyclic aromatic hydrocarbons and food pyrolysate heterocyclic amines belong to this class. Mycotoxins, such as aflatoxin B$_1$, can adsorb to blue cotton (Table 3), although its structure is not perfectly planar. It is therefore expected that blue cotton is useful in extracting these types of mutagenic compounds from complex mixtures such as food.

For extraction of mutagenic compounds from Difco Bacto beef extract, the sample was first diluted by dissolving in water, and the solution was then treated with blue cotton.[25] In this study of beef extract mutagens, several cycles of blue-cotton adsorption were performed, i.e., a blue-cotton extract was again dissolved in water, and the solution was treated with a new batch of blue cotton. With this repetition of the cycle, an efficient concentration of the mutagenic component was achieved with only a small loss of total activity (Table 4). By further purification with HPLC, the presence of MeIQx and IQ in the beef extract was demonstrated.

TABLE 4
Efficiency of Repetitive Blue-Cotton Processing in Purification
of Mutagens in Beef Extract[25]

	Sample	Blue cotton used	Material obtained	
			Weight (mg)	Mutagenicity[a]
Cycle 1	10 g Difco Bacto beef extract in 100 ml H_2O	0.8 g × 3[b]	15.6	150,400 (100%)
Cycle 2	15.6 mg in 100 ml H_2O	0.4 g × 2	0.3	118,200 (78%)
Cycle 3	0.3 mg in 50 ml H_2O	0.2 g × 2	0.1	101,800 (68%)

[a] The mutagenicity was assayed for small fractions of the samples using *S. typhimurium* TA98 with metabolic activation. The numbers represent those of revertants corresponding to the total material.
[b] This expression means that the blue-cotton adsorption was done three times, with 0.8 g fresh cotton at each adsorption.

More recently, Sugimura and co-workers applied the blue-cotton adsorption in the first step of quantifying IQ and MeIQx in beef extract.[8] They were able to achieve a 10^4-fold concentration of MeIQx-containing fraction by the use of two cycles of blue-cotton extraction, with an overall recovery of 71% for MeIQx. By a similar procedure, 4,8-Me$_2$IQx in beef extract was quantified.[47] These workers also succeeded in detecting Trp-P-2 in blue-cotton extract of broiled beef.[73]

By use of the blue-cotton extraction, MeIQx was identified as a major mutagenic component of *katsuobushi*, as already mentioned.[2] MeIQx and 4,8-Me$_2$IQx were also found in other Japanese smoked, dried fish products.[35] Since raw bonito does not contain these mutagens, investigations were initiated to determine which stage of processing is responsible for mutagen formation. It turned out that the process called *baikan*, in which the fish is dried at 80 to 120°C for a period of several days, is responsible.[50] Consistent with this finding, heating various fish at 100°C for 48 hr, which does not result in charring of the fish, can give rise to formation of MeIQx and 4,8-Me$_2$IQx.[53] In these studies, the blue-cotton method was used extensively.

Blue cotton has been utilized by many researchers as a regular step for preparing partially purified samples of food pyrolysate heterocyclic amines to be submitted to HPLC analysis. Thus, Edmonds et al.[48] quantified IQ and MeIQ in broiled salmon, Aeschbacher and co-workers[39] analyzed IQ and MeIQx in cooked beef products, and Manabe and co-workers[54] detected Glu-P-1 and Glu-P-2 in Worcestershire sauce. Recently, Zhang et al.[55] analyzed Chinese cooked fish and meat and detected IQ, MeIQ, MeIQx, 4,8-Me$_2$IQx, and PhIP.

Blue cotton has been used in studies on the mechanism of heterocyclic amine formation.[29,30,74,75] It has also been effectively applied in investigations of the metabolites of heterocyclic amines.[62,76,77]

Trp-P-1 and Trp-P-2 were found in dialysis fluid of patients with uremia.[64] In this study, a large volume of the dialysis fluid (40 l) was subjected to the blue-cotton extraction, and the mutagens in the extract were quantified by HPLC. High recoveries (60 to 70%) were noted both for Trp-P-1 and Trp-P-2. Likewise, MeIQx has been detected in the blue-cotton extract of dialysis fluid of uremic patients.[61] Glu-P-1 and Glu-P-2 are found in the blue-cotton extracts of human plasma.[63] Furthermore, Trp-P-1 and Trp-P-2 can be detected in normal human plasma[78] and urine.[79] The presence of heterocyclic amines in urine has been correlated to the diet ingested by human subjects.[79] An interesting finding is that Glu-P-1 and Glu-P-2 are present in human cataractous lens.[65]

TABLE 5
Amounts of Heterocyclic Amines in Cooked Foods[a]

Sample	Amount (ng/g cooked food)							
	IQ	MeIQx	4,8-DiMeIQx	PhIP	Trp-P-1	Trp-P-2	AαC	MeAαC
Broiled beef	0.19	2.11		15.7	0.21	0.25	1.20	
Fried ground beef		0.64	0.12	0.56	0.19	0.21		
Broiled chicken		2.33	0.81	38.1	0.12	0.18	0.21	
Broiled mutton		1.01	0.67	42.5		0.15	2.50	0.19
Food-grade beef extract		3.10[b]		3.62				
Bacteriological-grade beef extract	41.6[b]	58.7[b]	10.0[c]					

[a] Unless otherwise indicated, data represent unpublished results obtained in the National Cancer Center Research Institute, Tokyo.
[b] Reference 48.
[c] Reference 47.

V. HETEROCYCLIC AMINE CONTENTS IN FOOD

With a set of standardized isolation procedure, Sugimura and co-workers have systematically surveyed the amounts of heterocyclic amines in cooked foods. As shown in Table 5, IQ, MeIQx, and several other heterocyclic amines are present in cooked meat at low parts per billion levels, while PhIP is found at significantly higher levels.[6,46,47,84]

VI. ABBREVIATIONS

Trp-P-1, 3-amino-1,4-dimethyl-5*H*-pyrido[4,3-*b*]indole; Trp-P-2, 3-amino-1-methyl-5*H*-pyrido[4,3-*b*]indole; Glu-P-1, 2-amino-6-methyldipyrido[1,2-*a:3′,2′-d*]imidazole; Glu-P-2, 2-aminodipyrido[1,2-*a:3′,2′-d*]imidazole; AαC, 2-amino-9*H*-pyrido[2,3-*b*]indole; MeAαC, 2-amino-3-methyl-9*H*-pyrido[2,3-*b*]indole; IQ, 2-amino-3-methylimidazo[4,5-*f*]quinoline; MeIQ, 2-amino-3,4-dimethylimidazo[4,5-*f*]quinoline; MeIQx, 2-amino-3,8-dimethylimidazo[4,5-*f*]-quinoxaline; 4,8-DiMeIQx, 2-amino-3,4,8-trimethylimidazo[4,5-*f*]-quinoxaline; 7,8-DiMeIQx,2-amino-3,7,8-trimethylimidazo[4,5-*f*]-quinoxaline; PhIP, 2-amino-1-methyl-6-phenylimidazo[4,5-*b*]pyridine; Phe-P-1, 2-amino-5-phenylpyridine; Lys-P-1, 3,4-cyclopentenopyrido[3,2-*a*]carbazole.

REFERENCES

1. **Ames, B. N., McCann, J., and Yamasaki, E.,** Methods for detecting carcinogens and mutagens with the Salmonella/mammalian-microsome mutagenicity test, *Mutat. Res.,* 31, 347, 1975.
2. **Kikugawa, K., Kato, T., and Hayatsu, H.,** The presence of 2-amino-3,8-dimethylimidazo[4,5-*f*]quinoxaline in smoked dry bonito (Katsuobushi), *Jpn. J. Cancer Res. (Gann),* 77, 99, 1986.
3. **Sugimura, T., Kawachi, T., Nagao, M., Yahagi, T., Seino, Y., Okamoto, T., Shudo, K., Kosuge, T., Tsuji, K., Wakabayashi, K., Iitaka, T., and Itai, A.,** Mutagenic principle(s) in tryptophan and phenylalanine pyrolysis products, *Proc. Jpn. Acad.,* 53, 58, 1977.
4. **Hayatsu, H., Inoue, K., Ohta, H., Namba, T., Togawa, K., Hayatsu, T., Makita, M., and Wataya, Y.,** Inhibition of the mutagenicity of cooked-beef basic fraction by its acidic fraction, *Mutat. Res.,* 91, 437, 1981.

5. **Hatch, F. T., Felton, J. S., Stuermer, D. H., and Bjeldanes, L. F.,** Identification of mutagens from the cooking of food, in *Chemical Mutagens,* Vol. 9, de Serres, F. J. Ed., Plenum Press, New York, 1984, 111.

6. **Sugimura, T., Sato, S., and Wakabayashi, K.,** Mutagens/carcinogens in pyrolysates of amino acids and proteins and in cooked foods: heterocyclic aromatic amines, in *Chemical Induction of Cancer,* Vol. IIIC, Woo, Y.-T., Lai, D. Y., Arcos, J. C., and Argus, M. F., Eds., Academic Press, San Diego, 1988, 681.

7. **Wakabayashi, K. and Sugimura, T.,** to be published.

8. **Takahashi, M., Wakabayashi, K., Nagao, M., Yamamoto, M., Masui, T., Goto, T., Kinae, N., Tomita, I., and Sugimura, T.,** Quantification of 2-amino-3-methylimidazo[4,5-*f*]quinoline (IQ) and 2-amino-3,8-dimethylimidazo[4,5-*f*]quinoxaline (MeIQx) in beef extracts by liquid chromatography with electrochemical detection (LCEC), *Carcinogenesis,* 6, 1195, 1985.

9. **Yamaizumi, Z., Kasai, H., Nishimura, S., Edmonds, C. G., and McCloskey, J. A.,** Stable isotope dilution quantification of mutagens in cooked foods by combined liquid chromatography-thermospray mass spectrometry, *Mutat. Res.,* 173, 1, 1986.

10. **Murray, S., Gooderham, N. J., Boobis, A. R., and Davies, D. S.,** Measurement of MeIQx and DiMeIQx in fried beef by capillary column gas chromatography electron capture negative ion chemical ionisation mass spectrometry, *Carcinogenesis,* 9, 321, 1988.

11. **Tsuda, M., Negishi, C., Makino, R., Sato, S., Yamaizumi, Z., Hirayama, T., and Sugimura, T.,** Use of nitrite and hypochlorite treatments in determination of the contributions of IQ-type and non-IQ-type heterocyclic amines to the mutagenicities in crude pyrolyzed materials, *Mutat. Res.,* 147, 335, 1985.

12. **Kimura, T., Nakayama, T., Kurosaki, Y., Suzuki, Y., Arimoto, S., and Hayatsu, H.,** Absorption of 3-amino-1-methyl-5*H*-pyrido[4,3-*b*]indole, a mutagen-carcinogen present in tryptophan pyrolysate, from the gastrointestinal tract in the rat, *Jpn. J. Cancer Res. (Gann),* 76, 272, 1985.

13. **Kosuge, T., Tsuji, K., Wakabayashi, K., Okamoto, T., Shudo, K., Iitaka, Y., Itai, A., Sugimura, T., Kawachi, T., Nagao, M., Yahagi, T., and Seino, Y.,** Isolation and structure studies of mutagenic principles in amino acid pyrolysates, *Chem. Pharm. Bull.,* 26, 611, 1978.

14. **Yamamoto, T., Tsuji, K., Kosuge, T., Okamoto, T., Shudo, K., Takeda, K., Iitaka, Y., Yamaguchi, K., Seino, Y., Yahagi, T., Nagao, M., and Sugimura, T.,** Isolation and structure determination of mutagenic substances in L-glutamic acid pyrolysate, *Proc. Jpn. Acad.,* 54 (Ser. B), 248, 1978.

15. **Wakabayashi, K., Tsuji, K., Kosuge, T., Takeda, K., Yamaguchi, K., Shudo, K., Iitaka, Y., Okamoto, T., Yahagi, T., Nagao, M., and Sugimura, T.,** Isolation and structure determination of a mutagenic substance in L-lysine pyrolysate, *Proc. Jpn. Acad.,* 54 (Ser. B), 569, 1978.

16. **Yoshida, D., Matsumoto, T., Yoshimura, R., and Matsuzaki, T.,** Mutagenicity of amino-α-carbolines in pyrolysis products of soybean globulin, *Biochem. Biophys. Res. Commun.,* 83, 915, 1978.

17. **Yoshida, D., Nishigata, H., and Matsumoto, T.,** Pyrolytic yields of 2-amino-9*H*-pyrido[2,3-*b*]indole and 3-amino-1-methyl-5*H*-pyrido[4,3-*b*]indole as mutagens from proteins, *Agric. Biol. Chem.,* 43, 1769, 1979.

18. **Kasai, H., Yamaizumi, Z., Wakabayashi, K., Nagao, M., Sugimura, T., Yokoyama, S., Miyazawa, T., Spingarn, N. E., Weisburger, J. H., and Nishimura, S.,** Potent novel mutagens produced by broiling fish under normal conditions, *Proc. Jpn. Acad.,* 56 (Ser. B), 278, 1980.

19. **Yamaizumi, Z., Shiomi, T., Kasai, H., Nishimura, S., Takahashi, Y., Nagao, M., and Sugimura, T.,** Detection of potent mutagens, Trp-P-1 and Trp-P-2, in broiled fish, *Cancer Lett.,* 9, 75, 1980.

20. **Yokota, M., Narita, K., Kosuge, T., Wakabayashi, K., Nagao, M., Sugimura, T., Yamaguchi, K., Shudo, K., Iitaka, Y., and Okamoto, T.,** A potent mutagen isolated from a pyrolysate of L-ornithine, *Chem. Pharm. Bull.,* 29, 1473, 1981.

21. **Kasai, H., Yamaizumi, Z., Shiomi, T., Yokoyama, S., Miyazawa, T., Wakabayashi, K., Nagao, M., Sugimura, T., and Nishimura, S.,** Structure of a potent mutagen isolated from fried beef, *Chem. Lett.,* 485, 1981.

22. **Tada, M., Saeki, H., and Oikawa, A.,** The identification of 3-amino-9*H*-pyrido[3,4-*b*]indole derivatives in L-tryptophan pyrolysates, *Bull. Chem. Soc. Jpn.,* 56, 1450, 1983.

23. **Tada, M., Saeki, H., and Oikawa, A.,** Presence of 3-amino-β-carbolines (3-amino-9*H*-pyrido[3,4-*b*]indoles), effectors in the induction of sister-chromatid exchanges, in protein pyrolysates, *Mutat. Res.,* 121, 81, 1983.

24. **Grivas, S., Nyhammar, T., Olsson, K., and Jägerstad, M.,** Formation of a new mutagenic DiMeIQx compound in a model system by heating creatinine, alanine and fructose, *Mutat. Res.,* 151, 177, 1985.

25. **Hayatsu, H., Matsui, Y., Ohara, Y., Oka, T., and Hayatsu, T.,** Characterization of mutagenic fractions in beef extract and in cooked ground beef. Use of blue-cotton for efficient extraction, *Gann,* 74, 472, 1983.

26. **Yamazaki, E. and Ames, B. N.,** Concentration of mutagens from urine by adsorption with the nonpolar resin XAD-2. Cigarette smokers have mutagenic urine, *Proc. Natl. Acad. Sci. U.S.A.,* 74, 3555, 1977.

27. **Bjeldanes, L. F., Grose, K. R., Davis, P. H., Stuermer, D. H., Healy, S. K., and Felton, J. S.,** An XAD-2 resin method for efficient extraction of mutagens from fried ground beef, *Mutat. Res.,* 105, 43, 1982.

28. **Felton, J. S., Knize, M. G., Wood, C., Wuebbles, B. J., Healy, S. K., Stuermer, D. H., Bjeldanes, L. F., Kimble, B. J., and Hatch, F. T.**, Isolation and characterization of new mutagens from fried ground beef, *Carcinogenesis,* 5, 95, 1984.

29. **Jägerstad, M., Olsson, K., Grivas, S., Negishi, C., Wakabayashi, K., Tsuda, M., Sato, S., and Sugimura, T.**, Formation of 2-amino-3,8-dimethylimidazo[4,5-*f*]quinoxaline in a model system by heating creatinine, glycine and glucose, *Mutat. Res.,* 126, 239, 1984.

30. **Negishi, C., Wakabayashi, K., Tsuda, M., Sato, S., Sugimura, T., Saito, H., Maeda, M., and Jägerstad, M.**, Formation of 2-amino-3,7,8-trimethylimidazo[4,5-*f*]quinoxaline, a new mutagen, by heating a mixture of creatinine, glucose and glycine, *Mutat. Res.,* 140, 55, 1984.

31. **Knize, M. G., Andresen, B. D., Healy, S. K., Shen, N. H., Lewis, P. R., Bjeldanes, L. F., Hatch, F. T., and Felton, J. S.**, Effects of temperature, patty thickness and fat content on the production of mutagens in fried ground beef, *Food Chem. Toxicol.,* 23, 1035, 1985.

32. **Kikugawa, K., Kato, T., and Hayatsu, H.**, Mutagenicity of smoked, dried bonito products, *Mutat. Res.,* 158, 35, 1985.

33. **Felton, J. S., Knize, M. G., Shen, N. H., Lewis, P. R., Andresen, B. D., Happe, J., and Hatch, F. T.**, The isolation and identification of a new mutagen from fried ground beef: 2-amino-1-methyl-6-phenylimidazo[4,5-*b*]pyridine (PhIP), *Carcinogenesis,* 7, 1081, 1986.

34. **Grose, K. R., Grant, J. L., Bjeldanes, L. F., Andresen, B. D., Healy, S. K., Lewis, P. R., Felton, J. S., and Hatch, F. T.**, Isolation of the carcinogen IQ from fried egg patties, *J. Agric. Food. Chem.,* 34, 201, 1986.

35. **Kato, T., Kikugawa, K., and Hayatsu, H.**, Occurrence of the mutagens 2-amino-3,8-dimethylimidazo[4,5-*f*]quinoxaline (MeIQx) and 2-amino-3,4,8-trimethylimidazo[4,5-*f*]quinoxaline (4,8-Me₂IQx) in some Japanese smoked, dried fish products, *J. Agric. Food Chem.,* 34, 810, 1986.

36. **Knize, M. G., Happe, J. A., Healy, S. K., and Felton, J. S.**, Identification of the mutagenic quinoxaline isomers from fried ground beef, *Mutat. Res.,* 178, 25, 1987.

37. **Shioya, M., Wakabayashi, K., Sato, S., Nagao, M., and Sugimura, T.**, Formation of a mutagen, 2-amino-1-methyl-6-phenylimidazo[4,5-*b*]pyridine (PhIP) in cooked beef, by heating a mixture containing creatinine, phenylalanine and glucose, *Mutat. Res.,* 191, 133, 1987.

38. **Becher, G., Knize, M. G., Nes, I. F., and Felton, J. S.**, Isolation and identification of mutagens from a fried Norwegian meat product, *Carcinogenesis,* 9, 247, 1988.

39. **Turesky, R. J., Bur, H., Huynh-Ba, T., Aeschbacher, H. U., and Milon, H.**, Analysis of mutagenic heterocyclic amines in cooked beef products by high-performance liquid chromatography in combination with mass spectrometry, *Food Chem. Toxicol.,* 26, 501, 1988.

40. **Kikugawa, K., Kato, T., and Takahashi, S.**, Possible presence of 2-amino-3,4-dimethylimidazo[4,5-*f*]quinoline and other heterocyclic amine-like mutagens in roasted coffee beans, *J. Agric. Food Chem.,* 37, 881, 1989.

41. **Yamaguchi, K., Zenda, H., Shudo, K., Kosuge, T., Okamoto, T., and Sugimura, T.**, Presence of 2-aminopyrido[1,2-*a:3',2'-d*]imidazole in casein pyrolysate, *Gann,* 70, 848, 1979.

42. **Yamaguchi, K., Shudo, K., Okamoto, T., Sugimura, T., and Kosuge, T.**, Presence of 2-aminodipyrido[1,2-*a:3',2'-d*]imidazole in broiled cuttle-fish, *Gann,* 71, 743, 1980.

43. **Yamaguchi, K., Shudo, K., Okamoto, T., Sugimura, T., and Kosuge, T.**, Presence of 3-amino-1,4-dimethyl-5*H*-pyrido[4,3-*b*]indole in broiled beef, *Gann,* 71, 745, 1980.

44. **Hayatsu, H., Oka, T., Wakata, A., Ohara, Y., Hayatsu, T., Kobayashi, H., and Arimoto, S.**, Adsorption of mutagens to cotton bearing covalently bound trisulfo-copper-phthalocyanine, *Mutat. Res.,* 119, 233, 1983.

45. **Yoshida, D., Saito, Y., and Mizusaki, S.**, Isolation of 2-amino-3-methylimidazo[4,5-*f*]quinoline as mutagen from the heated product of a mixture of creatine and proline, *Agric. Biol. Chem.,* 48, 241, 1984.

46. **Hirose, M., Wakabayashi, K., Grivas, S., De Flora, S., Arakawa, N., Nagao, M., and Sugimura, T.**, Formation of a nitro derivative of 2-amino-3,4-dimethylimidazo[4,5-*f*]quinoline by photo-irradiation, *Carcinogenesis,* 11, 869, 1990.

47. **Takahashi, M., Wakabayashi, K., Nagao, M., Yamaizumi, Z., Sato, S., Kinae, N., Tomita, I., and Sugimura, T.**, Identification and quantification of 2-amino-3,4,8-trimethylimidazo[4,5-*f*]quinoxaline (4,8-DiMeIQx) in beef extract, *Carcinogenesis,* 6, 1537, 1985.

48. **Edmonds, C. G., Sethi, S. K., Yamaizumi, Z., Kasai, H., Nishimura, S., and McCloskey, J. A.**, Analysis of mutagens from cooked foods by directly combined liquid chromatography-mass spectrometry, *Environ. Health Perspect.,* 67, 35, 1986.

49. **Taylor, R. T., Fultz, E., and Knize, M.**, Mutagen formation in a model beef supernatant fraction. IV. Properties of the system, *Environ. Health Perspect.,* 67, 59, 1986.

50. **Kikugawa, K., Kato, T., and Hayatsu, H.**, The presence of 2-amino-3,8-dimethylimidazo[4,5-*f*]quinoxaline in smoked dry bonito (Katsuobushi), *Jpn. J. Cancer Res. (Gann),* 77, 99, 1986.

51. **Yamaizumi, Z., Kasai, H., Nishimura, S., Edmonds, C. G., and McCloskey, J. A.,** Stable isotope dilution quantification of mutagens in cooked foods by combined liquid chromatography-thermospray mass spectrometry, *Mutat. Res.,* 173, 1, 1986.

52. **Yamashita, M., Wakabayashi, K., Nagao, M., Sato, S., Yamaizumi, Z., Takahashi, M., Kinae, N., Tomita, I., and Sugimura, T.,** Detection of 2-amino-3-methylimidazo[4,5-*f*]quinoline in cigarette smoke condensate, *Jpn. J. Cancer Res. (Gann),* 77, 419, 1986.

53. **Kikugawa, K. and Kato, K.,** Formation of mutagens, 2-amino-3,8-dimethylimidazo[4,5-*f*]quinoxaline (MeIQx) and 2-amino-3,4,8-trimethylimidazo[4,5-*f*]quinoxaline (4,8-DiMeIQx), in heated fish meats, *Mutat. Res.,* 179, 5, 1987.

54. **Manabe, S., Kanai, Y., Yanagisawa, H., Tohyama, K., Ishikawa, S., Kitagawa, Y., and Wada, O.,** Detection of carcinogenic glutamic acid pyrolysis products in Worcestershire sauce by high-performance liquid chromatography, *Environ. Mol. Mutagen.,* 11, 379, 1988.

55. **Zhang, X.-M., Wakabayashi, K., Liu, Z.-C., Sugimura, T., and Nagao, M.,** Mutagenic and carcinogenic heterocyclic amines in Chinese cooked foods, *Mutat. Res.,* 201, 181, 1988.

56. **Yamamoto, M., Ishikawa, M., Masui, T., Nukaya, H., Tsuji, K., and Kosuge, T.,** Analysis of mutagenic and carcinogenic compounds in oil of charred egg yolk, *J. Food Hyg. Soc. Jpn.,* 30, 146, 1989.

57. **Gross, G. A., Philippossian, G., and Aeschbacher, H. U.,** An efficient and convenient method for the purification of mutagenic heterocyclic amines in heated meat products, *Carcinogenesis,* 10, 1175, 1989.

58. **Alexander, J., Wallin, H., Holme, J. A., and Becher, G.,** 4-(2-Amino-1-methylimidazo[4,5-*b*]pyrid-6-yl)phenyl sulfate — a major metabolite of the food mutagen 2-amino-1-methyl-6-phenylimidazo[4,5-*b*]pyridine (PhIP) in the rat, *Carcinogenesis,* 10, 1543, 1989.

59. **Holme, J. A., Wallin, H., Brunborg, G., Soderlund, E. J., Hongslo, J. K., and Alexander, J.,** Genotoxicity of the food mutagen 2-amino-1-methyl-6-phenylimidazo[4,5-*b*]pyridine (PhIP) : formation of 2-hydroxyamino-PhIP, a directly acting genotoxic metabolite, *Carcinogenesis,* 10, 1389, 1989.

60. **Wallin, H., Holme, J. A., Becher, G., and Alexander, J.,** Metabolism of the food carcinogen 2-amino-3,8-dimethylimidazo[4,5-*f*]quinoxaline in isolated rat liver cells, *Carcinogenesis,* 10, 1277, 1989.

61. **Yanagisawa, H., Manabe, S., Kitagawa, Y., Ishikawa, S., Nakajima, K., and Wada, O.,** Presence of 2-amino-3,8-dimethylimidazo[4,5-*f*]quinoxaline (MeIQx) in dialysate from patients with uremia, *Biochem. Biophys. Res. Commun.,* 138, 1084, 1986.

62. **Hayatsu, H., Kasai, H., Yokoyama, S., Miyazawa, T., Yamaizumi, Z., Sato, S., Nishimura, S., Arimoto, S., Hayatsu, T., and Ohara, Y.,** Mutagenic metabolites in urine and feces of rats fed with 2-amino-3,8-dimethylimidazo[4,5-*f*]quinoxaline, a carcinogenic mutagen present in cooked meat, *Cancer Res.,* 47, 791, 1987.

63. **Manabe, S., Yanagisawa, H., Ishikawa, S., Kitagawa, Y., Kanai, Y., and Wada, O.,** Accumulation of 2-amino-6-methyldipyrido[1,2-*a:3',2'-d*]imidazole and 2-aminodipyrido-[1,2-*a:3',2'-d*]imidazole, carcinogenic glutamic acid pyrolysis products, in plasma of patients with uremia, *Cancer Res.,* 47, 6150, 1987.

64. **Manabe, S., Yanagisawa, H., Guo, S.-B., Abe, S., Ishikawa, S., and Wada, O.,** Detection of Trp-P-1 and Trp-P-2, carcinogenic tryptophan pyrolysis products, in dialysis fluid of patients with uremia, *Mutat. Res.,* 179, 33, 1987.

65. **Manabe, S., Yanagisawa, H., Kanai, Y., and Wada, O.,** Presence of carcinogenic glutamic acid pyrolysis products in human cataractous lens, *Opthal. Res.,* 20, 20, 1988.

66. **Knize, M. G., Övervik, E., Midtvedt, T., Turteltaub, K. W., Happe, J. A., Gustafsson, J.-A., and Felton, J. S.,** The metabolism of 4,8-DiMeIQx in conventional and germ-free rats, *Carcinogenesis,* 10, 1479, 1989.

67. **Murray, S., Gooderham, J. J., Boobis, A. R., and Davies, D. S.,** Detection and measurement of MeIQx in human urine after ingestion of a cooked meat meal, *Carcinogenesis,* 10, 763, 1989.

68. **Sjodin, P., Wallin, H., Alexander, J., and Jägerstad, M.,** Disposition and metabolism of the food mutagen 2-amino-3,8-dimethylimidazo[4,5-*f*]quinoxaline (MeIQx) in rats, *Carcinogenesis,* 10, 1269, 1989.

69. **Hiramoto, K., Negishi, T., Namba, T., Katsu, T., and Hayatsu, H.,** Superoxide dismutase-mediated reversible conversion of 3-hydroxyamino-1-methyl-5*H*-pyrido[4,3-*b*]indole, the N-hydroxy derivative of Trp-P-2, into its nitroso derivative, *Carcinogenesis,* 9, 2003, 1988.

70. **Groopman, J. D., Donahue, P. R., Zhu, J., Chen, J., and Wogan, G. N.,** Aflatoxin metabolism in humans: Detection of metabolites and nucleic acid adducts in urine by affinity chromatography, *Proc. Natl. Acad. Sci. U.S.A.,* 82, 6492, 1985.

71. **Stanker, L. H., Watkins, B., Roger, N., and Vanderlaan, M.,** Monoclonal antibodies for dioxin: antibody characterization and assay development, *Toxicology,* 45, 229, 1987.

72. **Hayatsu, H.,** Blue cotton — broad possibility in assessing mutagens/carcinogens in the environment, in *Advances in Mutagenesis Research,* Vol. 1, Obe, G., Ed., Springer-Verlag, Berlin, 1990, 1.

73. **Takahashi, M., Wakabayashi, K., Nagao, M., Tomita, I., and Sugimura, T.,** Quantification of mutagenic/carcinogenic heterocyclic amines in cooked foods, *Mutat. Res.,* 147, 275, 1985.

74. **Negishi, C., Wakabayashi, K., Tsuda, M., Saito, H., Maeda, M., Sato, S., Sugimura, T., Jägerstad, M., Muramatsu, M., and Matsushima, T.,** Formation of MeIQx and a new mutagen DiMeIQx on heating mixtures of creatinine, glucose and amino acids, *Environ. Mutagen. Res. Commun.,* 6, 129, 1984.

75. **Negishi, C., Wakabayashi, K., Yamaizumi, Z., Saito, H., Sugimura, T., and Jägerstad, M.,** Identification of 4,8-DiMeIQx, a new mutagen, *Mutat. Res.,* 147, 267, 1985.

76. **Bashir, W., Kingston, D. G. I., Carman, R. J., Van Tassell, R. L., and Wilkins, T. D.,** Anaerobic metabolism of 2-amino-3-methyl-3H-imidzo[4,5-*f*]quinoline (IQ) by human fecal flora, *Mutat. Res.,* 190, 187, 1987.

77. **Inamasu, T., Luks, H. J., and Weisburger, J. H.,** Comparison of XAD-2 column and blue cotton batch techniques for isolation of metabolites of 2-amino-3-methylimidazo[4,5-*f*]quinoline, *Jpn. J. Cancer Res. (Gann),* 79, 42, 1988.

78. **Manabe, S. and Wada, O.,** Analysis of human plasma as an exposure level monitor for carcinogenic tryptophan pyrolysis products, *Mutat. Res.,* 209, 33, 1988.

79. **Ushiyama, H., Wakabayashi, K., and Sugimura, T.,** in preparation.

80. **Kasai, H., Yamaizumi, A., Nishimura, S., Wakabayashi, K., Nagao, M., Sugimura, T., Spingarn, N. E., Weisburger, J. H., Yokoyama, S., and Miyazawa, T.,** A potent mutagen in broiled fish. I. 2-Amino-3-methyl-3H-imidazo[4,5-*f*]quinoline, *J. Chem. Soc. Perkin 1,* p. 2290, 1981.

81. **Turesky, R. J., Forster, C. M., Aeschbacher, H. U., Wurzner, H. P., Skipper, P. L., Trudel, L. J., and Tannenbaum, S. R.,** Purification of the food-borne carcinogens 2-amino-3-methylimidazo[4,5-*f*]quinoline and 2-amino-3,8-dimethylimidazo[4,5-*f*]quinoxaline in heated meat products by immunoaffinity chromatography, *Carcinogenesis,* 10, 151, 1989.

82. **Vanderlaan, M., Watkins, B. E., Hwang, M., Knize, M. G., and Felton, J. S.,** Monoclonal antibodies for the immunoassay of mutagenic compounds produced by cooking beef, *Carcinogenesis,* 9, 153, 1988.

83. **Vanderlaan, M., Watkins, B. E., Hwang, M., Knize, M. G., and Felton, J. S.,** Monoclonal antibodies to 2-amino-1-methyl-6-phenylimidazo[4,5-*b*]pyridine (PhIP) and their use in the analysis of well-done fried beef, *Carcinogenesis,* 10, 2215, 1989.

84. **Takahashi, M., Wakabayashi, K., and Sugimura, T.,** in preparation.

Chapter 6.2

MYCOTOXINS

Yoshio Ueno

TABLE OF CONTENTS

I. INTRODUCTION

A variety of fungal metabolites are toxic to animals. After the finding that aflatoxin B_1 (AFB_1) is a potent mutagen and carcinogen, and is associated with the development of primary liver cancer in several endemic areas, the carcinogenicity and commodities have become of great concern as an environmental hazard.[1-4]

In this chapter, the author summarizes the current information on the genotoxicity, food contamination, analysis and metabolism of genotoxic mycotoxins, particularly of AFB_1, AFM_1, sterigmatocystin (ST), ochratoxin A (OTA), and emodin (EM), and related anthraquinones. Recent reviews are available for the methods of detection.[5,6]

II. EVALUATION OF GENOTOXICITY OF MYCOTOXINS

Many fungal species are present in cereals as plant parasites and contaminants, and they are often detected in food products. Furthermore, some fungi are used in fermentation and the food and pharmaceutical industries. The evaluation of genotoxicities of fungi and their metabolites was carried out by employing several short-term tests with microbes, cultured mammalian cells, and others, along with the long-term carcinogenicity tests.

First, *Rec*-assay with recombination-deficient mutant cells of *Bacillus subtilis* was introduced for the evaluation of 30 mycotoxins and 5 chemically modified toxins produced by *Penicillium*, *Aspergillus*, *Fusarium*, and other species. The data revealed that AFB_1, ST, patulin, penicillic acid, and others were positive in this assay.[7]

The Ames test was often used for the detection of mutagenicity of the toxic fungal metabolites. Dihydrobisfuranoid compounds such as AFB_1, AFG_1, ST, 5,6-dimethoxy-ST, versicolorin A, and austocystin A and D are highly mutagenic in the presence of the activation system, while no activity was found in two carcinogens luteoskyrin (LS) and OTA,[8-11] and the mutagenic acitivity appeared to be related to the terminal dihydrobisfuran ring and not to the anthraquinone moiety. Screening of mutagenic compounds in the natural and synthetic 90 anthraquinones revealed that 35% of 9,10-anthraquinoid compounds were mutagenic.[12] The mutagenicity of 41 fungal isolates representing common plant pathogens and food crop contaminants was evaluated by the Ames test. The culture extracts of several fungi such as *Botrytis cineria*, *Ceratocystis fimbriata*, and others were positive, along with the other known AFB_1-producers.[13]

SOS chromotest[14] also demonstrated the genotoxic activity of the mycotoxins and fungal cultures, but two carcinogens, LS and OTA, were negative in this test either in the presence or absence of the activation system.

The primary culture/DNA repair test using rat and mouse hepatocytes was applied for the evaluation of the genotoxicity of mycotoxins.[15] Among 28 mycotoxins and related compounds tested, several compounds of unknown carcinogenic potential (such as 5,6-dimethoxyl-ST, versicolorin A, averufin xanthomegnin, luteosporin, and chrysazin), as well as the carcinogenic mycotoxins AFB_1, ST, LS, and OTA were positive in this assay. The initiator tRNA acceptance assay was also introduced for the screening of carcinogens in 20 mycotoxins.[16] With the exception of citrinin, all carcinogenic compounds were positive, and five (OTA, patulin, penicillic acid, T-2 toxin, and zearalenone) that were negative in the Ames test were positive in this assay.

Table 1 summarizes the relationship between the carcinogenicity and genotoxicity of fungal toxins.

TABLE 1
Carcinogenicity and Short-Term Genotoxicity of Mycotoxins

Mycotoxins	Carcinogenicity	Rec assay[7]	SOS test[14]	Ames test[8,9]	USD test[16]	tRNA assay[14]
Aflatoxin B$_1$	+[17]	+	+	+	+	+
Aflatoxin G$_1$	+[17]	+	+	+		+
Aflatoxin M$_1$	+[18]			+		
Sterigmatocystin	+[19]	–	+	+	+	+
Versicolorin A	+[20]		+	+	+	
Luteoskyrin	+[21]	+	–	–	+	
Rugulosin	+[22]	+		–	–	
Skyrin	n.d.			–	–	
Chrysazin	n.d.		–		+	
Chrysophanol	n.d.		–		–	
Emodin	n.d.		–	+	–	
Penicillic acid	+[23]	–	–	+	+	+
Patulin	+[23]	+	–		+	+
PR-toxin	n.d.	–	–		+	+
Ochratoxin A	+[24,25]	–	–	+	+	+
Citrinin	+[26]	–	–		–	–
Rubratoxin B	n.d.	+	–		–	–
T-2 toxin	n.d.	–	–		+	+
Zearalenone	n.d.	+	–		+	+
Botryodiploidin	n.d.			+[27]		
Fusarin C	n.d.			+[28]		

Note: n.d. = no data available.

III. AFLATOXINS

A. CONTAMINATION OF AGRICULTURAL COMMODITIES AND FOODS BY AFLATOXINS

The aflatoxins (Figure 1) are unavoidable contaminants in a variety of foods consumed by humans. The aflatoxins of major concern are AFB$_1$, AFB$_2$, AFG$_1$, AFG$_2$, and AFM$_1$, because they are carcinogenic in experimental animals. The first four aflatoxins are often simultaneously present in food. AFB$_1$ is usually predominant. The acute toxicity and carcinogenic potential of AFB$_1$ are the highest. AFM$_1$, a metabolite of of AFB$_1$, is excreted in the milk of dairy cattle and other mammals that have consumed feeds contaminated with AFB$_1$.

Foods in which significant amounts of aflatoxins have been found include corn and corn products, peanuts and peanut products, Brazil nuts, pistachio nuts, almonds, pecans, filberts, and walnuts.[5,29]

Japan imports about 100,000 tons of peanuts and nearly 200,000 tons of other kinds of nuts every year.[30] Inspection for the possible presence of aflatoxins in Spanish-type raw shelled peanuts, imported during 1972 to 1987, revealed that 1.7% of a total 9450 samples were contaminated with over 10 ppb of aflatoxins and that 3.3% were contaminated at levels less than 10 ppb.[30] Of 610 bean samples, 1.8% were positive for aflatoxins with a range of 2 to 36 ppb.[31] Imported pistachios are sometimes contaminated with a high level of aflatoxins. Of 54 samples, 6 were positive and 1 sample contained 1830, 183, 32.9, and 13.9 ppb of AFB$_1$, AFB$_2$, AFM$_1$, and aflatoxicol I, respectively.[31] The presence of 1 ppb of AFM$_1$ was found in imported natural cheese.[31]

B. BIOTRANSFORMATION AND ITS MODIFICATION

AFB$_1$ administered to animals is metabolized into several derivatives.[32-33] Hepatic re-

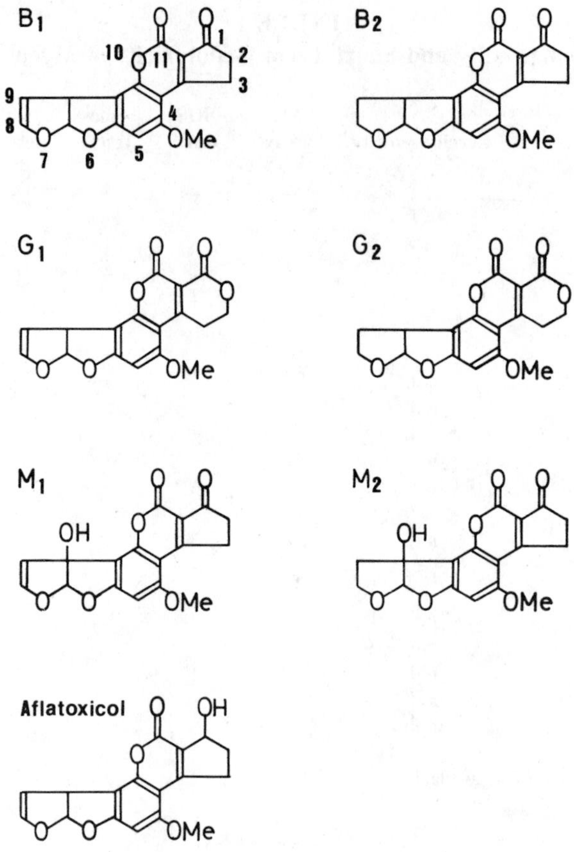

FIGURE 1.   Aflatoxins.

ductase in cytosol catalyzes the reduction of AFB_1 into aflatoxicol. The mixed function oxidase system in the microsomes catalyzes the epoxidation of the terminal double bond of the bisfuran ring, and the resulting AFB_1-8,9-oxide is highly reactive and covalently binds with macromolecules such as DNA, RNA, and proteins. The activated AFB_1 conjugates with glutathione (GSH) by cytosolic GSH-S-transferases and is hydrolized by microsomal epoxide hydrase.

Other microsome-dependent reactions are the hydroxylation of AFB_1 into AFM_1 and AFQ_1 and demethylation into AFP_1. The biological activity of these metabolites, except AFB_1-oxide, is far less than that of the parent AFB_1.

Several food additives and naturally occurring compounds are reported to modify the carcinogenic potential of AFB_1 *in vivo* as well as *in vitro*. Administration of β-naphthoflavone in rainbow trout induced hepatic microsomal AFB_1-hydroxylase.[34] This treatment also resulted in the reduction of AFB_1 carcinogenicity. In rats, phenolic antioxidants such as ethoxyquin, BHT and BHA, and an antischistosomal agent, dithiothione, reduced covalent binding of AFB_1 to DNA. This was due to the induction of the detoxification systems such as epoxide hydrase and GSH-S-tranferase.[35] Such alteration is closely associated with the reduction of AFB_1-induced hepatocarcinogenesis.[36] Ellagic acid, a plant phenol present in various fruits and nuts, depressed both the mutagenicity of AFB_1 in the Ames assay and the formation of AFB_1-DNA adduct in cultured explants of rat trachea and human tracheobronchus.[37] Indole-3-carbinol, a naturally occurring anticarcinogen found in cruciferous vegetables, depressed AFB_1-DNA binding and increased the formation of AFM_1 in trout.[38,39]

Genes for cytochrome P-450, which preferentially metabolizes AFB_1 to AFM_1, were isolated from hepatic microsomes of Ah-responsive mice. Analysis of the mRNA produced from the genes revealed that AFB_1-4-hydroxylase and aryl hydrocarbon hydroxylase, both regulated by the Ah locus, are the products of two separate genes.[46] Further experiment demonstrated that cytochrome P_3-450 cDNA encodes AFB_1-4-hydroxylase.[41]

C. CHEMICAL METHODS FOR DETECTION

Tiny amounts of AFB_1, AFM_1 and related compounds can give a blue or green fluorescence under UV, and this property allows the quantitative estimation of aflatoxins by thin-layer chromatography (TLC) and high-performance liquid chromatography (HPLC), as adopted in the official detection methods of the U.S.[42] Precoated silica gel plates are commonly employed in TLC analysis. The following solvent systems are recommended for AFB_1, B_2, G_1, and G_2, benzene-MeOH-HOAc (90:5:5), ether-MeOH-H_2O (96:3:1), $CHCl_3$-acetone-H_2O (88:12:1.5), and $CHCl_3$-isopropanol (99:1); for AFM_1, ether-MeOH-H_2O (95:4:1) and ether-hexane-MeOH-H_2O (85:10:4:1); for AFB_2a, $CHCl_3$-acetone-isopropanol (85:12.5:2.5); for AFM_1 derivatives, $CHCl_3$-MeOH-HOAc-H_2O (46:4:1:0.4).

For the extraction of AFB_1 in peanuts or meal, the sample is blended with methanol-H_2O (55:45) in the presence of hexane and NaCl. The aqueous methanol layer is separated from the hexane layer and extracted with $CHCl_3$. The $CHCl_3$ layer is evaporated into dryness under a gentle stream of N_2 gas, and the residue is dissolved in a small amount of benzene-CH_3CN for spotting onto a thin-layer chromatography (TLC) plate.

Clean-up procedures with minicolumn packed with silica gel or Florisil are essential for the analysis. Occasionally, blue fluorescent spots appear which can easily be mistaken for AFB_1. Therefore, chemical confirmation of identity of a toxin from all positive samples is essential. Treatment of AFB_1 by acetic anhydride-HCl gives a spot at Rf ca. 10% of that of untreated toxin. Trifluoroacetic acid (TFA) anhydride converts AFB_1 and AFG_1 into AFB_2a and AFG_2a, respectively, which give blue fluorescence at Rf ca. $1/4$ of that of AFB_1 and AFG_1. This TFA anhydride method is also applicable for the confirmation of AFM_1.

HPLC is common for the analysis of aflatoxins in food and biological fluids. For example, cottonseed products are extracted with acetone-H_2O and $Pb(OAc)_2$ or $Zn(OAc)_2$ is added and the filtrate is extracted with CH_2Cl_2. The CH_2Cl_2 layer is cleaned-up over a silica gel column, and 10 μl of the extracts containing 0.5 to 1.0 μg AFB_1/ml is charged on the HPLC column (μ-Porasil 300 × 4 mm; Partisil-10 250 × 4.6 mm), and the LC is developed with H_2O-saturated $CHCl_3$-cyclohexane-CH_3CN (25:7.5:1.0).

Liquid milk, powdered milk, or cheese is suspended in water and deproteinized by an addition of $Zn(SO_4)_2$ and NaOH. After extraction with $CHCl_3$, HPLC analysis is carried out with a reversed-phase column (C_{18}) and the solvent is composed from CH_3CN-methanol-H_2O (20:5:75). This method allows the estimation of a level of 10 ng/l AFM_1.[43]

A survey of AFM_1 with HPLC in dairy products marketed in Italy in 1984 revealed that 19.5, 26.5, and 53.5% of the French, German, and Dutch cheese samples, respectively, were over the detection limit (5 ng/kg) for AFM_1.[44]

Tandem mass spectrometry (MS/MS) was applied for the confirmation of purified toxins or detection of 10-ppb levels of toxins in matrix without a complicated clean-up procedure.[45]

D. IMMUNOASSAY OF AFLATOXINS, DNA, AND ALBUMIN ADDUCTS

With the development of several sensitive, specific, and simplified enzyme-linked immunosorbent assay (ELISA) and radioimmunoassay methods, it is now possible to monitor AFB_1 and its metabolites in food, body fluids, and excreta and to evaluate AFB_1 as a causal toxicant for primary liver cancer.[46,47] Several ELISA methods were proposed for AFB_1.[48-52] By combination of the one-step extraction method and ELISA, a level of 1 ppb of AFB_1 in butter peanuts was reproducibly estimated.[56] ELISA for AFM_1[53,54] and AFB_1-adducted DNA[55,56] was also proposed.

The immunoaffinity concentration of human urine can lead to the estimation of AFM_1

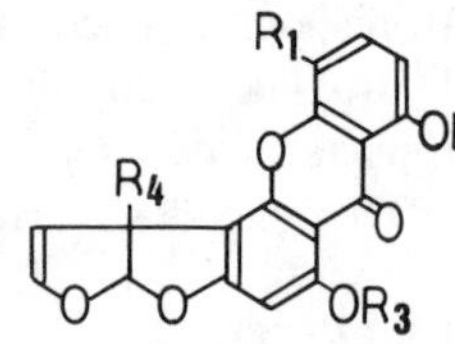

R₁	R₂	R₃	R₄	
H	H	CH₃	H	**Sterigmatocystin**
H	CH	CH₃	H	*O*-**Methylsterigmatocystin**
OCH₃	H	CH₃	H	**5-Methoxysterigmatocystin**
H	H	H	H	**Demethylsterigmatocystin**
H	CH₃	CH₃	OH	**Aspertoxin**

FIGURE 2.   Sterigmatocystins.

and AFB$_1$-guanine in AFB$_1$-exposed persons. An epidemiological survey with HPLC and ELISA revealed a good correlation between total dietary AFB$_1$ intake and total AFM$_1$ excretion.[54]

Since there is a good relationship between AFB$_1$ binding to serum albumin and the content of hepatic AFB$_1$-DNA adducts in rats exposed acutely or chronically to AFB$_1$,[57,58] AFB$_1$-albumin adducts in serum are expected to serve as an index for exposure of humans to AFB$_1$. A high incidence of primary liver cancer has been reported in Fushui province in the Guangxi region, China.[59] The survey in Guangxi revealed highly significant correlations among AFB$_1$-albumin adducts in serum, AFM$_1$ excretion, and AFB$_1$ intake.[60] From the slope of the regression line for the adduct level as a function of intake, it was calculated that 1.4 to 2.3% of ingested AFB$_1$ becomes covalently bound to serum albumin. This value is very similar to that observed when rats are administered AFB$_1$.[61]

In summary, (1) the immunoaffinity columns with monoclonal antibodies as the affinity ligands that recognize AFB$_1$, AFM$_1$, or AFB$_1$-modified DNA adducts are extremely useful for concentrating aflatoxin-containing fractions from biological materials; (2) further fractionation with HPLC and other techniques allows monitoring human exposure to AFB$_1$; and (3) the AFB$_1$-albumin adduct level in serum is also useful for the dosimetry of human exposure to AFB$_1$.[61,62]

IV. STERIGMATOCYSTIN

The genotoxicity of ST (Figure 2) is based on the epoxidation of the terminal double bond of the bisfuran ring, followed by a covalent binding to guanine residues of DNA at N-7, in a similar manner to AFB$_1$. ^{32}P-Postlabeling analysis also revealed the formation and persistence of ST-DNA adducts in the liver of rats given ST.[63]

The natural occurrence of ST in green coffee beans and pecans was reported, but the frequency was far less than that of aflatoxins,[29] although levels of several parts per million of ST were found in samples of feed associated with incidences of livestock poisoning.[64]

This mycotoxin was first isolated from *Aspergillus versicolor*, and subsequent survey demonstrated that various fungi belonging to the genera *Emericella*, *Chaetomium*, *Biopolaris*, and *Penicillium* are ST producers.

TLC analysis was recommended as an official method in the U.S.[42] Development of sample-charged silica gel plates with benzene-HOAc-MeOH (90:5:5), spraying of 20% ethanolic aluminum chloride solution, and heating at 80°C gives a yellow fluorescent spot of ST under UV light. Confirmation of ST can be carried out by derivatization into the acetate (acetic anhydride in pyridine) or into the TFA (TFA anhydride in benzene).

Acetylation followed by HPLC analysis has been recommended for the estimation of

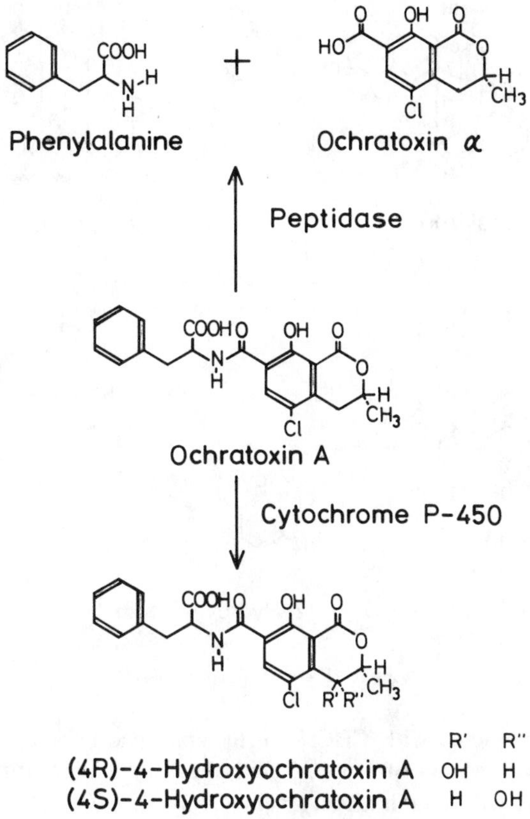

FIGURE 3. Ochratoxin A.

ST in barley.[65] HPLC methodology in conjunction with alkylphenone retention indices and UV-VIS spectra (diode array detection) was introduced for the multimycotoxin analysis in fungal metabolites including ST.[66]

An ELISA method was applied for the detection of ST in barley.[67]

V. OCHRATOXINS

OTA (Figure 3) (7-carboxy-5-chloro-8-hydroxy-3,4-dihydro-3-R-methyl-isocoumarin amide of L-β-phenylalanine) and related compounds are produced by *A. ochraceus*, *Penicillium viridicatum*, and other species.

OTA is nephrotoxic, and several epidemiological surveys revealed the association between OTA level in feeds and porcine nephropathy in several European countries.[3] The association of OTA levels in food and human serum with tumor incidence in urinary system was reported in Bulgaria.[68]

TLC analysis of OTA was recommended by IARC;[69] the analysis is carried out as follows. Foods are extracted with $CHCl_3$ in acidic aqueous suspensions, and the extract is cleaned-up over a column containing diatomaceous earth impregnated with a basic aqueous solution. After developing the plates charged with the cleaned-up samples and exposing the plates to ammonia, the fluorescent OTA spots are detectable within a limit of a few micrograms per kilogram.

HPLC procedures have been developed for OTA in foods and serum, with limits of detection in the range of 1 to 12 μg/kg.[70]

ELISA methods were applied for the detection of OTA in cereals[71] and animal tissues.[72]

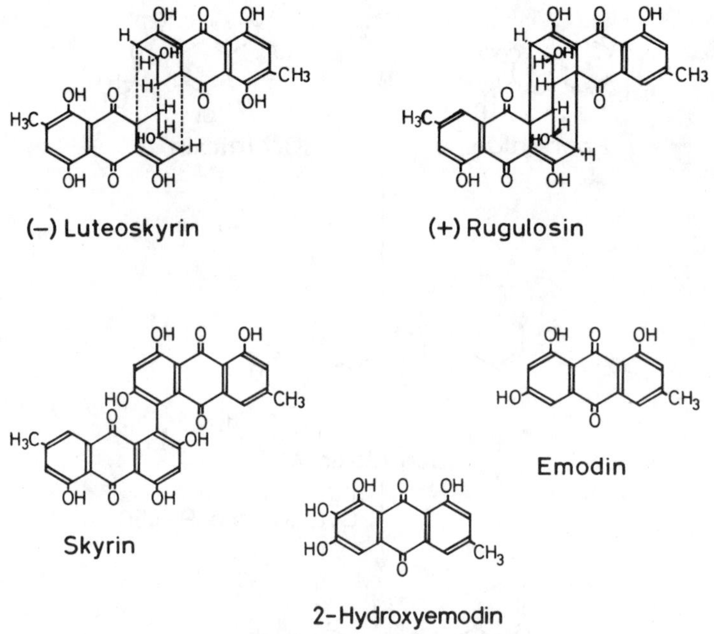

FIGURE 4.   Anthraquinones.

Chicken meat was extracted with $CHCl_3$-5% metaphosphoric acid (1:1), and OTA was assayed by ELISA with a limit of 1 ppb.[72] With the combination of the $CHCl_3$ extraction procedure and ELISA with a high-affinity monoclonal anti-OTA antibody, OTA in swine serum was estimated at a limit of 0.1 ppb.[74]

Confirmation of OTA is achieved by chemical methylation of OTA into OTA methyl ester, which can be detected by HPLC, or by enzymatic cleavage of the amide bond of OTA by carboxypeptidase, which results in the disappearance of OTA in the HPLC profile.

OTA is biotransformed into several metabolites (Figure 3). Carboxypeptidase hydrolizes OTA into OTA-α and phenylalanine. The microsomal cytochrome P-450 system of PCB-induced rats, particularly P-448 type, hydroxylates OTA at C-4 position into 4(S)- and 4(R)-hydroxy-OTA.[74] Both OTA-α and 4-OH-OTA are detected in the urine and blood of animals exposed to OTA. No significant toxicity is observed with OTA-α, while 4-OH-OTA possesses toxicity comparable to the parent OTA.

Exposure of farm animals and humans to OTA is a serious problem in Europe. Ubiqitous occurrence of this renal carcinogen in agricultural commodities and feeds has been noted.[3] Also, residues of OTA have been detected in the kidney and blood of swine.[3] Occurrence of OTA in swine blood was also reported in other countries such as Canada[75] and Japan.[73] As for human exposure, current survey in Germany has revealed that 96 human blood samples collected from a blood bank contained on the average 1.5 and 2.3 ppb of OTA in 1986 and 1987, respectively (HPLC detection limit, 0.1 ppb).[76] OTA was also detected from human milk samples in Germany.[77]

VI. EMODIN AND RELATED ANTHRAQUINONES

Many species of fungi and plants produce various derivatives of monomeric and dimeric polyhydroxy anthraquinones (Figure 4).[78] Among these anthraquinoid compounds, several monomeric anthraquinones such as EM (1,3,8-trihydroxy-6-methyl-anthraquinone) and related derivatives are constituents of rhubarb, an oriental medicine for diarrhea. They exhibit mutagenicity in the Ames/microsomes test (see Table 1) and cultured tumor cells.[79] Furthermore, dimeric anthraquinones such as LS and rugulosin are hepatotoxic and hepatocarcinogenic in mice[21,22] (see Table 1).

EM itself is not active in the Ames assay. This pigment is converted into active forms by microsomes, and 2-hydroxy-EM was isolated as an active principal.[80,81] Cytochrome P-450, particularly of P-448 form, catalyzes this biotransformation.[74,81] 2-Hydroxy-EM causes strand breaks in circular phage DNA.[83] Since a strong electron-spin resonance is observed for a mixture of 2-hydroxy-EM and DNA, free radicals derived from the anthraquinone may participate in this DNA damage.[83]

HPLC-ECD analysis has revealed the elevation of 8-hydroxydeoxyguanine contents in the hepatic DNA of mice administered with LS.[84] Since several hydroxy-radical-forming agents selectively hydroxylate deoxyguanosine residues of DNA at C-8 in *in vivo* and *in vitro* systems, it was presumed that the anthraquinones cause genotoxicity by their potential for hydroxy radical formation.

These anthraquinones show characteristic colors in solutions and on TLC plates, and this physical property is applied for identification and quantitation in TLC.[85] Since some anthraquinones bind with divalent cations such as Ca^{++} and Mg^{++}, silica gel used in column chromatography and TLC is usually treated with oxalic acid before use. Spraying of TLC plates with a methanolic $MgCl_2$ solution is used for identification of the anthraquinones. A method for TLC analysis of LS and related anthraquinones in rice grains was established with the solvent system composed from acetone-hexane-H_2O (5:5:3.5, upper layer). The detection limit is 0.05 μg per spot or 20-μg/kg samples.[85] HPLC with ODS column was applied for the quantitation of LS in rice grains with a limit of detection at 10 ppb.[86]

REFERENCES

1. **Ueno, Y.,** The toxicology of mycotoxins, *CRC Crit. Rev. Toxicol.,* 14(2), 99, 1985.
2. *Aflatoxins, IARC Monographs on the Evaluation of Carcinogenic Risks to Humans,* Suppl. 7, International Agency for Research on Cancer, Lyon, France, 1987, 83.
3. **Krogh, P., Ed.,** *Mycotoxins in Food,* Academic Press, London, 1987.
4. **Ueno, Y.,** Mycotoxins, in *Toxicological Aspects of Food,* Miller, K., Ed., Elsevier, Amsterdam, 1987, 139.
5. **Jelinek, C. F., Pohland, A., and Wood, G.,** Worldwide occurrence of mycotoxins in foods and feeds — an update, *J. Assoc. Off. Anal. Chem.,* 72, 223, 1989.
6. **Scott, P. M.,** Mycotoxins, *J. Assoc. Off. Anal. Chem.,* 71, 70, 1988.
7. **Ueno, Y. and Kubota, K.,** DNA-attacking ability of carcinogenic mycotoxins in recombination-deficient mutant cells of *Bacillus subtilis, Cancer Res.,* 36, 445, 1978.
8. **Ueno, Y., Kubota, K., Ito, T., and Nakamura, Y.,** Mutagenicity of carcinogenic mycotoxins in *Salmonella typhimurium, Cancer Res.,* 38, 536, 1978.
9. **Wehner, F. C., Theil, P. G., van Rensburg, S. J., and Demasius, P. C.,** Mutagenicity to *Salmonella typhimurium* of some *Aspergillus* and *Penicillium* mycotoxins, *Mutat. Res.,* 58, 193, 1978.
10. **Stark, A. A.,** Mutagenicity and carcinogenicity of mycotoxins: DNA binding as a possible mode of action, *Annu. Rev. Microbiol.,* 34, 235, 1980.
11. **Wong, J. J. and Hsieh, D. P. H.,** Mutagenicity of aflatoxins related to their metabolism and carcinogenic potential, *Proc. Natl. Acad. Sci. U.S.A.,* 73, 2241, 1976.
12. **Brown, J. and Dietrich, P. S.,** Mutagenicity of anthraquinone and benzanthrone derivatives in the *Salmonella*/microsome test: activation of anthraquinone glycosides by enzymatic extracts of rat cecal bacteria, *Mutat. Res.,* 66, 9, 1979.
13. **Bjeldanes, F. L., Chang, G. W., and Thomson, S. V.,** Detection of mutagens produced by fungi with the *Salmonella typhimurium* assay, *Appl. Environ. Microbiol.,* 35, 1150, 1978.
14. **Ueno, Y., Abe, K., and Sugiura, Y.,** Genotoxicity of mycotoxins evaluated by SOS microplates, in preparation.

15. **Mori, H., Kawai, K., Ohbayashi, F., Kuniyasu, T., Yamazaki, M., Hamasaki, T., and Williams, G. M.,** Genotoxicity of a variety of mycotoxins in the hepatocyte primary culture/DNA repair test using rat and mouse hepatocytes, *Cancer Res.,* 44, 2918, 1984.

16. **Hradec, J. and Vesely, D.,** The initiator tRNA acceptance assay as a short-term for carcinogenes. IV. Results with 20 mycotoxins, *Carcinogenesis,* 10, 213, 1989.

17. **Wogan, G. N., Edwards, G. S., and Newberne, P. M.,** Structure-activity relationships in toxicity and carcinogenicity of aflatoxins and analogs, *Cancer Res.,* 31, 1936, 1971.

18. **Hsieh, D. F. D. and Ruebner, B. H.,** An assessment of cancer risk from aflatoxins B_1 and M_1, in *Toxigenic Fungi—Their Toxins and Health Hazard,* Kurata, H. and Ueno, Y., Eds., Elsevier, Amsterdam, 1984, 332.

19. **Purchase, I. F. H. and van der Watt, J. J.,** Carcinogenicity of sterigamatocystin to rats, *Food Cosmet. Toxicol.,* 8, 289, 1969.

20. **Hendricks, J. D., Sinnhuber, R. O., Wales, J. H., Stack, M. E., and Hsieh, D. P. H.,** Hepatocarcinogenicity of sterigmatocystin and versicolorin A to rainbow trout *Salmo gairdneri* embryos, *J. Natl. Cancer Inst.,* 64, 1503, 1980.

21. **Uraguchi, K., Saito, M., Noguchi, Y., Takahashi, K., Enomoto, M., and Tatsuno, T.,** Chronic toxicity and carcinogenicity in mice of purified mycotoxins, luteoskyrin and cyclochlorotine, *Food Cosmet. Toxicol.,* 10, 193, 1972.

22. **Ueno, Y., Sato, N., Ito, T., Ueno, I., Enomoto, M., and Tsunoda, H.,** Chronic toxity and hepatocarcinogenicity of (+) rugulosin, anthraquinoid mycotoxin from *Penicillium* species: preliminary surveys in mice, *J. Toxicol. Sci.,* 5, 295, 1980.

23. **Dickens, F. and Jones, H. E. H.,** Carcinogenic activity of a series of reactive lactones and related substances, *Br. J. Cancer,* 15, 85, 1961.

24. **Kanisawa, M. and Suzuki, S.,** Induction of renal and hepatic tumors in mice by ochratoxin A, a mycotoxin, *Gann,* 69, 599, 1978.

25. **Bendele, A. M., Carlton, W. W., Krogh, P., and Lillehoj, E. B.,** Ochratoxin A, carcinogenesis in the (C57BL/6J × C3H), F_1 mouse, *J. Natl. Cancer Inst.,* 75, 733, 1985.

26. **Arai, M. and Hibino, T.,** Tumorigenicity of citrinin in male F344 rats, *Cancer Lett.,* 17, 281, 1983.

27. **Moule, Y., Decloitre, F., and Hamon, G.,** Mutagenicity of the mycotoxin botryodiploidin in the *Salmonella typhimurium*/microsomal activation test, *Environ. Mutagen.,* 3, 287, 1981.

28. **Wiebe, L. and Bjeldanes, L. F.,** Fusarin C, a mutagen from *Fusarium moniliforme* grown in corn, *J. Food Sci.,* 46, 1424, 1981.

29. **Pohland, A. E. and Wood, G. E.,** Occurrence of mycotoxins in food, in *Mycotoxins in Food,* Krogh, P., Ed., Academic Press, London, 1987, 35.

30. **Aibara, K. and Maeda, K.,** The regulation for aflatoxin in Japan and the current situation of aflatoxin contamination of food imported from throughout the world, in *Mycotoxins and Phycotoxins '88,* Natori, S., Hashimoto, K., and Ueno, Y., Eds., Elsevier, Amsterdam, 1989, 243.

31. **Nishijima, N.,** Survey for mycotoxins in commercial foods, in *Toxigenic Fungi—Their Toxins and Health Hazard,* Kurata, H. and Ueno, Y., Eds. Kodansha Ltd., Tokyo and Elsevier, Amsterdam, 1984, 172.

32. **Wong, Z. A. and Hsieh, D. H. P.,** The comparative metabolism and toxicokinetics of aflatoxin B_1 in the monkey, rat and mouse, *Toxicol. Appl. Pharmacol.,* 55, 115, 1980.

33. **Ueno, Y., Tashiro, F., and Nakaki, H.,** Mechanism of metabolic activation of aflatoxin B_1, *Gann,* 30, 111, 1985.

34. **Gurtoo, H. L., Koser, P. L., Bansal, S. K., Fox, H. W., Sharma, S. D., Mulhern, A. I., and Pavelic, Z. P.,** Inhibition of aflatoxin B_1-hepatocarcinogenesis in rats by β-naphthoflavone, *Carcinogenesis,* 6, 675, 1983.

35. **Kensler, T. W., Egner, P. A., Trush, M. A., Bueding, E., and Groopman, J. D.,** Modification of aflatoxin B_1 binding to DNA *in vivo* in rats fed phenolic antioxidants, ethoxyquin and a dithiothione, *Carcinogenesis,* 6, 759, 1985.

36. **Williams, G. M., Tanaka, T., and Maeura, Y.,** Dose-inhibition of aflatoxin B_1 induced hepatocarcinogenesis by the phenolic antioxidants, butylated hydroxyanisole and butylated hydroxytoluene, *Carcinogenesis,* 7, 1043, 1986.

37. **Mandal, S., Ahuja, A., Shivapurkar, N. M., Cheng, S.-J., Groopman, J. D., and Stone, G. D.,** Inhibition of aflatoxin B_1 mutagenesis in *Salmonella typhimurium* and DNA damage in cultured rat and human tracheobronchial tissues by ellagic acid, *Carcinogenesis,* 8, 1651, 1987.

38. **Goeger, D. E., Shelton, D. W., Hendricks, J. D., and Bailey, G. S.,** Mechanism of anti-carcinogenesis by indole-3-carbinol: effect on the distribution and metabolism of aflatoxin B_1 in rainbow trout, *Carcinogenesis,* 7, 2025, 1986.

39. **Dashwood, R. H., Arbogast, D. N., Fong, A. T., Pereira, C., Hendricks, J. D., and Bailey, G. S.,** Quantitative inter-relationships between aflatoxin B_1 carcinogen dose, indole-3-carbinol anti-carcinogen dose, target organ DNA adduction and final tumor response, *Carcinogenesis,* 10, 175, 1989.

40. **Koser, P., Faletto, M. B., Maccubbin, A. E., and Gurtoo, H. L.,** The genetics of aflatoxin B_1 metabolism. Association of the induction of aflatoxin B_1-4-hydroxylase with the transcriptional activation of cytochrome P_3-450 gene, *J. Biol. Chem.,* 263, 12584, 1988.

41. **Faletto, M. B., Koser, P. L., Battulas, N., Townsend, G. K., Maccubbin, A. E., Gelboin, H. V., and Gurtoo, H. L.,** Cytochrome P_3-450 cDNA encodes aflatoxin B_1-4-hydroxylase, *J. Biol. Chem.,* 263, 12187, 1988.

42. **Stoloff, L. and Scott, P. M.,** Natural poisons, in *Official Methods of Analysis: Association of Official Analytical Chemists,* Williams, S., Ed., AOAC, Virginia, 1984, 477.

43. **Chambon, P., Dano, S. D., Chambon, R., and Geahchan, A.,** Rapid determination of aflatoxin M_1 in milk and dairy products by high-performance liquid chromatography, *J. Chromatogr.,* 259, 372, 1983.

44. **Piva, G., Pietri, A., Galazzi, L., and Curto, O.,** Aflatoxin M_1 occurrence in dairy products marketed in Italy, *Food Addit. Contamin.,* 5, 133, 1987.

45. **Plattner, R. D., Bennett, G. A., and Stubblefield, R. D.,** Identification of aflatoxins by quadrupole mass spectrometry/mass spectrometry, *J. Assoc. Off. Anal. Chem.,* 67, 734, 1984.

46. **Chu, F. S.,** Immunochemical studies on mycotoxins, in *Toxigenic Fungi—Their Toxins and Health Hazard,* Kurata, H. and Ueno, Y., Eds., Kodansha Ltd., Tokyo and Elsevier, Amsterdam, 1984, 234.

47. **Garner, R. C.,** Monitoring aflatoxin exposure at a macromolecular level in man with immunological methods, in *Mycotoxins and Phycotoxins '88,* Natori, S., Hashimoto, K., and Ueno, Y., Eds., Elsevier, Amsterdam, 1989, 29.

48. **Sun, T., Wu, Y., and Wu, S.,** Monoclonal antibody against aflatoxin B_1 and its application, *Clin. J. Oncol.,* 5, 401, 1983.

49. **Ram, B. P. and Hart, L. P.,** Enzyme-linked immunosorbent assay of aflatoxin B_1 in naturally contaminated corn and cottonseed, *J. Assoc. Off. Anal. Chem.,* 69, 904, 1986.

50. **Kawamura, O., Nagayama, S., Sato, S., Ohtani, K., Ueno, I., and Ueno, Y.,** A monoclonal antibody-based enzyme-linked immunosorbent assay of aflatoxin B_1 in peanut products, *Mycotoxin Res.,* 4, 75, 1988.

51. **Mortimer, D. N., Shepherd, M. J., Gilbert, J., and Morgan, M. R. A.,** A survey of the occurrence of aflatoxin B_1 in peanut butters by enzyme-linked immunosorbent assay, *Food Addit. Contamin.,* 5, 127, 1987.

52. **Wilkinson, A. P., Denning, D. W., and Morgan, M. R. A.,** An ELISA method for the rapid and simple determination of aflatoxin in human serum, *Food Addit. Contamin.,* 5, 609, 1988.

53. **Goopman, J. D., Trubel, L. J., Donahue, P. R., Marshak-Rothstein, A., and Wogan, G. N.,** High affinity monoclonal antibodies for aflatoxins and their application to solid-phase immunoassays, *Proc. Natl. Acad. Sci. U.S.A.,* 81, 7728, 1985.

54. **Zhu, J.-Q., Zhang, L.-S., Hu, X., Chen, J.-S., Xu, Y.-C., Fremy, J., and Chu, F. S.,** Correlation of dietary aflatoxin B_1 level with excretion of aflatoxin M_1 in human urine, *Cancer Res.,* 47, 1848, 1987.

55. **Haugen, A., Groopman, J. D., Hsu, I. C., Goodrich, G. R., Wogan, G. N., and Harris, C. C.,** Monoclonal antibody to aflatoxin B_1-modified DNA detected by enzyme immunoassay, *Proc. Natl. Acad. Sci. U.S.A.,* 78, 4124, 1981.

56. **Groopman, J. D., Haugen, A., Goodrich, G. R., and Harris, C. C.,** Quantitation of aflatoxin B_1-modified DNA using monoclonal antibodies, *Cancer Res.,* 42, 3120, 1982.

57. **Wild, C. P., Garner, R. C., Montesano, R., and Tursi, F.,** Aflatoxin B_1 binding to plasma albumin and liver DNA upon chronic administration to rats, *Carcinogenesis,* 7, 853, 1986.

58. **Sabbioni, G., Skipper, P. L., Buchi, G., and Tannenbaum, S. R.,** Isolation and characterization of the major serum albumin adduct formed by aflatoxin B_1 *in vivo* in rats, *Carcinogenesis,* 8, 819, 1987.

59. **Yeh, F.-S., Mo, C.-C., and Yen, R.-C.,** Risk factors for hepatocellular carcinoma in Guangxi, People's Republic of China, *Natl. Cancer Inst. Monogr.,* 69, 47, 1985.

60. **Gan, L.-S., Skipper, P. L., Peng, X., Groopman, J. D., Chen, J.-S., Wogan, G. N., and Tannenbaum, S. R.,** Serum albumin adducts in the molecular epidemiology of aflatoxin carcinogenesis: correlation with aflatoxin B_1 intake and urinary excretion of aflatoxin M_1, *Carcinogenesis,* 9, 1323, 1988.

61. **Groopman, J. D., Donahue, P. R., Zhu, J., Chen, J., and Wogan, G. N.,** Aflatoxin metabolism in humans: detection of metabolites and nuclei acid adducts in urine by affinity chromatography, *Proc. Natl. Acad. Sci, U.S.A.,* 82, 6672, 1985.

62. **Groopman, J. D. and Nonahue, K. F.,** Aflatoxin, a human carcinogen: determination in foods and biological samples by monoclonal antibody affinity chromatography, *J. Assoc. Off. Anal. Chem.,* 71, 861, 1988.

63. **Reddy, M. V., Irvin, T. R., and Randerath, K.,** Formation and persistence of sterigmatocystin-DNA adducts in rat liver determined via ^{32}P-postlabeling analysis, *Mutat. Res.,* 152, 85, 1985.

64. **Vesonder, R. E. and Horn, B. W.,** Sterigmatocystin in dairy cattle feed contaminated with *Aspergillus versicolor, Appl. Environ. Microbiol.,* 49, 234, 1985.

65. **Abramson, D. and Thorsteinson, T.,** Determination of sterigmatocystin in barley by acetylation and liquid chromatography, *J. Assoc. Off. Anal. Chem.,* 72, 342, 1989.

66. **Frisvad, J. C. and Thrane, U.**, Standard high-performance chromatography of 182 mycotoxins and other fungal metabolites based on alkylphenone retention indices and UV-VIS spectra (Diode Array Detection), *J. Chromatogr.* 404, 195, 1987.

67. **Morgan, M. R. A., Kang, A. S., and Chan, H. W. S.**, Production of antisera against sterigmatocystin hemiacetal and its potential for use in an enzyme-linked immunosorbent assay for sterigmatocystin in barley, *J. Sci. Food Agric.*, 7, 873, 1986.

68. **Petkova-Bocharova, T., Chernozemsky, I. N., and Castegnaro, M.**, Ochratoxin A in human blood in relation to Balkan endemic nephropathy and urinary system tumours in Bulgaria, *Food Addit. Contamin.*, 5, 299, 1988.

69. **Nesheim, S.**, Thin-layer chromatographic determination of ochratoxin A in foodstuffs, in *Environmental Carcinogens Selected Methods of Analysis*, Vol. 5, Egan, H., Ed., International Agency for Research on Cancer, Lyon, France, 1982, 255.

70. **Josefsson, E. and Moller, T.**, High pressure liquid chromatographic determination of ochratoxin A and zearalenone in cereals, *J. Assoc. Off. Anal. Chem.*, 62, 1165, 1979.

71. **Morgan, M. R. A., McNerney, R., and Chan, H. W.-S.**, Enzyme-immunosorbent assay of ochratoxin A in barley, *J. Assoc. Off. Anal. Chem.*, 66, 1481, 1983.

72. **Kawamura, O., Sato, S., Kajii, H., Nagayama, S., Ohtani, K., Chiba, J., and Ueno, Y.**, A sensitive enzyme-linked immunosorbent assay of ochratoxin A based on monoclonal antibodies, *Toxicon*, 27, 887, 1989.

73. **Sato, S., Kawamura, O., and Ueno, Y.**, ELISA and HPLC analysis of ochratoxin A in the blood of swine, *J. Agric. Immunol.*, in preparation.

74. **Ueno, Y.**, Biotransformation of mycotoxins in the reconstituted cytochrome P-450 system, *Proc. Jpn. Assoc. Mycotoxicol.*, 22, 28, 1985.

75. **Marquardt, R., Frohlich, A. A., Sreemannayana, O., Abramson, D., and Bernatsky, A.**, Ochratoxin A in blood from slaughter pigs in western Canada, *Can. J. Vet. Res.*, 52, 185, 1988.

76. **Hald, B.**, Human exposure to ochratoxin A, in *Mycotoxins and Phycotoxins '88*, Natori, S., Hashimoto, K., and Ueno, Y., Eds., Elsevier, Amsterdam, 1989, 57.

77. **Gareis, M., Martbauer, E., Bauer, J., and Gedek, B.**, Determination of ochratoxin A in human milk, *Z. Lebens, Unters. Forsch.*, 186, 114, 1988.

78. **Ueno, Y.**, Hydroxyanthaquinones (skyrine), in *Mycotoxins—Production, Isolation, Separation and Purification*, Betina, V., E., Elsevier, Amsterdam, 1984, 329.

79. **Morita, H., Umeda, M., Masuda, T., and Ueno, Y.**, Cytotoxic and mutagenic effects of emodin on cultured mouse carcinoma EM3A cells, *Mutat. Res.*, 204, 329, 1988.

80. **Masuda, T. and Ueno, Y.**, Microsomal transformation of emodin into a direct mutagen, *Mutat. Res.*, 125, 135, 1983.

81. **Masuda, T., Haraikawa, K., Marooka, N., Nakano, S., and Ueno, Y.**, 2-Hydroxyemodin, an active metabolite of emodin in the hepatic microsomes of rats, *Mutat. Res.*, 149, 327, 1985.

82. **Tanaka, H., Haraikawa, K., Morooka, N., and Ueno, Y.**, Mechanism of metabolic activation of emodin, an anthraquinoid mycotoxin, *Mutat. Res.*, 164, 283, 1986.

83. **Kodama, M., Kamioka, Y., Nakayama, T., Nagata, C., Marooka, N., and Ueno, Y.**, Generation of free radicals and hydrogen peroxide from 2-hydroxyemodin, a direct acting mutagen, and DNA strand breaks by active oxygen, *Toxicol. Lett.*, 37, 149, 1987.

84. **Miyasaka, N., Kato, T., Masuda, T., and Ueno, Y.**, Formation of 8-hydroxydeoxyguanine moiety in DNA by luteoskyrin, a bis-anthraquinoid hepatocarcinogen, in mice, in preparation.

85. **Ueno, Y.**, Thin-layer chromatographic determination of luteoskyrin in rice grains, in *Environmental Carcinogens Selected Methods of Analysis*, Vol. 5, Egan, H., Ed., International Agency for Research on Cancer, Lyon, France, 1982, 450.

86. **Nakagawa, T., Kawamura, T., Fujimoto, Y., and Tatsuno, T.**, Determination of luteoskyrin in grains by high performance liquid chromatography (HPLC) (in Japanese), *Proc. Jpn. Assoc. Mycotoxicol.*, 18, 31, 1983.

Chapter 7

FATE OF INGESTED MUTAGENS

Chapter 7.1

ACTIVATION OF FOOD MUTAGENS

Yasushi Yamazoe and Shogo Ozawa

TABLE OF CONTENTS

I. INTRODUCTION

More than 15 different mutagenic heterocyclic arylamines have been identified as products of heating food.[1,2] Also, food can sometimes be contaminated with mutagenic mycotoxins (see Chapter 6.2). These mutagens are in general indirectly acting; they require conversion into activated forms in order to become mutagenic to cells. In the biological setting, this activation is carried out enzymatically. In this chapter, we discuss the pathway and the enzymes involved in biological activation. Considerable knowledge has been accumulated concerning the metabolic activation, on the basis mostly of studies using *in vitro* systems. Whole-animal *in vivo* studies, on the other hand, produce information as to the entire fate of these mutagens. This aspect will be the major subject of Chapter 7.2 by Alexander and Wallin.

II. ACTIVATION OF MUTAGENIC PYROLYSATES OF FOOD AND OTHER NATURAL PRODUCTS

Most of the food pyrolysis mutagens bear heteroaromatic rings and a primary amino group in their molecules and can be divided into two subgroups, α-aminopyridines and α-amino-*N*-methylimidazoles. We discuss here the metabolic activation of these two groups of pyrolysate amines one by one. Polycyclic aromatic hydrocarbons and their amino and nitro derivatives, some of which are also formed by pyrolysis of food or related materials, are also dealt with.

A. ACTIVATION OF α-AMINOPYRIDINES
1. Amino-γ-Carbolines: Trp-P-1 and Trp-P-2

Trp-P-2 is metabolized *in vitro* into at least four derivatives.[3] Two of the metabolites are mutagenic to *Salmonella typhimurium* TA98 without addition of S9-mix. Structures of these active metabolites were identified as N-hydroxy-Trp-P-2 and NO-Trp-P-2.[4] The latter is considered to be produced through the microsomal or superoxide dismutase-catalyzed oxidation of N-hydroxy-Trp-P-2.[3,5] Furthermore, the rate of the formation of N-hydroxy-Trp-P-2 in hepatic microsomes is shown to correlate well with the number of revertants inducible in *S. typhimurium* TA98.[6]

Trp-P-1 is also oxidized *in vitro* to a mutagenic derivative.[7] The compound formed is likely to be the 2-hydroxyamino derivative, although its chemical structure is not yet firmly established.

Thus, *N*-hydroxylation is considered to be the main route in the mutagenic activation of both Trp-P-1 and Trp-P-2 (Figure 1).

2. Amino-aza-δ-Carbolines: Glu-P-1 and Glu-P-2

Both Glu-P-1 and Glu-P-2 (Figure 2) are converted *in vitro* into at least three metabolites.[8] Similar to Trp-P-2, N-hydroxy and nitroso derivatives of Glu-P-1 were the main mutagenic metabolites formed on incubating Glu-P-1 with hepatic microsomes from PCB-treated rats. Human livers also catalyze the *N*-hydroxylation of Glu-P-1, and the average rate of activation is somewhat higher than that in rat livers, although capacities for *N*-hydroxylation vary among individual human subjects.[9]

3. Aminopyridine: Phe-P-1

From the pyrolysates of phenylalanine and phenylalanine-containing mixtures, two heterocyclic arylamines are isolated as potent mutagens to *S. typhimurium* TA98. One has an α-aminopyrido structure (Phe-P-1), while the other belongs to aminoimidazoles (PhIP). Phe-P-1 is structurally related to a human carcinogen, 4-aminobiphenyl (Figure 3). Similar to 4-aminobiphenyl, Phe-P-1 was converted in rat hepatocytes to the oxidized metabolites that

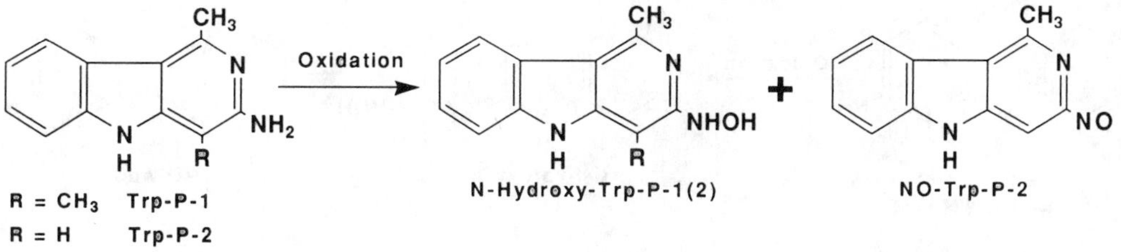

FIGURE 1. Metabolic activation of Trp-P-1 and Trp-P-2.

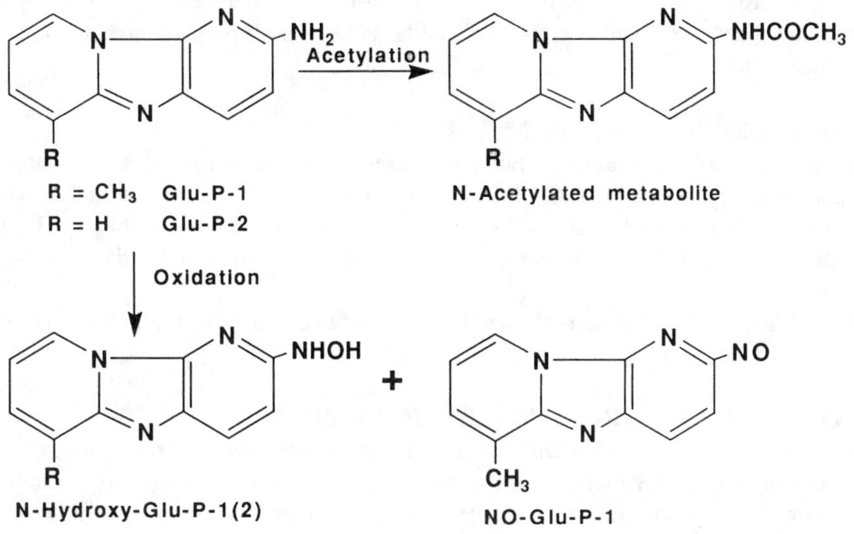

FIGURE 2. Metabolism of Glu-P-1 and Glu-P-2. *N*-Hydroxy-Glu-P-1 and NO-Glu-P-1 are detected in microsomal reactions, but only *N*-acetylated metabolite of Glu-P-1 is identified *in vivo*, except for unchanged Glu-P-1. Formation of *N*-hydroxy-Glu-P-2 is tentative.

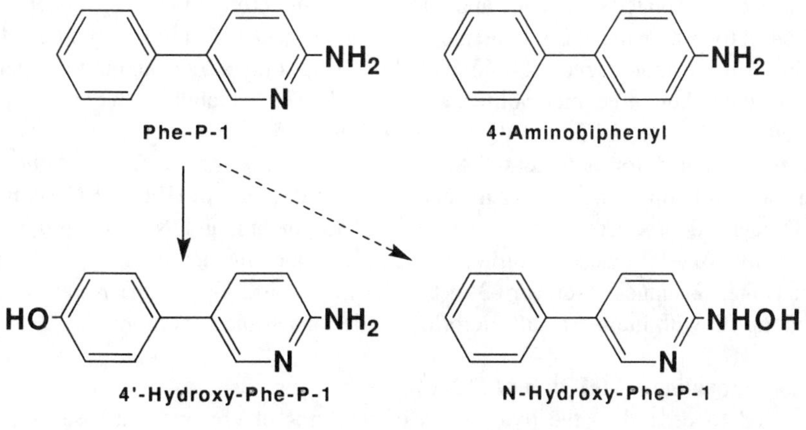

FIGURE 3. Structures of Phe-P-1 and its metabolites. Formation of *N*-hydroxy-Phe-P-1 is not yet reported, but it is suggested from the binding of microsomal metabolite of Phe-P-1 to DNA.

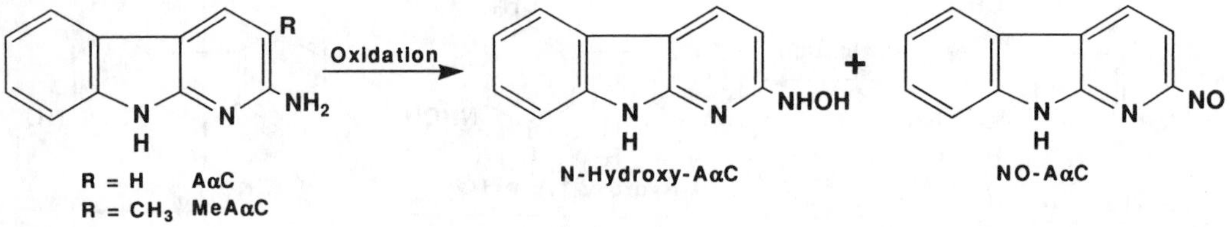

FIGURE 4.   Structures of AαC and MeAαC and the activation pathway of AαC. In microsomal reacting systems, metabolites other than the *N*-oxidized derivatives are detected, but not yet identified.

include 4′-hydroxy-Phe-P-1.[10] Formation of the *N*-hydroxy derivative has not yet been demonstrated. However, synthetic N-hydroxy-Phe-P-1 has been prepared as a possible mutagenic metabolite.

4. Amino-α-Carbolines: AαC and MeAαC

AαC and MeAαC are present in fairly high amounts in the pyrolysates of tryptophan-containing food and in the condensate of cigarette smoke. In the hepatic microsomal activating system, AαC is metabolized to five different derivatives, which are separable by HPLC.[11] Among them, *N*-hydroxy- and nitroso-derivatives were identified as the active metabolites (Figure 4).

Although amino-β-carboline is shown to be present in casein pyrolysate, its metabolism is not yet well understood.[12]

B. ACTIVATION OF AMINO-*N*-METHYLIMIDAZOLES

Several mutagenic amino-*N*-methylimidazoles have been isolated from cooked foods. These compounds require metabolic activation to exert genotoxic effects. They commonly bear the 1-methyl-2-aminoimidazole moiety. Studies on the structure-activity relationship suggest the requirement of the 1-methyl-2-aminoimidazole moiety for the maximal effect of aminoimidazoles.[13-15]

1. Aminoquinolines: IQ and MeIQ

From the pyrolysates of protein-rich foods, two mutagenic aminoquinolines, IQ and MeIQ, have been isolated. Both IQ and MeIQ are indirectly active; they undergo *N*-hydroxylation before showing the mutagenic effect and the modification of biological molecules.[16] In the urine and/or feces of animals fed [^{14}C]-IQ, only a trace amount of unchanged IQ was detected, but three metabolites were found. The metabolites were identified as demethylated IQ, N-acetyl-IQ, and IQ sulfamate (Figure 5).[17-20] The total amounts of these metabolites accounted for only less than $^1/_3$ of the excreted radioactivity. Although the structures are not known, three polar metabolites were detected in HPLC.[21] N-Hydroxy-IQ and its O-acylated products bind to macromolecules including DNA and proteins.[22] In addition, N-hydroxy-IQ reacts with glutathione to form a conjugate *in vitro*.[23] A quinolone-type metabolite, 2-amino-3,6-dihydro-3-methyl-7*H*-imidazo[4,5-*f*]quinoline-7-one, is formed on incubating IQ with human fecal microflora under anaerobic conditions.[24]

2. Aminoquinoxalines: MeIQx and DiMeIQx

In cooked foods and in the pyrolysates of mixtures of creatine, glucose, and amino acids, three aminoquinoxalines have been isolated and identified as MeIQx and two regio-isomers of DiMeIQx. The metabolic fate of MeIQx is shown in Figure 6. MeIQx is oxidized to four metabolites, 4(or 5)-hydroxy-, 8-hydroxymethyl, 3-demethyl, and 2-*N*-hydroxy derivatives.[25,26] Some oxidized metabolites are further transformed into the glucuronide, sulfate, and to a small extent the acetate, although considerable portions of MeIQx are excreted as nonoxidized forms. Whereas no direct evidence is obtained for the formation of N-hydroxy-

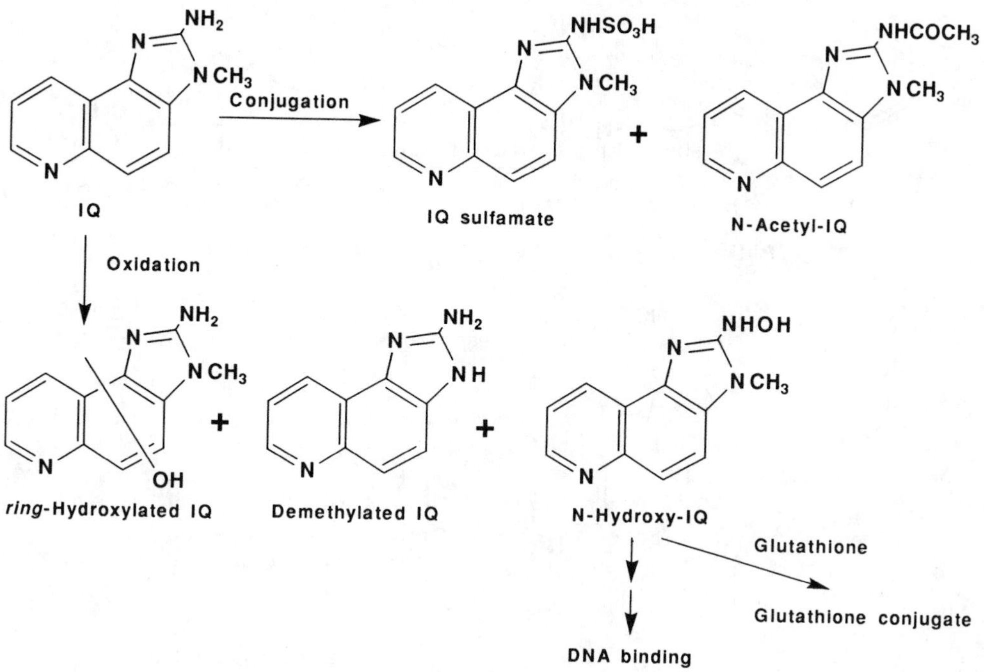

FIGURE 5.   Metabolism of IQ.

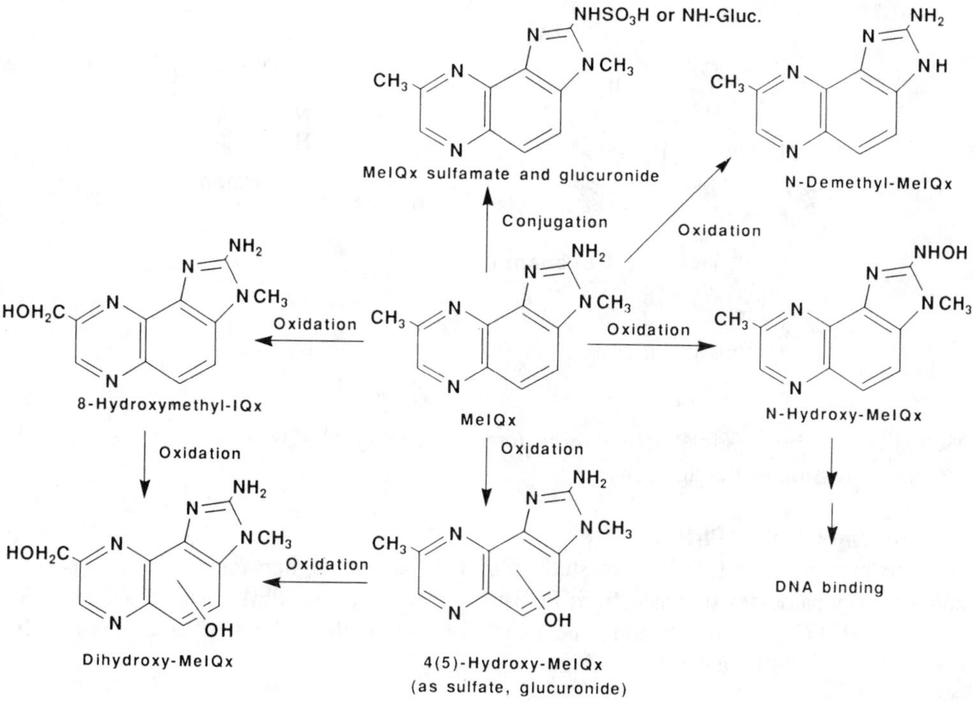

FIGURE 6.   Metabolism of MeIQx. gluc: glucuronide

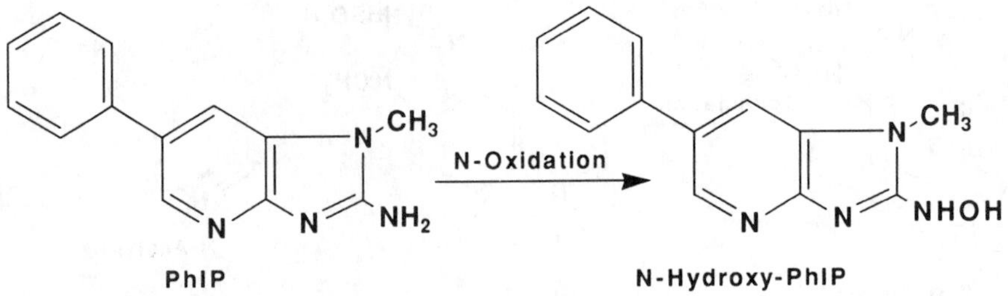

FIGURE 7.   Metabolic activation of PhIP.

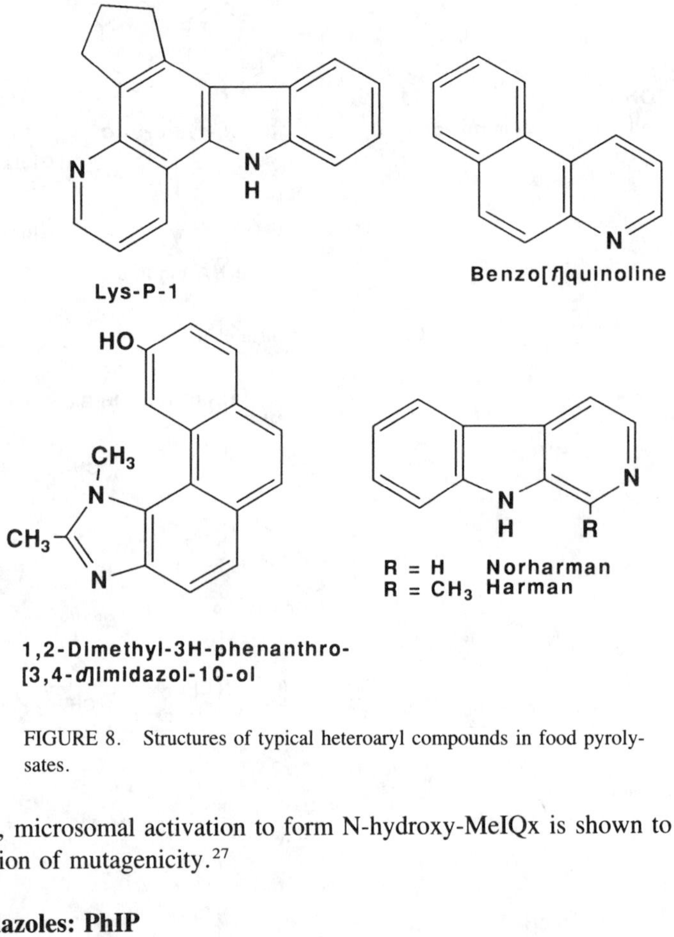

FIGURE 8.   Structures of typical heteroaryl compounds in food pyrolysates.

MeIQx *in vivo*, microsomal activation to form N-hydroxy-MeIQx is shown to be essential for the expression of mutagenicity.[27]

3. Aminoimidazoles: PhIP

The metabolism of PhIP was studied *in vitro* using rat hepatocytes.[28] In a system containing hepatocytes obtained from PCB-treated rats, several PhIP metabolites were detected in HPLC analysis. Among the metabolites, N-hydroxy-PhIP is considered to be responsible for the mutagenicity[29] (Figure 7).

C. POLYCYCLIC HETEROARYL COMPOUNDS

Several polycyclic heteroaryl compounds that are present in the pyrolysates of foods and in the condensate of cigarette smoke are mutagenic to *Salmonella*. The compounds listed in Figure 8 are not directly active; they require enzymatic transformation to exhibit genotoxic activity.

Benzo[*f*]quinoline, which is detected in soybean pyrolysate and cigarette smoke, is mutagenic to *Salmonella* in the presence of S9-mix.[30,31] This compound is oxidized into several phenolic, dihydrodiol, and *N*-oxidized derivatives by microsomal enzymes. Among them, benzo[*f*]quinoline 7,8-diol is the most mutagenic, as detected with *Salmonella* in the presence of S9-mix which has been prepared from PCB-treated rats. Thus, similar to the activation of homocyclic analogues of aryl compounds, the bay-region epoxide-diol, benzo[*f*]quinoline 7,8-diol-9,10-oxide, it is likely to be an ultimate form for causing mutagenicity.

From opium pyrolysates, several hydroxyphenanthrene compounds were isolated and shown as mutagenic to *Salmonella*.[32] These chemicals were metabolized into their phenols and dihydrodiols by rat liver 9000 g supernatant, a fact suggesting that arene oxides are the ultimate active metabolites.

β-Carbolines norharman and harman are non- or weakly mutagenic in the presence of S9-mix, but have a comutagenic activity to enhance the mutagenicity of other heterocyclic amines. Aniline is not mutagenic to *Salmonella* even in the S9-mix-containing system, but shows a mutagenic effect in the presence of norharman.[33,34] In a microsomal reaction, harman was metabolized into phenolic derivatives,[35] but none of the metabolites had the comutagenic effect. This finding suggests that an unchanged or unstable metabolite of norharman is responsible for the comutagenic effect.

D. *N*-SUBSTITUTED ARYL COMPOUNDS

1-Nitropyrene is known to be produced in diesel exhaust and is also detected in the neutral extract of grilled chicken.[36] Rats given [G-³H]-1-nitropyrene orally excreted more than 80% of the radioactivity into feces and urine, with a ratio of 3 to 1.[37] Several phenols and two dihydrodiols were detected in the excreta (Figure 9).[37,38] A nitro-reduction product, 1-aminopyrene and the *N*-acetylated derivative were also identified as minor metabolites. The reduction intermediate, N-hydroxy derivative, is probably the active form that causes mutagenesis. 1-Nitropyrene phenols and dihydrodiols are mostly conjugated as the glucuronides and sulfates. Considerable portions of the arene oxide-derived metabolites are also excreted as the conjugated forms with glutathione, cysteine, and mercapturic acid.

A human bladder carcinogen, 2-naphthylamine, is also known to be formed by the pyrolysis of glutamic acid and leucine,[39] although the existence of the compound in cooked foods is unclear. 2-Naphthylamine is metabolized by cytochrome P-450 or prostaglandin H synthase to the active intermediate, 2-N-hydroxyaminonaphthalene and 2-amino-1-naphthol.[40]

III. ACTIVATION OF OTHER FOOD-CONTAMINATING MUTAGENS

Cooked food or raw materials are shown to be contaminated occasionally with mycotoxins such as aflatoxin B_1. In addition, mutagenicity is also detected in food additives and spices. However, the metabolic fate of these chemicals is mostly unclear, except for a few mycotoxins[41] and nitrofurans.[42,43]

IV. ENZYMES INVOLVED IN THE ACTIVATION OF FOOD MUTAGENS

A. OXIDATION AND REDUCTION
1. Cytochrome P-450

Most food mutagens are not active by themselves and show their mutagenicity in the presence of S9-mix. In *Salmonella* mutagenesis test, the capacities of S9-mix for activation are localized mainly in the microsomal fraction, as shown by studying subcellular fractions

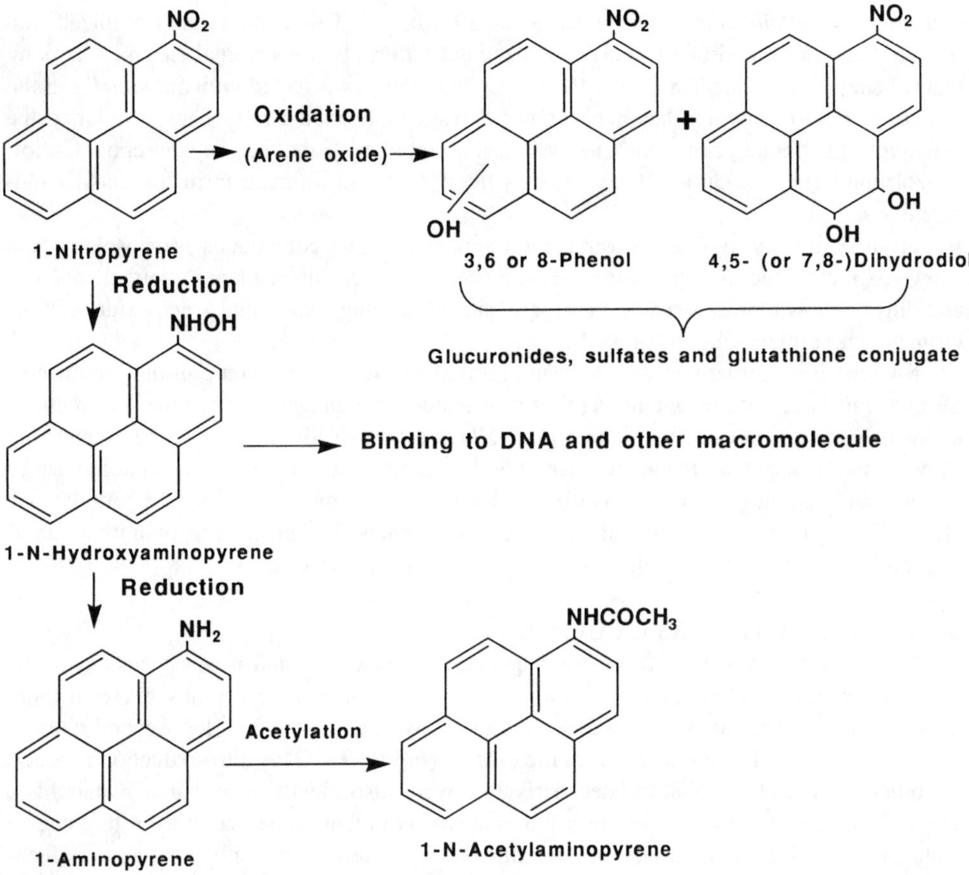

FIGURE 9.   Metabolic pathway of 1-nitropyrene. Although not yet proven, covalent binding of arene oxide derivatives of 1-nitropyrene to DNA is also possible.

of rat liver homogenates. The microsomal enzymes show catalysis in the presence of NADPH as the cofactor.

As described above, heterocyclic amines are mostly activated by *N*-hydroxylation of their primary amino group. Cytochrome P-450 and flavin monooxygenase are the enzymes responsible for many reactions taking place in hepatic microsomes.[44] However, no clear evidence has been presented about the involvement of flavin monooxygenase in the activation of heterocyclic amines, although this capacity of flavin monooxygenase has not yet been studied with a purified preparation.[45] Consistent with the data obtained from hepatic microsomes of typical cytochrome P-450 inducer-treated rats, the numbers of heterocyclic amine-induced revertants of *Salmonella* were higher with the use of 3-methylcholanthrene-inducible rat cytochrome P-450s than with phenobarbital-inducible or -noninducible P-450s. Two major phenobarbital-inducible forms, P-450b and P-450e, showed similar activation capacities for both aminopyridines and aminoimidazoles. A 3-methylcholanthrene-inducible form, P-450d (P-448-H), showed the highest catalysis in the mutagenic activation of all seven arylamines examined.[27,46,47] Glu-P-1, AαC, and IQ are transformed into their mutagenic metabolites at more than 10 times higher rates by P-450d than by P-450c (P-448-L), in spite of the fact that P-450c shows a high activity in the hydroxylation of benzo(*a*)pyrene and that it mediates the activation of Trp-P-2, IQ, and MeIQx with 6 to 15 times higher rates than P-450b and P-450e.[47,48] A similar difference between two P-448 isozymes was also observed in systems of HeLa cells infected with recombinant vaccinia virus that contained either P₁-450 (a P-

450c orthologue in mice) or P_3-450 (a P-450d orthologue) gene, with respect to their capacity to activate arylamines including IQ, DiMeIQx, and PhIP.[49] As described in Table 1, a major constitutive form in rats, P-450-male,[50] can catalyze the activation of IQ and MeIQx at rates similar to those of P-450c.[26] Although P-450d has the higher capacities for the mutagenic activation of arylamines, content in untreated rat liver is limited. Therefore, the activation of food-derived arylamines in rats is considered to be catalyzed both by P-450d and P-450-male. Glu-P-1, IQ, and MeIQx were activated to their mutagenic metabolites at slightly higher rates in liver microsomes of humans than of rats.[9] The rate of the human-liver microsomal activation for Glu-P-1 was shown to correlate with the content of immunoreactive P-448-H (P-450d), which was detectable by Western blotting using anti-rat P-448-H.[9] These results suggest that the activation of food-derived arylamines in human livers is catalyzed mainly by a form of P-450 similar to P-450d of rats.[9,51] For the activation of polycyclic aryl compounds, P-450c (or the orthologue in other species), rather than P-450d, is suggested to be involved in the formation of reactive intermediates such as arene oxides.[52] A mycotoxin, aflatoxin B_1, is activated through the formation of the epoxide.[41] Several forms of cytochrome P-450 have been suggested to mediate the activation. Mizokami et al.[53] isolated a novel form of P-450, P-450AFB, which had an extremely high capacity for the mutagenic activation of aflatoxin B_1, from the livers of 3-methylcholanthrene-treated hamsters. The form is inducible by the treatment of hamsters with 3-methylcholanthrene-type inducers, but the isolated cDNA sequence shows a rather close similarity to the gene sequence of P-450a (testosterone 7α-hydroxylase).[54]

2. Peroxidase

Food-derived arylamines have been shown to undergo peroxidase-mediate oxidations. Trp-P-1, Trp-P-2, and Glu-P-1 are oxidized to their nonmutagenic products by various mammalian peroxidases that include myeloperoxidase, lactoperoxidase, and catalase.[55,56] Prostaglandin H synthase from microsomes of ram seminal vesicles is shown to oxidize Trp-P-2, Glu-P-1, IQ, and MeIQ.[57,58] Involvement of lipoxygenase in Glu-P-1 activation is also suggested.[59] Kadlubar et al.[60,61] examined human extrahepatic tissues and showed that microsomes of human colon and bladder can mediate the DNA binding of Glu-P-1, Trp-P-2, and IQ and that the binding reaction requires the presence of arachidonic acid. In this experiment, the rate of Glu-P-1 binding to externally added DNA was higher than that of Trp-P-2 and IQ and was similar to that of a human bladder carcinogen, 4-aminobiphenyl. Although the significance of mammalian peroxidases on the activation of food-derived arylamines is still not well understood, these peroxidases may have a role in this action within the extrahepatic tissues. In addition, myoglobin and hemoglobin are shown to oxidize various heterocyclic amines.[62] Thus, these hemoproteins may contribute as trappers to decrease the amounts of arylamines accessible to cellular DNA.

B. ACYLATION AND CONJUGATION
1. Acetylation
a. N-Acetylation

Similar to 2-aminofluorene, Trp-P-1, Glu-P-1, IQ, and MeIQx are shown to be excreted in the rat as the N-acetylated forms.[63-65,20] However, the amounts excreted were less than a few percent of the total excretions, which is consistent with the low N-acetylating activities of rat liver cytosol for food-derived arylamines.[66] The rates of rat cytosolic N-acetylation of these food-derived arylamines were 20 to 3000 times lower than that of 2-aminofluorene.[66] Thus, in contrast to 2-aminofluorene, N-acetylation may have only a limited significance in the metabolic fate of food-derived arylamines. Among food-derived arylamines, Glu-P-1 is most rapidly N-acetylated and IQ and MeIQx are only slowly N-acetylated.[66]

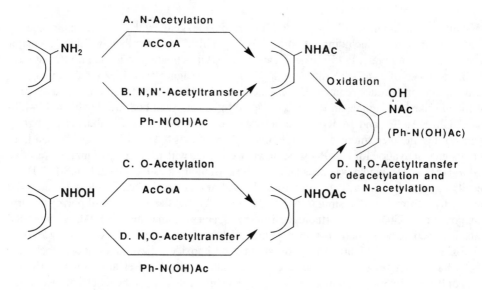

FIGURE 10.   Metabolic fates of arylamines and *N*-hydroxyarylamines through acetylation. Hepatic enzymes in rats catalyze all four reactions, A to D, while *Salmonella* enzymes mediate only the reactions A and C.

b. O-Acetylation

Considerable evidence has been accumulated regarding the role of cytosolic acetyltransferase in the activation of *N*-hydroxyarylamines through direct *O*-acetylation.[47,67,71] *N*-Hydroxylated intermediates of Trp-P-2, Glu-P-1, and IQ are shown *in vitro* to bind to DNA through the enzymatic transformation into their putative *O*-acetylated (*N*-acetoxy) intermediates.[22,70,72] Among two hamster acetyltransferases AT-I and AT-II that have been purifed to homogeneity,[73] AT-I catalyzes the *O*-acetylations of heterocyclic *N*-hydroxylamines, but AT-II does not, although both enzymes can mediate the *O*-acetylation of *N*-hydroxy-2-aminofluorene.[73] In the metabolism of arylamines, four different types of acetylation, *N*-acetylation, *N,N'*-acetyltransfer, *O*-acetylation, and *N,O*-acetyltransfer, have been reported to occur in mammalian cytosol (Figure 10).[73,74] *N,O*-Acetyltransfer of *N*-hydroxyarylacetamides (hydroxamic acids) into *N*-acetoxy derivatives is considered to be a main activating pathway of homocyclic arylamines.[75] However, no N-acetyl-N-hydroxy-Glu-P-1 formation from *N*-hydroxy-Glu-P-1 can be detected with use of mammalian cytosols.[70] This finding suggests a minor role of *N,O*-acetyltransfer in the activation of food-derived arylamines.

To *S. typhimurium* TA98, *N*-hydroxyarylamines are directly mutagenic without the aid of externally added S9-mix.[76] However, some *N*-hydroxyarylamines are activated to *O*-acetylated forms by the action of an internal acetyltransferase.[68,77] Purified *Salmonella* acetyltransferase has no detectable *N,O*-acetyltransferase activity, but catalyzed *O*-acetylation of various *N*-hydroxyarylamines.[78] With several food derived arylamines, different mutagenic responses in Salmonella were observed between TA98 and the acetyltransferase-deficient mutant TA981, 8-dnp$_6$.[79] These results indicated an essential role of internal *O*-acetylation in causing the mutagenic responses in bacteria, although *N*-hydroxyarylamines such as 1-*N*-hydroxyaminopyrene do not require acetyltransferase-mediated activation to exert the *Salmonella* mutagenicity. Substrate specificities of *Salmonella* and mammalian acetyltransferases differ. N-Hydroxy-IQ is a good substrate for bacterial, but not rat liver acetyltransferase,[22] which could explain at least partly the difference in the genotoxic effect of IQ in bacteria and mammalian cells.[80-83] A gene of a Salmonella acetyltransferase has recently been cloned, and the plasmid (pYG219) carrying the gene has been introduced in TA98 and TA100 strains.[84] The new strains showed two orders of magnitude greater capacity in the activation of N-hydroxy-Glu-P-1 and N-hydroxy-2-aminofluorene.

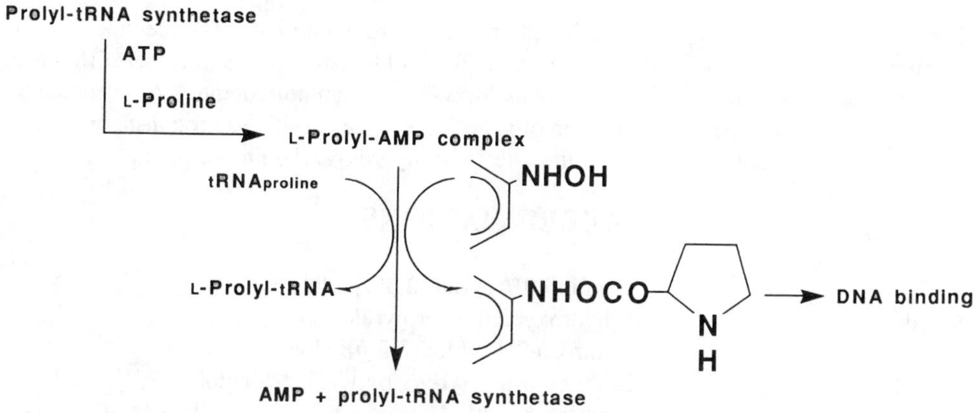

Prolyl-tRNA synthetase

FIGURE 11. Mechanism of tRNA synthetase-mediated aminoacylation of N-hydroxyarylamines.

2. Aminoacylation

As with 4-hydroxyaminoquinoline-1-oxide,[85] purified seryl-tRNA synthetase from yeast mediated the activation of N-hydroxy-Trp-P-2 in the presence of serine and ATP.[86] In rat liver cytosol, the binding of N-hydroxy-Trp-P-2 to DNA was most efficient in the presence of proline, instead of serine.[67,87] The reaction is catalyzed by prolyl-tRNA synthetase and is considered to proceed as shown in Figure 11. In a similar system, N-hydroxy-Glu-P-1 and N-hydroxy-IQ were poor substrates for this enzyme.[22]

3. Glucuronidation

Glucuronide conjugates were detected in the urine and bile of animals given IQ, MeIQ, or MeIQx.[26,88,89] In addition, glucuronidase-sensitive conjugates were also observed in the urine of rats treated with Glu-P-1 or Trp-P-1.[61,62] These conjugates were the N-glucuronides of unchanged aminoimidazoles and the O-glucuronides of their phenolic metabolites.

4. Glutathione Conjugation

Glutathione is known to react with various electrophilic intermediates through reduction and conjugation.[90] Addition of glutathione is reported to reduce the number of revertants inducible in the *Salmonella* mutagenesis test with Trp-P-1.[91] Furthermore, a depletion of glutathione by treatment of liver with diethylmaleate enhanced the binding of Trp-P-2 to hepatocyte DNA.[92] Similar enhancement was also observed with IQ.[45] These results suggest the protective role of glutathione in the activation of food-derived arylamines. However, a glutathione conjugate of N-hydroxy-Trp-P-2, the chemical structure of which is unclear, is shown to have higher mutagenicity than N-hydroxy-Trp-P-2.[93] Umemoto et al.[94] have recently determined the structures of two adducts formed in the reaction of NO-Glu-P-1 with glutathione: sulfinamide and N-hydroxy-sulfonamide. Until now, the significance of glutathione S-transferase on the metabolism of N-hydroxyarylamines and arylnitrates has not been clear. Also, characteristics of its isozymes have not been studied in detail, although isozymes containing subunit 1 or 2 are reported to have a high activity in aflatoxin metabolism.[95]

5. Sulfation

Studies *in vivo* indicate that sulfamate formation (N-sulfo-conjugation) is a main metabolic pathway of aminoimidazoles. Although a sulfamate was detected in animals treated with 2-naphthylamine, the amount was very small among the total metabolites excreted.[96]

In a cytosolic system containing 3-phosphoadenosine 5'-phosphosulfate (PAPS), N-hydroxy-Glu-P-1 and N-hydroxy-IQ were activated to their O-sulfates, which bind to calf thymus DNA.[69,79,97] PAPS-dependent sulfation of N-hydroxy-Glu-P-1 was three times higher with hepatic cytosols of male rats as compared to those of female rats.[98] Similar sex difference was also observed with the sulfation of *N*-hydroxy-2-acetylaminofluorene.[99] A recent study indicates that the sex-specific expression of growth hormone-sensitive sulfotransferase is the cause of the observable difference in the rates of *N*-hydroxyarylamine sulfation.[98]

ABBREVIATIONS

AαC	2-amino-9*H*-pyrido[2,3-*b*]indole
N-hydroxy-AαC	2-hydroxyamino-9*H*-pyrido[2,3-*b*]indole
NO-AαC	2-nitroso-9*H*-pyrido[2,3-*b*]indole
MeAαC	2-amino-3-methyl-9*H*-pyrido[2,3-*b*]indole
Glu-P-1	2-amino-6-methyldipyrido[1,2-*a*:3′,2′-*d*]imidazole
N-hydroxy-Glu-P-1	2-hydroxyamino-6-methyldipyrido[1,2-*a*:3′,2′-*d*]imidazole
NO-Glu-P-1	2-nitroso-6-methyldipyrido[1,2-*a*:3′,2′-*d*]imidazole
Glu-P-2	2-aminodipyrido[1,2-*a*:3′,2′-*d*]imidazole
N-hydroxy-Glu-P-2	2-hydroxyaminodipyrido[1,2-*a*:3′,2′-*d*]imidazole
IQ	2-amino-3-methylimidazo[4,5-*f*]quinoline
N-acetyl-IQ	2-acetylamino-3-methylimidazo[4,5-*f*]quinoline
N-hydroxy-IQ	2-hydroxyamino-3-methylimidazo[4,5-*f*]quinoline
IQ-sulfamate	^{2}*N*-sulfonyloxy derivative of 2-amino-3-methylimidazo[4,5-*f*]quinoline
demethylated IQ	2-aminoimidazo[4,5-*f*]quinoline
MeIQ	2-amino-3,4-dimethylimidazo[4,5-*f*]quinoline
MeIQx	2-amino-3,8-dimethylimidazo[4,5-*f*]quinoxaline
N-hydroxy-MeIQx	2-hydroxyamino-3,8-dimethylimidazo[4,5-*f*]quinoxaline
8-hydroxymethyl-MeIQx	2-amino-3-methyl-8-hydroxymethylimidazo[4,5-*f*]quinoxaline
N-demethyl-MeIQx	2-amino-8-methylimidazo[4,5-*f*]quinoxaline
MeIQx sulfamate	^{2}*N*-sulfonyloxy derivative of 2-amino-3,8-dimethylimidazo[4,5-*f*]quinoxaline
4,8-DiMeIQx	2-amino-3,4,8-trimethylimidazo[4,5-*f*]quinoxaline
Phe-P-1	2-amino-5-phenylpyridine
N-hydroxy-Phe-P-1	2-hydroxyamino-5-phenylpyridine
PhIP	2-amino-1-methyl-6-phenylimidazo[4,5-*b*]pyridine
Trp-P-1	3-amino-1,4-dimethyl-5*H*-pyrido[4,3-*b*]indole
N-hydroxy-Trp-P-1	3-hydroxyamino-1,4-dimethyl-5*H*-pyrido[4,3-*b*]indole
Trp-P-2	3-amino-1-methyl-5*H*-pyrido[4,3-*b*]indole
N-hydroxy-Trp-P-2	3-hydroxyamino-1-methyl-5*H*-pyrido[4,3-*b*]indole
NO-Trp-P-2	3-nitroso-1-methyl-5*H*-pyrido[4,3-*b*]indole

REFERENCES

1. **Sugimura, T. and Sato, S.,** Mutagens-carcinogens in foods, *Cancer Res.,* 43, 2415, 1983.
2. **Sugimura, T.,** Past, present, and future of mutagens in cooked foods, *Environ. Health Perspect.,* 67, 5, 1986.
3. **Yamazoe, Y., Ishii, K., Kamataki, T., Kato, R., and Sugimura, T.,** Isolation and characterization of active metabolites of tryptophan-pyrolysate mutagen, Trp-P-2, formed by rat liver microsomes, *Chem. Biol. Interact.,* 30, 125, 1980.

4. **Yamazoe, Y., Ishii, K., Kamataki, T., and Kato, R.,** Structural elucidation of a mutagenic metabolite of 3-amino-1-methyl-5H-pyrido[4,3-b]indole, *Drug. Metab. Dispos.,* 9, 292, 1981.
5. **Hiramoto, K., Negishi, K., Namba, T., Katsu, T., and Hayatsu, H.,** Superoxide dismutase-mediated reversible conversion of 3-hydroxyamino-1-methyl-5H-pyrido[4,3-b]indole, the N-hydroxy derivative of Trp-P-2, into its nitroso derivative, *Carcinogenesis,* 9, 2003, 1988.
6. **Yamazoe, Y., Kamataki, T., and Kato, R.,** Species difference in N-hydroxylation of a tryptophan pyrolysis product in relation to mutagenic activation, *Cancer Res.,* 41, 4518, 1981.
7. **Ishii, K., Yamazoe, Y., Kamataki, T., and Kato, R.,** Metabolic activation of mutagenic tryptophan pyrolysis products by rat liver microsomes, *Cancer Res.,* 40, 2596, 1980.
8. **Ishii, K., Yamazoe, Y., Kamataki, T., and Kato, R.,** Metabolic activation of glutamic acid pyrolysis products, 2-amino-6-methyldipyrido[1,2-a:3',2'-d]imidazole and 2-aminodipyrido[1,2-a:3',2'-d]imidazole, by purified cytochrome P-450, *Chem. Biol. Interact.,* 38, 1, 1981.
9. **Yamazoe, Y., Abu-Zeid, M., Yamauchi, K., and Kato, R.,** Metabolic activation of pyrolysate arylamines by human liver microsomes: possible involvement of a P-448-H type cytochrome P-450, *Jpn. J. Cancer Res. (Gann),* 79, 1159, 1988.
10. **Stavenuiter, J. F. C., Verrips-Kroon, M., Bos, E. J., and Westra, J. G.,** Syntheses of 5-phenyl-2-pyridinamine, a possibly carcinogenic pyrolysis product of phenylalanine, and some of its putative metabolites, *Carcinogenesis,* 6, 13, 1985.
11. **Niwa, T., Yamazoe, Y., and Kato, R.,** Metabolic activation of 2-amino-9H-pyrido[2,3-b]indole by rat-liver microsomes, *Mutat. Res.,* 95, 159, 1982.
12. **Tada, M., Saeki, H., and Oikawa, A.,** Presence of 3-amino-β-carbolines (3-amino-9H-pyrido[3,4-b]indoles), effectors in the induction of sister chromatid exchanges, in protein pyrolysates, *Mutat. Res.,* 121, 81, 1983.
13. **Nagao, M., Wakabayashi, K., Kasai, H., Nishimura, S., and Sugimura, T.,** Effect of methyl substitution on mutagenicity of 2-amino-3-methylimidazo[4,5-f]quinoline, isolated from broiled sardine, *Carcinogenesis,* 2, 1147, 1981.
14. **Grivas, S. and Jägerstad, M.,** Mutagenicity of some synthetic quinolines and quinoxalines related to IQ, MeIQ or MeIQx in Ames test, *Mutat. Res.,* 137, 29, 1984.
15. **Kaiser, G., Harnasch, D., King, M. T., and Wild, D.,** Chemical structure and mutagenic activity of aminoimidazoquinolines and aminonaphthimidazoles related to 2-amino-3-methylimidazo[4,5-f]quinoline, *Chem. Biol. Interact.,* 57, 97, 1986.
16. **Yamazoe, Y., Shimada, M., Kamataki, T., and Kato, R.,** Microsomal activation of 2-amino-3-methylimidazo[4,5-f]quinoline, a pyrolysate of sardine and beef extracts, to a mutagenic intermediate, *Cancer Res.,* 43, 5768, 1983.
17. **Turesky, R. J., Skipper, P. L., Tannenbaum, S. R., Coles, B., and Ketterer, B.,** Sulfamate formation is a major route for detoxication of 2-amino-3-methylimidazo[4,5-f]quinoline in the rat, *Carcinogenesis,* 7, 1483, 1986.
18. **Barnes, W. S. and Weisburger, J. H.,** Fate of the food mutagen 2-amino-3-methylimidazo[4,5-f]quinoline (IQ) in Sprague-Dawley rats, *Mutat. Res.,* 156, 83, 1985.
19. **Størmer, F. C., Alexander, J., and Becher, G.,** Fluorometric detection of 2-amino-3-methylimidazo[4,5-f]quinoline, 2-amino-3,4-dimethylimidazo[4,5-f] quinoline and their N-acetylated metabolites excreted by the rat, *Carcinogenesis,* 8, 1277, 1987.
20. **Peleran, J. C., Rao, D., and Bories, G. F.,** Identification of the cooked food mutagen 2-amino-3-methylimidazo[4,5-f]quinoline (IQ) and its N-acetylated and 3-N-demethylated metabolites in rat urine, *Toxicology,* 43, 193, 1987.
21. **Inamasu, T., Luks, H. J., and Weisburger, J. H.,** Comparison of XAD-2 column and blue cotton batch techniques for isolation of metabolites of 2-amino-3-methylimidazo[4,5-f]quinoline, *Jpn. J. Cancer Res. (Gann),* 79, 42, 1988.
22. **Yamazoe, Y., Abu-Zeid, M., Dawei, G., Staiano, N., and Kato, R.,** Enzymatic acetylation and sulfation of N-hydroxyarylamines in bacteria and rat livers, *Carcinogenesis,* 10, 1675, 1989.
23. **Abu-Zeid, M., Ketterer, B., Yamazoe, Y., and Kato, R.,** unpublished data.
24. **Bashir, M., Kingston, D. G. I., Carman, R. J., van Tassell, R. L., and Wilkins, T. D.,** Anaerobic metabolism of 2-amino-3-methyl-3H-imidazo[4,5-f]quinoline (IQ) by human fecal flora, *Mutat. Res.,* 190, 187, 1987.
25. **Hayatsu, H., Kasai, H., Yokoyama, S., Miyazawa, T., Yamaizumi, Z., Sato, S., Nishimura, S., Arimoto, S., Hayatsu, T., and Ohara, Y.,** Mutagenic metabolites in urine and feces of rats fed with 2-amino-3,8-dimethylimidazo[4,5-f]quinoxaline, a carcinogenic mutagen present in cooked meat, *Cancer Res.,* 47, 791, 1987.
26. **Sjödin, P., Wallen, H., Alexander, J., and Jägerstad, M.,** Disposition and metabolism of the food mutagen 2-amino-3,8-dimethylimidazo[4,5-f]quinoxaline (MeIQx) in rats, *Carcinogenesis,* 10, 1269, 1989.
27. **Yamazoe, Y., Abu-Zeid, M., Manabe, S., Toyama, S., and Kato, R.,** Metabolic activation of a protein pyrolysate promutagen 2-amino-3,8-dimethylimidazo[4,5-f]quinoxaline (MeIQx) by rat liver microsomes and purified cytochrome P-450, *Carcinogenesis,* 9, 105, 1988.

28. **Holme, J. A., Wallin, H., Søderlund, E. J., Brunborg, G., Hongslo, J. K., and Alexander, J.,** Genotoxicity of the food mutagen 2-amino-1-methyl-6-phenylimidazo[4,5-*b*]pyridine (PhIP): formation of 2-hydroxyamino-PhIP, a directory acting genotoxic metabolite, *Carcinogenesis,* 10, 1389, 1989.

29. **McManus, M. E., Felton, J. S., Knize, M. G., Burgess, W., Roberts-Thomson, S., Pond, S. M., Stupans, I., and Veronese, M. E.,** Activation of the food-derived mutagen 2-amino-1-methyl-6-phenylimidazo[4,5-*b*]pyridine by rabbit and human liver microsomes and purified forms of cytochrome P-450, *Carcinogenesis,* 10, 357, 1989.

30. **Kosuge, T., Tsuji, K., Wakabayashi, K., Okamoto, T., Shudo, K., Iitaka, Y., Itai, A., Sugimura, T., Kawachi, T., Nagao, M., Yahagi, T., and Seino, Y.,** Isolation and structure studies of mutagenic principles in amino acid pyrolysate, *Chem. Pharm. Bull.,* 26, 611, 1978.

31. **Kandaswami, C., Kumar, S., Dubey, S. K., and Sikka, H. C.,** Metabolism of benzo[*f*]quinoline by rat liver microsomes, *Carcinogenesis,* 8, 1861, 1987.

32. **Friesen, M., O'Neill, I. K., Malaveille, C., Garren, L., Hautefeuille, A., and Bartsch, H.,** Substituted hydroxyphenanthrenes in opium pyrolysates implicated in oesophageal cancer in Iran: structures and *in vitro* metabolic activation of a novel class of mutagens, *Carcinogenesis,* 8, 1423, 1987.

33. **Nagao, M., Yahagi, T., Honda, M., Seino, Y., Matsushima, T., and Sugimura,T.,** Demonstration of mutagenicity of aniline and *o*-toluidine by norharman, *Proc. Jpn. Acad. Ser B.,* 34, 1977.

34. **Nagao, M., Yahagi, T., and Sugimura, T.,** Differences in effects of norharman with various classes of chemical mutagens and amount of S9, *Biochem. Biophys. Res. Commun.,* 83, 373, 1978.

35. **Tweedie, D. J., Prough, R. A., and Burke, M. D.,** Effects of induction on the metabolism and cytochrome P-450 binding of harman and other β-carbolines, *Xenobiotica,* 18, 785, 1988.

36. **Kinouchi, T., Tsutsui, H., and Ohnishi, Y.,** Detection of 1-nitropyrene in yakitori (grilled chicken), *Mutat. Res.,* 171, 105, 1986.

37. **Kinouchi, T., Morotomi, M., Mutai, M., Fifer, E. K., Beland, F. A., and Ohnishi, Y.,** Metabolism of 1-nitropyrene in germ-free and conventional rats, *Jpn. J. Cancer Res. (Gann),* 77, 356, 1986.

38. **El-Bayoumy, K. and Hecht, S. S.,** Identification and mutagenicity of metabolites of 1-nitropyrene formed by rat liver, *Cancer Res.,* 43, 3132, 1983.

39. **Masuda, Y., Mori, K., and Kuratsune, M.,** Studies on bladder carcinogens in the human environment. I. Naphthylamines produced by pyrolysis of amino acids, *Int. J. Cancer,* 2, 489, 1967.

40. **Yamazoe, Y., Miller, D. W., Weis, C. C., Dooley, K. L., Zenser, T. V., Beland, F. A., and Kadluber F. F.,** DNA adducts formed by *ring*-oxidation of the carcinogen 2-naphthylamine with prostaglandin H synthase *in vitro* and in the dog urothelium *in vivo, Carcinogenesis,* 6, 1379, 1985.

41. **Essigman, J. M., Croy, R. G., Bennett, R. A., and Wogan, G. N.,** Metabolic activation of aflatoxin B1: Patterns of DNA adduct formation, removal, and excretion in relation to carcinogenesis, *Drug Metab. Rev.,* 13, 581, 1982.

42. **Jonen, H. G.,** Reductive and oxidative metabolism of nitrofurantoin in rat liver, *Naunyn-Schmiedeberg's Arch. Parmacol.,* 315, 167, 1980.

43. **Mattammal, M. B., Zenser, T. V., and Davis, B. B.,** Anaerobic metabolism and nuclear binding of the carcinogen 2-amino-4-(5-nitro-2-furylthiazole (ANFT), *Carcinogenesis,* 3, 1339, 1982.

44. **Frederick, C. B., Mays, J. B., Ziegler, D. M., Guengerich, F. P., and Kadlubar, F. F.,** Cytochrome P-450- and flavin-containing monoxygenase-catalyzed formation of the carcinogen N-hydroxy-2-aminofluorene and its covalent binding to nuclear DNA, *Cancer Res.,* 42, 2671, 1982.

45. **Loretz, L. J. and Pariza, M. W.,** Effect of glutathione levels, sulfate levels, and metabolic inhibitors on covalent binding of 2-amino-3-methylimidazo[4,5-*f*]quinoline and 2-acetylaminofluorene to cell macromolecules in primary monolayer cultures of adult rat hepatocytes, *Carcinogenesis,* 5, 895, 1984.

46. **Kato, R., Kamataki, T., and Yamazoe, Y.,** N-Hydroxylation of carcinogenic and mutagenic aromatic amines, *Environ. Health Perspec.,* 49, 21, 1983.

47. **Kato, R. and Yamazoe, Y.,** Metabolic activation and covalent binding to nucleic acids of carcinogenic heterocyclic amines from cooked foods and aminoacid pyrolysates, *Jpn. J. Cancer Res. (Gann),* 78, 297, 1987.

48. **Kamataki, T., Maeda, K., Yamazoe, Y., Matsuda, N., Ishii, K., and Kato, R.,** A high-spin form of cytochrome P-450 highly purified from polychlorinated biphenyl-treated rats, *Mol. Pharmacol.,* 24, 146, 1983.

49. **Snyderwine, E. G. and Battula, N.,** Selective mutagenic activation by cytochrome P3-450 of carcinogenic arylamines found in foods, *J. Natl. Cancer Inst.,* 81, 223, 1989.

50. **Kamataki, T., Maeda, K., Yamazoe, Y., Nagai, T., and Kato, R.,** Sex-difference of cytochrome P-450 in the rat: purification, characterization, and quantitation of constitutive forms of cytochrome P-450 from liver microsomes of male and female rats, *Arch. Biochem. Biophys.,* 225, 758, 1983.

51. **McManus, M. E., Burgess, W., Stupans, I., Trainor, K. J., Fenech, M., Robson, R. A., Morley, A. A., and Snyderwine, E. G.,** Activation of the food-derived mutagen 2-amino-3-methylimidazo[4,5-*f*]quinoline by human-liver microsomes, *Mutat. Res.,* 204, 185, 1988.

52. **Conney, A. H.**, Induction of microsomal enzymes by foreign chemicals and carcinogenesis by polycyclic aromatic hydrocarbons: G. H. A. Clowes Memorial Lecture, *Cancer Res., 42*, 4875, 1982.

53. **Mizokami, K., Nohmi, T., Fukuhara, M., Takanaka, A., and Omori, Y.**, Purification and characterization of a form of cytochrome P-450 with high specificity for aflatoxin-B1 from 3-methylcholanthrene-treated hamster liver, *Biochem. Biophys. Res. Commun., 139*, 466, 1986.

54. **Fukuhara, M., Nagata, K., Mizokami, K., Yamazoe, Y., Takanaka, A., and Kato, R.**, Complete cDNA sequence of a major 3-methylcholanthrene-inducible cytochrome P-450 isozyme (P-450AFB) of syrian-hamsters with high activity toward aflatoxin-B1, *Biochem. Biophys. Res. Commun., 162*, 265, 1989.

55. **Yamada, M., Tsuda, M., Nagao, M., Mori, M., and Sugimura, T.**, Degradation of mutagens from pyrolysates of tryptophan, glutamic acid and globulin by myeloperoxidase, *Biochem. Biophys. Res. Commun., 90*, 769, 1979.

56. **Yagi, M.**, Oxidation of tryptophan-P-1 and P-2 by beef liver catalase-H_2O_2 intermediate: comparsion with horseradish peroxidase, *Cancer Biochem. Biophys., 4*, 105, 1979.

57. **Petry, T. W., Krauss, R. S., and Eling, T. E.**, Prostaglandin H synthase-mediated biocativation of the amino acid pyrolysate product Trp P-2, *Carcinogenesis, 7*, 1397, 1986.

58. **Wild, D. and Degen, G. H.**, Prostaglandin H synthase-dependent mutagenic activation of heterocyclic aromatic amines of the IQ-type, *Carcinogenesis, 8*, 541, 1987.

59. **Nemoto, N. and Takayama, S.**, Activation of 2-amino-6-methyldipyrido[1,2-*a*:3',2'-*d*]imidazole, a mutagenic pyrolysis product of glutamic acid, to bind to protein by NADPH-dependent and -independent enzyme systems, *Carcinogenesis, 5*, 653, 1984.

60. **Kadlubar, F. F., Yamazoe, Y., Lang, N. P., Chu, D. Z. J., and Beland, F. A.**, Carcinogen-DNA adduct formation as a predictor of metabolic activation pathways and reactive intermediates in benzidine carcinogens, in *Biological Reactive Intermediates*, Vol. III, Kocsis, J. J., Jollow, D. J., Witmer, C. M., Nelson, J. O., and Snyder, R., Eds., Plenum Press, New York, 1986, 537.

61. **Flammang, T. J., Yamazoe, Y., Benson, R. W., Roberts, D. W., Potter, D. W., Chu, D. W., Lang, D. Z. J., and Kadlubar, F. F.**, Arachidonic acid-dependent peroxidative activation of carcinogenic arylamines by extrahepatic human tissue microsomes, *Cancer Res., 49*, 1977, 1989.

62. **Arimoto, S., Ohara, Y., Hiramoto, K., and Hayatsu, H.**, Inhibitory effect of myoglobin and hemoglobin on the direct-acting mutagenicity of protein pyrolysate heterocyclic amine derivatives, *Mutat. Res., 192*, 253, 1987.

63. **Rafter, J. J. and Gustafsson, J.-Å.**, Metabolism of the dietary carcinogen Trp-P-1 in rats, *Carcinogenesis, 7*, 1291, 1986.

64. **Negishi, C., Umemoto, A., Rafter, J. J., Sato, S., and Sugimura, T.**, N-Acetyl derivative as the major active metabolite of 2-amino-6-methyldipyrido[1,2-*a*:3',2'-*d*]imidazole in rat bile, *Mutat. Res., 175*, 23, 1986.

65. **Kanai, Y., Manabe, S., and Wada, O.**, Detection of N-acetyl derivative of 2-amino-6-methyldipyrido[1,2-*a*:3',2'-*d*]imidazole (Glu-P-1) in Glu-P-1-injected rats, *Mutat. Res., 207*, 63, 1988.

66. **Shinohara, A., Yamazoe, Y., Saito, K., Kamataki, T., and Kato, R.**, Species differences in the N-acetylation by liver cytosol of mutagenic heterocyclic aromatic amines in protein pyrolysates, *Carcinogenesis, 5*, 683, 1984.

67. **Yamazoe, Y., Shimada, M., Kamataki, T., and Kato, R.**, Covalent binding of N-hydroxy-Trp-P-2 to DNA by cytosolic proline-dependent system, *Biochem. Biophys. Res. Commun., 107*, 165, 1982.

68. **Saito, K., Yamazoe, Y., Kamataki, T., and Kato, R.**, Mechanism of activation of proximate mutagens in Ames' tester strains: The acetyl CoA-dependent enzyme in *Salmonella typhimurium* TA98 deficient in TA98/1,8-DNP_6 catalyzes DNA binding as the cause of mutagenicity, *Biochem. Biophys. Res. Commun., 116*, 141, 1983.

69. **Shinohara, A., Saito, K., Yamazoe, Y., Kamataki, T., and Kato, R.**, DNA binding of N-hydroxy-Trp-P-2 and N-hydroxy-Glu-P-1 by acetyl-CoA dependent enzyme in mammalian liver cytosol, *Carcinogenesis, 6*, 305, 1985.

70. **Shinohara, A., Saito, K., Yamazoe, Y., Kamataki, T., and Kato, R.**, Acetyl coenzyme A-dependent activation of N-hydroxy derivative of carcinogenic arylamines: Mechanism of activation, species difference, tissue distribution and acetyl donor specificity, *Cancer Res., 46*, 4362, 1986.

71. **Flammang, T. J. and Kadlubar, F. F.**, Acetyl coenzyme A-dependent metabolic activation and N-hydroxy-3,2'-dimethyl-4-aminobiphenyl and several carcinogenic N-hydroxyarylamines in relation to tissue and species differences, other acetyldonors, and arylhydroxamic acid-dependent acyltransferases, *Carcinogenesis, 7*, 919, 1986.

72. **Snyderwine, E. G., Wirth, P. J., Roller, P. P., Adamson, R. H., Sato, S., and Thorgeirsson, S. S.**, Mutagenicity and in vitro covalent DNA binding of 2-hydroxyamino-3-methylimidazolo[4,5-*f*]quinoline, *Carcinogenesis, 9*, 411, 1988.

73. **Kato, R. and Yamazoe, Y.**, Further metabolic activations of mutagenic and carcinogenic N-hydroxyarylamine by conjugating enzymes, in *Xenobiotic Metabolism and Disposition*, Kato, R., Estabrook, R. W., and Cayen, M. N., Eds., Taylor & Francis, London, 1989, 383.

74. **Weber, W. W. and Hein, D. W.**, N-acetylation pharmacogenetics, *Pharm. Rev.*, 37, 25, 1985.

75. **Allaben, W. T. and King, C. M.**, The purification of rat liver arylhydroxamic acid, N,O-acyltransferase, *J. Biol. Chem.*, 259, 12128, 1984.

76. **Saito, K., Yamazoe, Y., Kamataki, T., and Kato, R.**, Syntheses of hydroxyamino, nitroso and nitro derivatives of Trp-P-2 and Glu-P-1, amino acid pyrolysate mutagens and their direct mutagenicities toward *Salmonella typhimurium* TA98 and TA98NR, *Carcinogenesis*, 4, 1547, 1983.

77. **McCoy, E. C., McCoy, G. D., and Rosenkranz, H. S.**, Esterification of arylhydroxylamines: evidence for a specific gene product in mutagenesis, *Biochem. Biophys. Res. Commun.*, 108, 1362, 1982.

78. **Saito, K., Shinohara, A., Kamataki, T., and Kato, R.**, Metabolic activation of mutagenic N-hydroxyarylamines by O-acetyltransferase in *Salmonella typhimurium* TA98, *Arch. Biochem. Biophys.*, 239, 286, 1985.

79. **Nagao, M., Fujita, Y., Wakabayashi, K., and Sugimura, T.**, Ultimate forms of mutgenic and carcinogenic heterocyclic amines produced by pyrolysis, *Biochem. Biophys. Res. Commun.*, 114, 626, 1983.

80. **Gayda, D. P. and Pariza, M. W.**, Activation of 2-amino-3-methylimidazo[4,5-*f*]quinoline and 2-aminofluorene for bacterial mutagenesis by primary monolayer cultures of adult rat hepatocytes, *Mutat. Res.*, 118, 7, 1983.

81. **Nakayasu, M., Nakasato, F., Sakamoto, H., Terada, M., and Sugimura, T.**, Mutagenic activity of heterocyclic amines in Chinese hamster lung cells with diphtheria toxin resistance as a marker, *Mutat. Res.*, 118, 91, 1983.

82. **Thompson, L. H., Carrano, A. V., Salazar, E., Felton, J. S., and Hatch, F. T.**, Comparative genotoxic effects of the cooked-food-related mutagens Trp-P-2 and IQ in bacteria and cultured mammalian cells, *Mutat. Res.*, 117, 243, 1983.

83. **Wild, D., Gocke, E., Harnasch, D., Kaiser, G., and King, M. T.**, Differential mutagenic activity of IQ (2-amino-3-methylimidazo[4,5-*f*]quinoline) in *Salmonella typhimurium* strains *in vitro* and *in vivo*, in Drosophila, and in mice, *Mutat. Res.*, 156, 93, 1985.

84. **Watanabe, M., Ishidate, M. J., and Nohmi, T.**, New modified strains of *Salmonella typhimurium* TA98, TA100, very sensitive to nitroarenes and aromatic amines by cloning of acetyltransferase gene, *Environ. Mol. Mutat.*, 14 (Suppl.) 15, 1989.

85. **Tada, M. and Tada, M.**, Seryl-tRNA synthetase and activation of the carcinogen 4-nitroquinoline 1-oxide, *Nature (London)*, 255, 510, 1975.

86. **Yamazoe, Y., Tada, M., Kamataki, T., and Kato, R.**, Enhancement of binding of N-hydroxy-Trp-P-2 to DNA by seryl-tRNA synthetase, *Biochem. Biophys. Res. Commun.*, 102, 432, 1981.

87. **Yamazoe, Y., Shimada, M., Shinohara, A., Saito, K., Kamataki, T., and Kato, R.**, Catalysis of the covalent binding of 3-hydroxyamino-1-methyl-5*H*-pyrido[4,3-*b*]indole to DNA by a L-proline- and adenosine triphosphate-dependent enzyme in rat hepatic cytosol, *Cancer Res.*, 45, 2495, 1985.

88. **Gooderham, N. J., Watson, D., Rice, J. C., Murray, S., Taylor, G. W., and Davis, D. S.**, Metabolism of the mutagen MeIQx in vivo: metabolite screening by liquid chromatography-thermospray mass spectrometry, *Biochem. Biophys. Res. Commun.*, 148, 1377, 1987.

89. **Turesky, R. J., Aeschbacher, H. U., Malnoë, A., and Würzner, H. P.**, Metabolism of the food-borne mutagen/carcinogen 2-amino-3,8-dimethylimidazo[4,5-f]quinoxaline in the rat: Assessment of billiary metabolites for genotoxicity, *Food Chem. Toxicol.*, 26, 105, 1988.

90. **Ketterer, B.**, Protective role of glutathione and glutathione transferases in mutagenesis and carcinogenesis, *Mutat. Res.*, 202, 343, 1988.

91. **de Waziers, I. and Decloître, F.**, Effect of glutathione and uridine-5'-diphosphoglucuronic acid on the mutagenicity of tryptophan pyrolysis products (Trp-P-1 and Trp-P-2) by rat-liver and -intestine S9 fraction, *Mutat. Res.*, 139, 15, 1984.

92. **Mita, S., Yamazoe, Y., Kamataki, T., and Kato, R.**, Metabolic activation of Trp-P-2, a tryptophan-pyrolysis mutagen, by isolated rat hepatocytes, *Biochem. Pharmacol.*, 32, 1179, 1983.

93. **Saito, K., Yamazoe, Y., Kamataki, T., and Kato, R.**, Activation and detoxication of N-hydroxy-Trp-P-2 by glutathione and glutathione transferases, *Carcinogenesis*, 4, 1551, 1983.

94. **Umemoto, A., Grivas, S., Yamaizumi, Z., Sato, S., and Sugimura, T.**, Non-enzymatic glutathione conjugation of 2-nitroso-6-methyldipyrido[1,2-*a*:3',2'-*d*]imidazole (NO-Glu-P-1) in vitro: N-hydroxy-sulfonamide, a new binding form of arylnitroso compounds and thiols, *Chem. Biol. Interact.*, 68, 57, 1988.

95. **Coles, B., Meyer, D. J., Ketterer, B., Stanton, C. A., and Garner, R. C.**, Studies on the detoxication of microsomally-activated aflatoxin B1 by glutathione and glutathione transferases in vitro, *Carcinogenesis*, 6, 693, 1985.

96. **Boyland, E., Mason, D., and Orr, S. F. D.**, The biochemistry of aromatic amines 2. The conversion of arylamines into arylsulphamic acids and arylamine-N-glucuronic acids, *Biochem. J.*, 65, 417, 1957.

97. **Snyderwine, E. G., Roller, P. P., Adamson, R. H., Sato, S., and Thorgeirsson, S. S.**, Reaction of N-hydroxylamine and N-acetoxy derivatives of 2-amino-3-methylimidazo[4,5-*f*]quinoline with DNA. Synthesis and identification on N-(deoxyguanosin-8-yl)-IQ, *Carcinogenesis*, 9, 1061, 1988.

98. **Yamazoe, Y., Manabe, S., Murayama, N., and Kato, R.,** Regulation of hepatic sulfotransferase catalyzing the activation of N-hydroxyarylamide and N-hydroxyarylamine by growth hormone, *Mol. Pharmacol.,* 32, 536, 1987.

99. **DeBaun, J. R., Miller, E. C., and Miller, J. A.,** N-Hydroxy-2-acetylaminofluorene sulfotransferase: its probable role in carcinogenesis and in protein-(methion-S-yl) binding in rat liver, *Cancer Res.,* 30, 577, 1970.

Chapter 7.2

METABOLIC FATE OF HETEROCYCLIC AMINES FROM COOKED FOOD

Jan Alexander and Håkan Wallin

TABLE OF CONTENTS

I. INTRODUCTION

Several highly mutagenic compounds have been isolated from the crust of fried or broiled meat and fish. All these mutagens are heterocyclic primary amines which can also be produced by amino acid pyrolyzation or by heating model systems composed of amino acids, creatine, and sugar.[1-3] On the basis of chemical structure the compounds might be divided into the pyridoimidazoles or the amino acid pyrolysates [i.e., 3-amino-1,4-dimethyl-5*H*-pyrido[4,3-*b*]indole (Trp-P-1), 3-amino-1-methyl-5*H*-pyrido[4,3-*b*]indole (Trp-P-2), 2-amino-6-methyl-dipyrido[1,2-*a*:3′,2′-*d*]imidazole] and the aminoimidazoazaarenes (AIA) [i.e., 2-amino-3-methylimidazo[4,5-*f*]quinoline (IQ), 2-amino-3,4-dimethylimidazo[4,5-*f*]quinoline (MeIQ)(the quinolines), 2-amino-3,8-dimethylimidazo[4,5-*f*]quinoxaline (MeIQx), 2-amino-3,4,8-trimethylimidazo[4,5-*f*]quinoxaline (diMeIQx) (the quinoxalines), and 2-amino-1-methyl-6-phenylimidazo[4,5-*b*]-pyridine (PhIP)] (Figure 1).[3] This presentation will cover the metabolic fate of these heterocyclic amines. We will, however, focus on the so-called AIA class of compounds because they are the most prevalent in the western diet. The amino acid pyrolysate compounds, which are less common in food, have, however, also been extensively studied, especially with regard to metabolic activation.[1,3-5]

Only very few studies contain data on ingested heterocyclic amines in humans. Conclusions must therefore be inferred from studies of heterocyclic amines in various animal model systems. Most studies on the metabolic fate of mutagenic heterocyclic amines have been performed with isolated enzymes, subcellular fractions, and cells *in vitro*. Much fewer studies have been published on whole animals. Furthermore, most of the *in vitro* studies have focused on metabolic activation and mechanisms of formation of reactive metabolites.[4,5] Less attention has been paid to detoxification pathways, which are also important for the fate and the possible harmful effects of these compounds. In addition to the general problems of extrapolation of *in vitro* and animal data to the human situation, most metabolic *in vivo* studies have been performed with single doses, which are much higher than those expected from the human exposure situation.

II. UPTAKE AND DISTRIBUTION

The absorption of heterocyclic amines from the gastrointestinal tract seems to be efficient and rapid. Using an *in situ* technique with intestinal loops in rats it was found that Trp-P-2 was not absorbed in the stomach at a pH below pH 3, whereas absorption occurred rapidly in the small intestine and colon when the pH was above 4.0.[6] This indicates transport across the intestinal wall of the unionized lipophilic form of the amine. Radiolabeled Trp-P-2 or Glu-P-1 given to rats by gavage was absorbed 1 to 2 h.[4] Studies with closed intestinal segments of mice showed that IQ was primarily taken up by the small intestine.[7] Sixty percent or more of a single dose of the heterocyclic amines given by gavage was recovered in urine and feces as metabolites within a few days, indicating efficient intestinal absorption.[4,7-15]

Heterocyclic amines bind to dietary fibers both *in vitro* and *in vivo*.[16-22] In comparison with MeIQ alone, the intrasanguinous host-mediated bacterial mutagenicity was reduced in rats when MeIQ was given combined with a fiber-rich diet.[21] Furthermore, the intestinal microflora increased the binding of heterocyclic amines to fecal components.[19,22] The binding of heterocyclic amines to fibers or other fecal material probably results in lower bioavailability of the compounds. In one experiment, however, dietary fiber did not reduce IQ genotoxicity in the colon.[23]

For humans the few existing data indicate a rapid uptake of the heterocyclic amines.[24] For example, some MeIQx ingested via cooked beef was excreted as unchanged compound in the urine the first 12 hr, but not thereafter.[24] This agrees with the increased mutagenicity

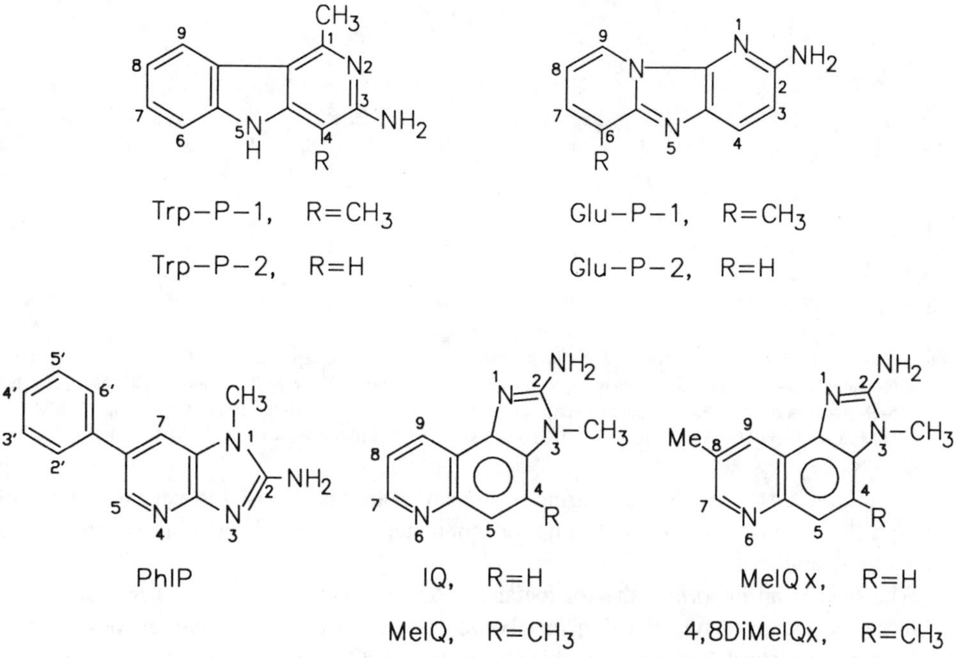

FIGURE 1. Chemical structures of compounds discussed in this chapter.

in urine of healthy humans that was detected within a day after intake of fried meat.[25,26] It is, however, impossible from these data to evaluate the efficiency of absorption. Although they contain no data on rate and efficiency, a few other studies demonstrate that the compounds are taken up by humans. Thus, heterocyclic amines have been detected in dialysis fluid of kidney patients.[27-30] Also, bile fluid and urine from humans were found to contain Glu-P-1 and a Glu-P-1 metabolite.[31]

Once absorbed, the compounds are distributed by the bloodstream to most parts of the body. This apparently is a rapid process as ^{14}C-Glu-P-1, ^{14}C-Trp-P-1, ^{14}C-IQ, and ^{14}C-MeIQ injected intravenously in rats and mice are cleared from the blood within minutes or a few hours.[32-34] As revealed by autoradiography, these heterocyclic amines (i.e., Glu-P-1, Trp-P-1, MeIQ, and IQ) are widely distributed throughout the body except for the central nervous system.[32-34] This indicates poor penetration of the blood brain barrier. At short survival times these compounds seemed to accumulate in excretory organs such as liver, bile, intestinal contents, kidney and urine, salivary glands, zymbal glands, and nasal mucosa. Also, endocrine tissue such as thyroid and adrenals contained high levels. All compounds had a strong affinity for melanin. Trp-P-1 was furthermore concentrated to lung tissue after β-naphthoflavone (BNF) pretreatment.[33] Trp-P-1, IQ, and MeIQ passed the placenta in mice, but no radioactivity was retained in fetal tissue except in fetal retinas containing melanin.[32-34] Apart from the liver (Trp-P-1, Glu-P-1, IQ, and MeIQ), kidney (IQ and MeIQ), and nasal mucosa (Trp-P-1 and Glu-P-1) where non-extractable radioactivty were found, radiolabeling left most tissues after 1 to 6 days.[32-34]

III. METABOLISM

Heterocyclic amines seem to be rapidly and extensively metabolized. After a single dose the major part of the dose is excreted within 1 to 3 days as metabolites and only a few percent of the parent compound is found in urine and feces.[4,8-15,35-39] With MeIQx and DiMeIQx up to 20% may be excreted unchanged in urine.[8,12-14] These studies were performed

TABLE 1

	Metabolites of AIA											
	1	2	3	4	5	6	7	8	9	10	11	12
IQ	+	(+)	(+)	+	+	−	−	+	+			
MeIQ	+	−	(+)	+	+	−	−	+	+			
MeIQx	+	(+)	(±)	+	+	(+)	−	+	+	+	+	
DiMeIQx	+	(+)	(+)	−	−	−	−	−	−	+		+
PhIP	+	+	−	?	+	+	+	+	+			

Note: Major + and minor(+) pathways for which evidence exists, or not reported − . 1: N-OH; 2: N-demethyl; 3: N-acetyl; 4: N-sulfamate; 5: N-glucuronide; 6: N-SO-glutathione; 7: N(OH)-glucuronide; 8: IQ, MeIQ, MeIQx: 5-O-glucuronide, PhIP: 4'-O-glucuronide; 9: IQ, MeIQ, MeIQx: 5-O-sulfate, PhIP: 4'-O-sulfate; 10: MeIQx, DiMeIQx: 8-CH$_2$OH; 11: MeIQx: 8-CH$_2$OH-5-O-sulfate; 12: DiMeIQx: 4-CH$_2$OH, N-demethyl.

with one large dose. The bile and urine of humans naturally exposed to very low doses of heterocyclic amines via food consumption contained very small amounts of unchanged heterocyclic amines.[31]

The liver is an important site for metabolic transformation, but also other tissues such as intestines, kidney, lung, and skin metabolize (i.e., activate) heterocyclic amines.[4-6,33,40-45] Furthermore, intestinal bacteria anaerobically converted IQ to 2-amino-3,6-dihydro-3-methyl-7H-imidazo[4,5-f]quinoline-7-one, which was directly mutagenic in *Salmonella typhimurium*.[46,47]

Most of our knowledge on the metabolism of heterocyclic amines is derived from studies either on subcellular fractions and enzymes or cellular model systems. The metabolic transformations reported for these compounds are direct conjugation reactions, oxidations, demethylations, and conjugation of oxidation products (Figure 2 and Table 1).

A. METABOLIC ACTIVATION

Oxidation reactions can be divided into those leading to metabolic activation and detoxification reactions. Heterocyclic amines are metabolically activated through hydroxylation of the exocyclic amino group. This has been excellently reviewed by Kato[4] in 1986 and Kato and Yamazoe[5] in 1987. Although this reaction can be catalyzed by prostaglandin synthetases, cytochromes P-450 are the dominating catalyzing enzymes.[4,5] The highest rate of N-hydroxylation for all heterocyclic amines is seen with rat liver P-450IA2 (i.e., P450d, P448 high spin).[4,5] Activity is also seen with P-450IA1; however, this varies greatly between different heterocyclic amines.[4,5,48,49] Toward PhIP and Trp-P-2 the N-hydroxylating rate of P-450IA1 relative to that of P-450IA2 is much higher (i.e., 0.5 and 0.33, respectively) in comparison with other heterocyclic amines (i.e., IQ = 0.076, Glu-P-1 = 0.032).[4,5,48] P-450IA2 also seems to be the most active enzyme in mice.[50] Rabbit liver P-450IA2 activates IQ 7.7 times more efficiently than P-450IA1,[51] whereas toward PhIP rabbit liver P-450IA1 is three times more active than P-450IA2.[52] Also, the rat liver male P-450 form activates heterocyclic amines.[4] In contrast, much lower activity has been seen with other P-450 forms (e.g., rat P-450IIB1, rabbit P-450IIB4, and rat and rabbit P-450IIE1).[4,5,48,49,51,52]

Homogenates from various tissues of rodents, as well as human and monkey liver microsomes, activate heterocyclic amines to bacterial mutagens.[4,5,40,43,44,52-56] Bioactivation of several heterocyclic amines was studied in a series of human liver microsomes which had been selected on the basis of characteristic levels of individual cytochrome P-450 enzymes with help of antibodies and also in reconstituted human P-450 systems.[56] Human P-450$_{PA}$ (low K$_m$ phenacetine-O-deethylase) (tentatively P-450IA2) appears to be the major human enzyme involved in the bioactivation of IQ, MeIQ, MeIQx, Glu-P-1, Glu-P-2, and Trp-P-

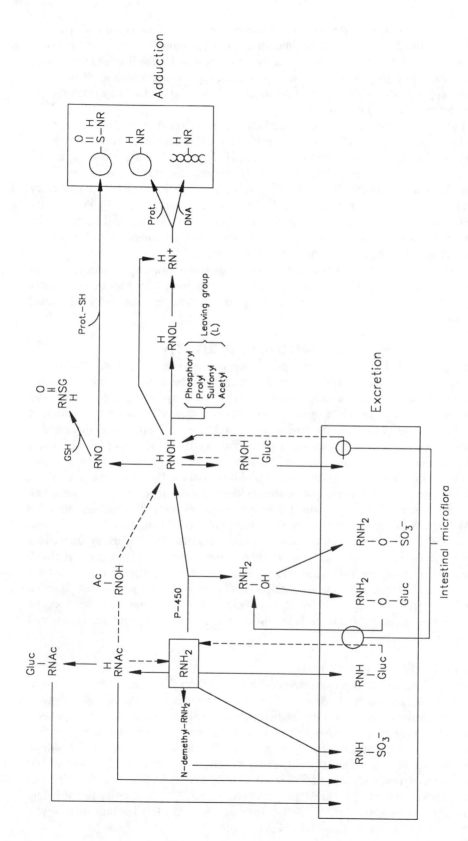

FIGURE 2. Scheme of suggested metabolic pathways for heterocyclic amines. Significance and role of pathways for individual amines are discussed in the text and Table 1. The dashed arrows indicate pathways not confirmed.

2, while more than one enzyme appears to contribute to the bioactivation of Trp-P-1. Apparently P-450IA2 is also the most important form metabolizing aromatic amines in humans *in vivo*.[57] The highest activation capacity is usually found in the liver and the activity in liver is induced by polycyclic aromatic hydrocarbon-type inducers such as PCB, methylcholanthrene, and BNF.[4,5,32,33,40,43,59-61] In humans P-450IA2 is induced by smoking and intake of charcoal-containing food.[56]

The liver P-450 enzymes are widely distributed to many extrahepatic tissues, but there is also reason to believe that P-450 enzymes specific to other tissues may activate heterocyclic amines in extrahepatic tissues (e.g., lung, intestine, kidney, etc.). For example, 2-amino-fluorene is effectively *N*-hydroxylated by a P-450 enzyme specific to rabbit lung.[18,22,62] Although Clara cells from rabbit lung cells effectively activate MeIQ and IQ,[40] the specificity of different P-450 forms has not been examined. Enzyme induction (e.g., by PCB, methylcholanthrene and BNF) occurs also in organs such as intestine, kidney, and lung.[40,43] The induction probably affects the metabolism and distribution of the compounds (this was for example demonstrated for Trp-P-1 after BNF pretreatment).[33,34]

The half-lives of the direct-acting mutagenic metabolites formed by hapatocytes varies from $^3/_4$ hr for IQ and MeIQ to 24 hr for PhIP.[58,60] Therefore, they may be sufficiently stable to be transported from the liver to other organs. However, this has not been studied in any detail.

B. ESTERIFICATION OF THE *N*-HYDROXY DERIVATIVE

Although the *N*-hydroxy derivatives or metabolically activated IQ, MeIQ, and MeIQx form protein and DNA adducts and are directly mutagenic in bacteria, it is well established that a second activation step, namely *O*-acetylation of the *N*-hydroxy group, is important for mutagenic events in *Salmonella typhimurium*. This and other esterification reactions have been studied by Kato, Yamazoe, and co-workers[60b] and have been extensively discussed in their review papers.[4,5] The esterification of the *N*-hydroxy IQ compounds probably is less efficient in mammalian cells, since most AIA compounds are much less genotoxic in mammalian cells.[4,5] In this respect, Trp-P-2 and PhIP differ from the IQ compounds since they, although less active in bacteria, potentially produce genotoxic effects in mammalian cells.[60,60a,60c] Apparently PhIP and Trp-P-2 do not require *O*-acetylation as, in contrast to IQ and MeIQ, they induce mutations in the *O*-acetyltransferase-lacking strain *S. typhimurium* TA98-1,6-DNP$_8$.[4,5,58,60,60a,60c] Esterification reactions suggested for *N*-hydroxy derivatives of heterocyclic amines are, besides *O*-acetylation, amino acylation by amino acyl tRNA synthetase (prolylation), sulfation, and phosphorylation.[5] It is not clear if these reactions are important and operate in mammalian cells. *In vitro* experiments suggest that there are considerable differences in the preferred esterification reaction among the heterocyclic amines.[4,5] There are also differences between species, at least regarding acetylation and amino acylation.[5]

C. ADDUCT FORMATION *IN VIVO*

Several groups have shown that heterocyclic amines form DNA and protein adducts *in vitro*.[4,5,61,62a-63] Again, much less data are found in the literature on the *in vivo* situation. However administration of IQ to rats results in covalent binding of IQ to serum proteins albumin and hemoglobin.[64] DNA-adduct formation in rats has been measured by ^{32}P-postlabeling for all the common heterocyclic amines.[65-68] A total level of 0.5 to 270 adducts per 10^7 nucleotide has been noted in the liver after exposure of rats to a single dose of 100 mg amine per kilogram body weight.[65] Several adducts were detected for each amine. As to species, monkeys and rats had similar patterns of IQ adducts.[69] By measuring radiolabeling of DNA after administration of ^{14}C-MeIQx to mice the "covalent binding index" was determined to be of a similar magnitude as that of other potent carcinogens.[70]

N-Oxidized forms of heterocyclic amines (e.g., Glu-P-1, Trp-P-2, MeIQx, and PhIP) react with sulfhydryl groups on proteins and with glutathion (GSH).[4,71,72] This has been demonstrated *in vivo* for IQ where a sulfinamide adduct form with albumin in exposed rats.[64] Several GSH derivatives of Trp-P-2 from rats were also characterized,[4] and the formation of a PhIP-GSH sulfinamide is catalyzed by GSH-transferase in hepatocytes.[86] Also, MeIQx forms a GSH derivative in hepatocytes.[74] The common pathway probably is the reaction of nitroso derivatives with GSH. This had, for example, recently been shown with NO-Glu-P-1, which reacts spontaneously with GSH to form both a sulfinamide and a sulfonamide.[72] NOH-Trp-P-2 is conjugated to GSH in a GSH transferase catalyzed reaction.[4]

D. CONJUGATION REACTIONS

Recently we discovered that a semistable *N*-glucuronide conjugate of 2-hydroxamino-PhIP was formed in rat hepatocytes.[71] This is the first time a semistable transportable metabolite of an activated heterocyclic amine has been described. This was a major derivative of PhIP in the bile of exposed rats. It is possible that this metabolite is hydrolyzed in the large intestine by bacterial β-glucuronidases. This will liberate the directly acting genotoxic metabolite 2-hydroxamino-PhIP, causing DNA damage in the cells of the intestinal wall. One might therefore suspect PhIP to be a colon carcinogen.

N-Acetylation of heterocyclic amines, which is a common and well-known reaction for aromatic amines, is, even if a great species variation exists, a minor pathway for the heterocyclic amines.[4,9,35-38,75] In the rat only a few percent of Glu-P-1,[9] IQ,[39] MeIQ,[38,39] and MeIQx[75] were found to be excreted as *N*-acetylated metabolites. The trace amounts of *N*-acetyl-Glu-P-1 observed in human bile and urine suggests that the reaction occurs in humans.[31] *N*-Acetylation does not represent a strict detoxification pathway as the compounds may be easily deacetylated and still exert genotoxic activity in bacteria and mammalian cells *in vitro*.[76] It is not known whether *N*-acetylated heterocyclic amines can be *N*-hydroxylated directly. However, the slightly greater DNA damaging effect of *N*-acetyl-IQ in comparison with IQ in isolated hepatocytes might indicate direct *N*-hydroxylation followed by *N,O*-acetyltransfer.[76]

E. DETOXIFICATION

The formation of water-soluble detoxified metabolites has been studied less. From animal studies, however, it can be inferred that many of the heterocyclic amines such as Glu-P-1, Trp-P-1, Trp-P-2, IQ, MeIQ, MeIQx, diMeIQx, and PhIP are rapidly converted to water-soluble products which are excreted in feces via bile and in urine.[4,8-14,36] (Figure 2).

The elucidation of metabolic detoxification pathways of heterocyclic amines is a complex task, but apart from studies in animals valuable information can again be collected from various *in vitro* model systems such as isolated cell fractions, enzymes, and cells. A very convenient way of studying metabolism of heterocyclic amines is to use isolated rat liver cells.[35,36,58,60,74,77] These cells contain both oxidative and conjugation enzyme systems. The metabolites are easily isolated from the incubation medium of the hepatocytes in comparison with complex matrices such as urine, bile, and feces. In this *in vitro* system the metabolism can be modulated by the use of enzyme inducers (e.g., P-450:PCB-pretreatment) or inhibitors (e.g., P-450:α-naphthoflavone (ANF), metyrapone; sulfotransferase:pentachlorophenol; glucuronyl transferase:galactosamine, and glutathione depletion by pretreating hepatocytes with diethylmaleate). Sulfate conjugates can be detected by using $^{35}SO_4{}^{2-}$. Isolated metabolites as well as incubates may be further characterized by treatment with β-glucuronidase and arylsulfatase. Metabolites may then be compared to those found *in vivo*. We have found the hepatocyte model system to be a good approximation of the *in vivo* situation for several amines.[12,35,36,74]

Detailed characterizaton of metabolic products has, to our knowledge, only been performed on IQ,[15,37-39,78] MeIQ,[35] MeIQx,[12-14,74] and PhIP.[36,60,71,87] Much less is known about the metabolites of Glu-P-1,[4,9] Trp-P-1,[10] Trp-P-2,[4] and diMeIQx[8] (Table 1).

Two major detoxification pathways in the rat for IQ, MeIQ, and MeIQx seem to be sulfate or glucuronyl conjugation of the exocyclic amino group.[12-14,35,74] The product of sulfation, the sulfamate derivative, is most prominent for IQ[15,35,77] for which this is the dominating metabolite in the rat. Also, a major part of MeIQx given to rats could be recovered as the sulfamate. The N-glucuronide of MeIQx was found both in treated rats and mice.[79] In hepatocytes isolated from uninduced rats IQ, MeIQx, and MeIQ were converted to sulfamates, with the IQ-sulfamate as a dominating metabolite, and to N-glucuronides.[35,74,77] In comparison with IQ a larger part of MeIQ was converted to the N-glucuronide.[35] For PhIP, only traces, if any, of the sulfamate could be observed with isolated hepatocytes.[36] IQ and MeIQx were N-acetylated at low rates by hepatocytes.[35,77] The rate was about 3 times larger for MeIQ.[43] The N-acetylated IQ and MeIQ were further transformed to N-glucuronyl derivatives by the hepatocytes.[35] With hepatocytes from PCB-pretreated animals, sulfates and glucuronides of oxidized metabolites were the principal metabolites, whereas much less of the compounds IQ, MeIQ, and MeIQx were directly conjugated to N-glucuronides and sulfamates.[35,74]

The oxidative pathways leading to detoxification are oxidations of exocyclic methyl groups and of the aromatic ring systems. Oxidative demethylation of the N-methyl group of IQ, MeIQx, and diMeIQx has been observed in rats and for PhIP in isolated hepatocytes.[8,38,75,78] This pathway seems, however, to be a minor pathway. More important are the oxidations occurring at the ring systems or at the C-methyl groups. Ring-hydroxylation has been observed at the 5 position for IQ[35,78] and MeIQ[35] (see Figure 1) and 5 or 4 position for MeIQx[12,13,75] (see Figure 1). The major detoxification product of PhIP[36] is oxidized in position 4 of the phenyl ring (see Figure 1). These reactions, which take place both in isolated hepatocytes and rats, can be induced by pretreatment of the animals with PCB before isolation of the heptocytes and are inhibited by P-450 inhibitor ANF.[35,36,74]

Although pretreatment of the animals with the P-450 inducer BNF increases the retention of Trp-P-1 in lung tissue, kidney cortex, and small intestine, such induction reduces the retention of Trp-P-1 and Glu-P-1 in the liver.[33,34] This indicates induction of ring-hydroxylations leading to detoxification and increased excretion of the compounds. This notion is supported by the fact that BNF pretreatment combined with the P-450 inhibitor 9-hydroxyyellipticine increases liver retention of Trp-P-1 and the overall body retention of Glu-P-1.[33,34] With diMeIQx, BNF pretreatment increased the fecal excretion of metabolites in rats.[8]

The 8-methyl group of MeIQx is oxidized in rats as well as *in vitro* with hepatocytes.[13,74,75] This metabolite can still be activated to a bacterial mutagen, but it is less mutagenic than the parent compound.[75] 8-Methyl oxidation as well as N-acetylation have also been reported for diMeIQx in rats.[8] With the exception of PhIP, only minor amounts, if any, of the unconjugated ring-hydroxylated compounds were found in hepatocyte incubations or *in vivo*.[11-13,35-39,43,74,77] In general they were rapidly converted to glucuronyl or sulfate conjugates.

The ring-hydroxylated glucuronide of MeIQx is a major metabolite in bile.[12] This metabolite, however, is absent in fecal material of MeIQx-exposed rats.[12] The glucuronyl conjugate must therefore be further metabolized in the intestine, probably by bacterial β-glucuronidase.

By mass spectroscopy of urinary metabolites it was determined that Trp-P-1 was metabolized to hydroxylated and acetylated derivatives in rats exposed to Trp-P-1.[10] Similarly, several uncharacterized ether-glucuronide derivatives of ring-hydroxylated Trp-P-2 were found in rat urine of exposed animals.[4] Glu-P-1 glucuronides of unknown structure have also been reported.[4]

Whereas the enzymology of mutagenic activation and N-hydroxylation has been extensively studied, little is known about the ring- and C-methyl hydroxylations. With PhIP we recently found that ring-hydroxylation in rat liver microsomal incubations were induced by BNF, PCB, and isosafrole pretreatment of the animals and that this reaction was inhibited by ANF.[48] No induction was seen with phenobarbital. With isolated P-450 enzymes, rat P-450IA1 had the largest ring-hydroxylating activity followed by rat P-450IA2 being 5 times less active. As mentioned previously, P-450IA2 N-hydroxylated PhIP at a rate twice that of P-450IA1.[48]

F. P-450 INDUCTION BY HETEROCYCLIC AMINES

Heterocyclic amines may induce their own metabolism by increasing P-450 levels. Trp-P-1 and Trp-P-2 given in the diet induce P-450IA type in both Ah and non-Ah responsive female mice. Much less induction was seen in males, and induction in females could be depressed by administration of testosterone.[80,81] Recently it was reported that intraperitoneally injected Trp-P-1, Trp-P-2, Glu-P-1, Glu-P-2, AAC, MeAAC, IQ, or MeIQx all induced hepatic cytochrome P-450IA enzymes, particularly P-450IA2.[82] The selective induction by IQ of the P-450IA family in liver and not P-450IIB has been reported.[83] MeIQx induced P-450IA enzymes in various rat organs, i.e., liver, kidney, and lung, whereas PhIP induced P-450IA in the intestine of exposed rats. Administration of a fried meat diet to rats induced P-450IA1- and 2-dependent activity in the intestine but no changes in liver P-450 levels.[22,84]

IV. EXCRETORY PATHWAYS

The heterocyclic amines are rapidly excreted from the body.[4,8-14] No or only negligible amounts are found in expired CO_2.[11,12] The dominating excretory pathways are via the bile to feces and urine. The two pathways seem to be about equally important, but this seems to vary somewhat between experiments and between compounds. For Trp-P-1 and Glu-P-1 fecal excretion seems to be more important.[4,10] Germ-free rats have a slower fecal excretion of heterocyclic amines (e.g., Trp-P-1 and diMeIQx) in comparison with normal rats and this may at least partly be due to a slower transition time of feces in the gut of the germ-free rats.[8,10] A considerable amount (20.5 and 63%) of fecal radioactivity was bound and nonextractable from fecal material of rats exposed to ^{14}C-Trp-P-1 and ^{14}C-DiMeIQx, respectively.[8,10] Using germ-free animals the amount decreased to 3.6 and 48.8, respectively.[8,10] Except for MeIQx and diMeIQx where about 20% of the unchanged compound was found in urine, only a few percent of the parent compound of other heterocyclic amines is excreted in urine.[8,12-14,37-39] Much less parent MeIQx was recovered in bile and feces.[12-14] The large fraction of unchanged compound excreted may be due to the administration of large doses, i.e., 20 mg/kg, which might saturate the metabolizing systems or the high polarity of MeIQx and diMeIQx. Only 2 to 5% of the unchanged MeIQx was recovered in the urine of humans exposed to MeIQx at much lower dose, i.e., 290 to 850 ng in one single meal of fried meat.[24]

A. SEX DIFFERENCES

Differences in metabolism between sexes has only been studied to a limited degree and seem (sex organs excluded) to be small. Similar autoradiographic distribution patterns in mice of both sexes have been reported for IQ, MeIQ, and Trp-P-1.[32,33] In rats, only small differences were seen between sexes for the total excretion of metabolites from IQ, MeIQ, and MeIQx via feces or urine.[11,12] Only for MeIQx have sex-specific patterns been studied. The female rats had a greater tendency to form the ring-hydroxylated O-glucuronide as compared to male rats. No exclusive metabolites were seen in either sex.[12] Sex-dependent differences in metabolic activation of heterocyclic amines by isolated hepatocytes have also been reported.[85]

V. CONCLUDING REMARKS

Heterocyclic amines are rapidly taken up and distributed via blood to most parts of the body. The liver seems to be a major site of metabolism. A minor fraction is bound to macromolecules (DNA, proteins) in the tissues or blood after being metabolized to reactive metabolites by *N*-hydroxylation. Metastable metabolites may be formed in the liver and transported to other organs (e.g., PhIP). The compounds are also metabolized in extrahepatic tissues (e.g., Trp-P-1 in lung) where they may bind locally. The heterocyclic amines are also rapidly metabolized to water-soluble detoxified conjugates either directly or via ring-hydroxylations. The metabolites leave the body via bile to feces, where glucuronides may be hydrolyzed, or via urine. Except for a small fraction covalently bound in tissues, virtually all of a single-dose heterocyclic amine leaves the body within 1 to 3 days.

ACKNOWLEDGMENTS

The authors wish to thank Lise Timm Haug for art work and Anne Lene Solbakken for excellent secretarial assistance. Håkan Wallin is a research fellow of the Norwegian Research Council for Science and Humanities.

REFERENCES

1. **Sugimura, T., Sato, S., and Wakayabashi, K.,** Chemical induction of cancer. Structural basis and biological mechamisms, in *Mutagens/Carcinogens in Pyrolysates of Amino Acids and Proteins in Cooked Food: Heterocyclic Amines,* Academic Press, San Diego, 1988, 681.
2. **Felton, J. S., Knize, M., Andresen, B. D., Bjeldanes, L. F., and Hatch, F. J.,** Identification of the mutagens in cooked beef., *Environ. Health Perspect.,* 67, 17, 1986.
3. **Alexander, J., Becher, G., and Busk, L.,** Cooked food mutagens—a general overview, *Vår Föda* (Natl. Food Administration, Sweden), 42, Suppl. 2, 9—59, 1989.
4. **Kato, R.** Metabolic activation of mutagenic heterocyclic aromatic amines from protein pyrolysates, *CRC Crit. Rev. Toxicol.,* 16, 307, 1986.
5. **Kato, R. and Yamazoe, Y.,** Metabolic activation and covalent binding to nucleic acids of carcinogenic heterocyclic amines from cooked foods and amino acid pyrolysates, *Jpn. J. Cancer Res.,* 78, 297, 1987.
6. **Kimura, T., Nakayama, T., Kurosaki, Y., Zuzuki, Y., Arimoto, S., and Hayatsu, H.,** Absorption of 3-amino-1-methyl-5H-pyrido[4,3-b]indole, a mutagen-carcinogen present in tryptophan pyrolysate, from the gastro-intestinal tract in the rat, *Jpn. J. Cancer Res.,* 76, 272, 1985.
7. **Alldrick, A. J. and Rowland, I. R.,** Distribution of radiolabeled [2-14C]IQ and MeIQx in the mouse, *Toxicol. Lett.,* 44, 183, 1988.
8. **Knize, M. G., Övervik, E., Midtvedt, T., Turteltaub, K. W., Happe, J. A., Gustafsson, J.-A., and Felton, J. S.,** The metabolism of 4,8-DiMeIQx in conventional and germ-free rats, *Carcinogenesis,* 10, 1479, 1988.
9. **Negishi, C., Umemoto, A., Rafter, J. J., Sato, S., and Sugimura, T.,** *N*-Acetyl derivative as the major active metabolite of 2-amino-6-methyldipyrido[1,2-a:3′,2′-d]imidazole in rat bile, *Mutat. Res.,* 175, 23, 1986.
10. **Rafter, J. J. and Gustafsson, J.-A.,** Metbolism of the dietary carcinogen TRP-P-1 in rats, *Carcinogenesis,* 7, 1291, 1986.
11. **Sjödin, P. and Jägerstad, M.,** A balance study of 14C-labeled 3H-imidazo[4,5-f]quinolin-2-amines (IQ and MeIQ) in rats, *Food Chem. Toxicol.,* 22, 207, 1984.
12. **Sjödin, P., Wallin, H., Alexander, J., and Jägerstad, M.,** Disposition and metabolism of the food mutagen 2-amino-3,8-dimethyl(4,5-f)quinoxaline (MeIQx) in rats, *Carcinogenesis,* 10, 1269, 1989.
13. **Turesky, R. J., Aeschbacher, H. U., Wurzner, H. P., Skipper, P. L., and Tannenbaum, S. R.,** Major routes of metabolism of the food-borne carcinogen 2-amino-3,8-dimethylimidazo[4,5-f]quinoxaline in the rat, *Carcinogenesis,* 9, 1043, 1988.
14. **Turesky, R. J., Aeschbacher, H. U., Malnoe, A., and Wurzner, H. P.,** Metabolism of the food-borne mutagen/carcinogen 2-amino-3,8-dimethylimidazo[4,5-f]quinoxaline in the rat: assessment of biliary metabolites for genotoxicity, *Food Chem. Toxicol.,* 26, 105, 1988.

153

15. **Turesky, R. J., Skipper, P. L., Tannenbaum, S. R., Coles, B., and Ketterer, B.**, Sulfamate formation is a major route for detoxification of 2-amino-3-methylimidazo[4,5-f]quinoline in the rat, *Carcinogenesis,* 7, 1483, 1986.

16. **Barnes, W. S., Maiello, J., and Weisburger, J. H.**, In vitro binding of the food mutagen 2-amino-3-methylimidazo[4,5-f]quinoline to dietary fibers, *J. Natl. Cancer Inst.,* 70, 757, 1983.

17. **Kada, T., Kato, M., Aikawa, K., and Kiriyama, S.**, Adsorption of pyrolysate mutagens by vegetable fibers, *Mutat. Res.,* 141, 149, 1984.

18. **Lindeskog, P., Överik, E., Nilsson, L., Nord, C. E., and Gustafsson, J.-A.**, Influence of fried meat and fiber on cytochrome P-450 mediated activity and excretion of mutagens in rats, *Mutat. Res.,* 204, 553, 1988.

19. **Morotomi, M. and Mutai, M.**, In vitro binding of potent mutagenic pyrolysates to intestinal bacteria, *J. Natl. Cancer Inst.,* 77, 195, 1986.

20. **Sjödin, P., Nyman, M., Asp, N.-G., and Jägerstad, M.**, Binding of carbon 14-labeled food mutagens (IQ, MeIQ, MeIQx) to dietary fiber in vitro, *J. Food Sci.,* 50, 1680, 1985.

21. **Howes, A. J., Rowland, I. R., Lake, B. G., and Alldrick, A. J.**, Effect of dietary fibre on the mutagenicity and distribution of 2-amino-3,4-dimethylimidazo[4,5-f]quinoline (MeIQ), *Mutat.Res.,* 210, 227, 1989.

22. **Övervik, E.**, *Formation and Biological Effects of Cooked-Food Mutagens,* Dissertation, Karolinska Institute, Stockholm, Sweden, 1989.

23. **Dolara, P., Caderni, G., Bianchini, F., and Tanganelli, E.**, Nuclear damage of colon epithelial cells by the food carcinogen 2-amino-3-methylimidazo[4,5-f]quinoline (IQ) is modulated by dietary lipids, *Mutat. Res.,* 175, 255, 1986.

24. **Murray, S., Gooderham, M. J., Boobis, A. R., and Davies, D. S.**, Detection and measurement of MeIQx in human urine after ingestion of a cooked meat meal, *Carcinogenesis,* 10, 763, 1989.

25. **Hayatsu, H., Hayatsu, T., and Ohara, Y.**, Mutagenicity of human urine caused by ingestion of fried ground beef, *Jpn. J. Cancer Res.,* 76, 445, 1985.

26. **Dolara, P., Caderni, G., Salvadori, M., and Lodovici, M.**, Urinary mutagens in human after fried pork and bacon meals, *Cancer Lett.,* 22, 275, 1984.

27. **Manabe, S., Yanagisawa, H., Guo, S. B., Abe, S., Ishikawa, S., and Wado, O.**, Detection of Trp-P-1 and Trp-P-2, carcinogenic tryptophan pyrolysis products, in dialysis fluid of patients with uremia, *Mutat. Res.,* 179, 33, 1987.

28. **Wakabashi, K., Shioya, M., Nagao, M., Sugimura, T.**, Detection of heterocyclic amines in dialysates from patients with uremia, *Proc. Annu. Meet. Jpn. Cancer Assoc.,* 46, 18, 1987.

29. **Yanagisawa, H., Manabe, S., and Wada, O.**, Detection of IQ-type heterocyclic amines in dialysis fluid of uremic patients treated by peritoneal dialysis, *Jpn. J. Nephrol.,* 29, 1153, 1987.

30. **Yanagisawa, H., Manage, S., Kitagawa, Y., Ishikawa, S., Nakajima, K., and Wada, O.**, Presence of 2-amino-3,8-dimethylimidazo[4,5-f]quinoxaline (MeIQx) in dialysate from patients with uremia, *Biochem. Biophys. Res. Commun.,* 138, 1084, 1986.

31. **Kanai, Y., Manabe, S., and Wada, O.**, In vitro and in vivo N-acetylation of carcinogenic glutamic acid pyrolysis products in humans, *Carcinogenesis,* 9, 2179, 1988.

32. **Bergman, K.**, Autoradiographic distribution of 14C-labeled 3H-imidazo[4,5-f]quinoline-2-amines in mice, *Cancer Res.,* 45, 1351, 1985.

33. **Brandt, I., Gustafsson, J.-A., and Rafter, J.**, Distribution of the carcinogenic tryptophan pyrolysis product Trp-P-1 in control, 9-hydroxyellipiticine and beta-naphthoflavone pretreated mice, *Carcinogenesis,* 4, 1291, 1983.

34. **Brandt, I., Kowalski, B., Gustafsson, J.-A., and Rafter, J.**, Tissue localisation of the carcinogenic glutamic acid pyrolysis product Glu-P-1 in control and beta-naphthoflavone-treated mice and rats, *Carcinogenesis,* 10, 1529, 1989.

35. **Alexander, J., Holme, J. A., Wallin, H., and Becher, G.**, Characterisation of metabolites of the food mutagens 2-amino-3-methylimidazo(4,5-f)quinoline (IQ) and 2-amino-3,4-dimethylimidazo(4,5-f)-quinoline (MeIQ) formed after incubation with isolated rat liver cells, *Chem. Biol. Interact.,* 72, 125, 1989.

36. **Alexander, J., Wallin, H., Holme, J. A., and Becher, G.**, 1-Methylimidazo(4,5-b)pyrid-6-yl) phenyl sulfate—a major metabolite of the food mutagen 2-amino-1-methylimidazo(4,5-f)pyridine (PhIP) in rat, *Carcinogenesis,* 10, 1543, 1989.

37. **Barnes, W. S. and Weisburger, J. H.**, Fate of the food mutagen 2-amino-3-methylimidazo[4,5-f]-quinoline (IQ) in Sprague-Dawley rats. I. Mutagens in the urine, *Mutat. Res.,* 156, 83, 1985.

38. **Peleran, J. C., Rao, D., and Bories, G. F.**, Identification of the cooked food mutagen 2-amino-3-methylimidazo[4,5-f]quinoline (IQ) and its N-acetylated and 3-N-demethylated metabolites in rat urine, *Toxicology,* 43, 193, 1987.

39. **Størmer, F. C., Alexander, J., and Becher, G.**, Fluorometric detection of 2-amino-2-methylimidazo[4,5-f]quinoline, 2-amino-3,4-dimethylimidazo[4,5-f]quinoline and their N-acetylated metabolites excreted by the rat, *Carcinogenesis,* 8, 1277, 1987.

40. **Aune, T. and Aune, K. T.,** Mutagenic activation of IQ and Me-IQ by liver and lung microsomes from rabbit and mouse, and with isolated lung cells from the rabbit, *Carcinogenesis,* 7, 273, 1986.

41. **de Waziers, I. and Decloitre, F.,** Effect of glutathione and uridine-5'-diphosphoglucuronic acid on the mutagenicity of tryptophan pyrolysis products (Trp-P-1 and Trp-P-2) by rat-liver and -intestine S9 fraction, *Mutat. Res.,* 139, 15, 1984.

42. **de Waziers, I. and Decloitre, F.,** Formation of mutagenic derivatives from tryptophan pyrolysis products (Trp P1 and Trp P2) by rat intestinal S9 fraction, *Mutat. Res.,* 119, 103, 1983.

43. **Holme, J. A., Alexander, J., and Dybing, E.,** Mutagenic activation of 2-amino-3-methylimidazo(4,5-f)-quinoline (IQ) and 2-amino-3,4-dimethylimidazo(4,5-f)-quinoline (MeIQ) by subcellular fractions and cells isolated from small intestine, kidney and liver of the rat, *Cell. Biol. Toxicol.,* 3, 51, 1987.

44. **Kawakubo, Y., Manabe, S., Yamazoe, Y., Nishikawa, T., and Kato, R.,** Properties of cutaneous acetyltransferase catalyzing N- and O-acetylation of carcinogenic arylamines and N-hydroxyarylamines, *Biochem., Pharmacol.,* 37, 265, 1988.

45. **Shinohara, A., Saito, K., Yamazoe, Y., Kamataki, T., and Kato, R.,** Acetyl coenzyme A dependent activation of N-hydroxy derivatives of carcinogenic arylamines: mechanism of activation species difference, tissue distribution, and acetyl donor specificity, *Cancer Res.,* 46, 4362, 1986.

46. **Bashir, M., Kingston, D., G., Carman, R. J., van Tassell, R. L., and Wilkins, T. D.,** Anaerobic metabolism of 2-amino-3-methyl-3H-imidazo[4,5-f]quinoline (IQ) by human fecal flora, *Mutat. Res.,* 190, 187, 1987.

47. **Carman, R. J., Van Tassell, R. L., Kingston, D. G., Bashir, M., and Wilkins, T. D.,** Conversion of IQ, a dietary pyrolysis carcinogen to a direct-acting mutagen by normal intestinal bacteria of humans, *Mutat. Res.,* 206, 335, 1988.

48. **Wallin, H., Mikalsen, A., Guengerich, F. P., Ingelman-Sundberg, M., Solberg, K. E., Rossland, O. J., and Alexander, J.,** Differential rates of metabolic activation and detoxication of the food mutagen 2-amino-1-methyl-6-phenylimidazo(4,5-b)pyridine by different cytochrome P-450 enzymes, *Carcinogenesis,* 11, 489—492, 1990.

49. **Yamazoe, Y., Abu-Zeid, M., Manabe, S., Toyama, S., and Kato, R.,** Metabolic activation of protein pyrolysate promutagen 2-amino-3,8-dimethylimidazo[4,5-f]quinoxaline by rat liver microsomes and purified cytochrome P-450, *Carcinogenesis,* 9, 105, 1988.

50. **Snyderwine, E. G. and Battula, N.,** Selective mutagenic activation by cytochrome P3-450 of carcinogenic arylamines found in food, *J. Natl. Cancer Inst.,* 81, 223, 1989.

51. **McManus, M. E., Burgess, W., Snyderwine, E., and Stupans, I.,** Specificity of rabbit cytochrome P-450 isozymes involved in the metabolic activation of food derived mutagen 2-amino-3-methylimidazo[4,5-f]quinoline, *Cancer Res.,* 48, 4513, 1988.

52. **McManus, M. E., Felton, J. S., Knize, M. G., Burgess, W. M., Roberts-Thomson, S., Pond, S. M., Stupans, I., and Veronese, M. E.,** Activation of the food-derived mutagen 2-amino-1-methyl-6-phenylimidazo[4,5-b]pyridine by rabbit and liver microsomes and purified forms of cytochrome P-450, *Carcinogenesis,* 10, 357, 1989.

53. **Ishida, Y., Negeshi, C., Umemoto, U. F., Y., Sato, S., Sugimura, T., Thorgeirsson, S. S., and Adamson, R. H.,** Activation of mutagenic and carcinogenic heterocyclic amines by S-9 from the liver of a rhesus monkey, *Toxicol. In Vitro,* 1, 45, 1986.

54. **Yamazoe, Y., Abu-Zeid, M., Yamauchi, K., and Kato, R.,** Metabolic activation of pyrolysate arylamines by human liver microsomes; possible involvement of a P-488-H type cytochrome P-450, *Jpn. J. Cancer Res.,* 79, 1159, 1988.

55. **Felton, J. S. and Healy, S. K.,** Activation of mutagens in cooked ground beef by human-liver microsomes, *Mutat. Res.,* 140, 61, 1984.

55a. **McManus, M. E., Burgess, W., Stupans, I., Trainor, K. J., Fenech, M., Robson, R. A., Morley, A. A., and Snyderwine, E. G.,** Activation of the food-derived mutagen 2-amino-3-methylimidazo[4,5-f]quinoline by human-liver microsomes, *Mutat. Res.,* 204, 185, 1988.

56. **Shimada, T., Iwasaki, M., Martin, M. V., and Guengeric, F. P.,** Human liver microsomal cytochrome P-450 enzymes involved in the bioactivation of procarcinogens detected by the umu gene response in salmonella typhimurium TA 1535/pSK1002, *Cancer Res.,* 49, 3218, 1989.

57. **Butler, M. A., Iwasaki, M., Guengerich, F. P., and Kadlubar, F. F.,** Human liver cytochrome P-450PA is primarily responsible for caffeine 3-demethylation and carcinogenic arylamine N-oxidation, *Environ. Mol. Mutagen.,* 14 (Suppl. 15), 33, 1989.

58. **Holme, J. A., Brunborg, G., Alexander, J., Trygg, B., and Bjornstad, C.,** Modulation of the mutagenic effects of 2-amino-3-methylimidazo[4,5-f]quinoline (IQ) and 2-amino-3,4-dimethylimidazo[4,5-f]quinoline (MeIQ) in bacteria with rat-liver 9000 × g supernatant or monolayers of rat hepatocytes as an activation system, *Mutat. Res.,* 197, 39, 1988.

59. **Holme, J. A., Hongslo, J. K., Søderlund, E., Brunborg, G., Christensen, T., Alexander, J., and Dybing, E.,** Comparative genotoxic effects of IQ and MeIQ in *Salmonella typhimurium* and cultured mammalian cells, *Mutat. Res.,* 187, 181, 1987.

60. **Holme, J. A., Wallin, H., Brunborg, G., Søderlund, E. J., Hongslo, J. K., and Alexander, J.,** Genotoxicity of the food mutagen 2-amino-1-methyl-6-phenyl(4,5-b)pyridine: formation of 2-hydrox-amino-PhIP, a directly acting genotoxic metabolite, *Carcinogenesis,* 10, 1389, 1989.

60a. **Thompson, L. H., Tucker, J. D., Stewart, S. A., Christensen, M. L., Salazar, E. P., Carrano, A. V., and Felton, J. S.,** Genotoxicity of compounds from cooked beef in repair-deficient CHO cells versus Salmonella mutagenicity, *Mutagenesis,* 2, 483, 1987.

60b. **Yamazoe, Y., Shimada, M., Shinohara, A., Saito, K., Kamataki, T., and Kato, R.,** Catalysis of the covalent binding of 3-hydroxyamino-1-methyl-5H-pyrido[4,3-b]indole to DNA by a L-proline- and adenosine triphosphate-dependent enzyme in rat hepatic cytosol, *Cancer Res.,* 45, 2495, 1985.

60c. **Thompson, L. H., Carrano, A. V., Salazar, E. P., Felton, J. S., and Hatch, F. T.,** Comparative genotoxic effects of cooked-food-mutagens Trp-P-2 and IQ in bacteria and cultured mammalian cells, *Mutat. Res.,* 117, 243, 1983.

61. **Wallin, H. and Alexander, J.,** A model system for studying covalent binding of food carcinogens MeIQx, MeIQ and IQ to DNA and protein, *IARC Sci. Publ.,* 89, 113, 1988.

62. **Robertson, I. C. G., Philpot, R. M., Zeiger, E., and Wolf, C. R.,** Specificity of rabbit pulmonary cytochrome P-450 in the activation of several aromatic amines and aflatoxin B1, *Mol. Pharmacol.,* 20, 662, 1981.

62a. **Snyderwine, E. G., Roller, P. P., Adamson, R. H., Sato, S., and Thorgeirsson, S. S.,** Reaction of N-hydroxylamine and N-acetoxy derivatives of 2-amino-3-methylimidazolo[4,5-f]quinoline with DNA. Synthesis and identification of N-(deoxyguanosin-8-yl)-IQ, *Carcinogenesis,* 9, 1061, 1988.

63. **Snyderwine, E. G., Wirth, P. J., Roller, P. P., Adamson, R. H., Sato, S., and Thorgeirsson, S. S.,** Mutagenicity and in vitro covalent DNA binding of 2-hydroxyamino-3-methylimidazolo[4,5-f]-quinoline, *Carcinogenesis,* 9, 411, 1988.

64. **Turesky, R. J., Skipper, P. L., and Tannenbaum, S. R.,** Binding of 2-amino-3-methylimidazo-[4,5-f]quinoline to hemoglobin and albumin in vivo in the rat. Identification of an adduct suitable for dosimetry, *Carcinogenesis,* 8, 1537, 1987.

65. **Yamashita, K., Umemoto, A., Grivas, S., Kato, S., Sato, S., and Sugimura, T.,** Heterocyclic amine-DNA adducts analysed by 32P-postlabeling method, *Nucleic Acids Symp. Ser.,* 111, 1988.

66. **Yamashita, K., Umemoto, A., Grivas, S., Kato, S., and Sugimura, T.,** In vitro reaction of hydroxyamino derivatives of MeIQx, Glu-P-1, and Trp-P1 with DNA: 32P-postlabeling of DNA adducts formed in vivo by the parent amine and in vitro by their hydroxamino derivatives, *Mutagenesis,* 3, 515, 1988.

67. **Schut, H. A. J., Putman, K. L., and Randerath, K.,** 32P-postlabeling analysis of DNA adducts in liver, small and large intestine of male Fisher-344 rats after intraperitoneal administration of 2-amino-3-methylimidazo(4,5-f)quinoline (IQ), in *Carcinogenic and Mutagenic Responses to Aromatic Amines and Nitroarenes,* Elsevier, New York, 1988.

68. **Schut, H. A. J., Putman, K. L., and Randerath, K.,** DNA adduct formation of the carcinogen 2-amino-3-methylimidazo[4,5-f]-quinoline in target tissues of the F-344 rat, *Cancer Lett.,* 41, 345, 1988.

69. **Snyderwine, E. G., Yamashita, K., Adamson, R. H., Sato, S., Nagao, M., Sugimura, T., and Thorgeirsson, S. S.,** Use of the 32P-postlabeling method to detect DNA adducts of 2-amino-3-methylimidazolo[4,5-f]quinoline (IQ) in monkeys fed IQ: identification of N-(deoxyguanosin-8-yl)-IQ adduct, *Carcinogenesis,* 9, 1739, 1988.

70. **Alldrick, A. J. and Lutz, W. K.,** Covalent binding of (2-14C)2-amino-3,8-dimethylimidazo(4,5-f)-quinoxaline (MeIQx) to mouse DNA in vivo, *Carcinogenesis,* 10, 1419, 1989.

71. **Alexander, J., Wallin, H., Holme, J. A., Brunborg, G., Søderlund, E. J., Becher, G., Mikalsen, A., and Hongslo, J. K.,** Metabolism of heterocyclic amines in cooked food, in Mutation and the environment, Part E. Environmental genotoxicity, risks and modulation, Mendelsohn, M. L. and Albertini, R. J., Eds., *Progr. Clin. Biol. Res.,* 340D, 159—168, 1990.

72. **Umemoto, A., Grivas, S., Yamaizumi, Z., Sato, S., and Surimura, T.,** Non-enzymatic glutathione conjugation of 2-nitroso-6-methyldipyrido [1,2-a:3',2'-d] imidazole (NO-Glu-P-1) in vitro: N-hydroxy-sulfonamide, a new binding form of arylnitroso compounds and fluids, *Chem. Biol. Interact.,* 68, 57, 1988.

73. **Umemoto, A., Monden, Y. Tsuda, M., Grivas, S., and Sugimura, T.,** Oxidation of the 2-hydroxyamino derivative of 2-amino-6-methyl-dipyrido[1,2-a:3',2'-d]imidazole (Glu-P-1) to its 2-nitroso form, an ultimate form reacting with hemoglobin thiol groups, *Biochem. Biophys. Res. Commun.,* 151, 1326, 1988.

74. **Wallin, H., Holme, J. A., Becher, G., and Alexander, J.,** Metabolism of the food carcinogen 2-amino-3,8-dimethyl(4,5-f)quinoxaline in isolated rat liver cells, *Carcinogenesis,* 10, 1277, 1989.

75. **Hayatsu, H., Kasai, H., Yokoyama, S., Miyazawa, T., Yamaizumi, Z., Sato, S., Nishimura, S., Arimoto, S., Hayatsu, T., and Ohara, Y.,** Mutagenic metabolites in urine and feces of rats fed with 2-amino-3,8-dimethylimidazo[4,5-f]quinoxaline, a carcinogenic mutagen present in cooked meat, *Cancer Res.,* 47, 791, 1987.

76. **Brunborg, G., Holme, J. A., Alexander, J., Becher, G., and Hongslo, J. K.,** Genotoxic activity of the N-acetylated metabolites of the food mutagens 2-amino-3-methylimidazo[4,5-f]quinoline (IQ) and 2-amino-3,4-dimethylimidazo[4,5-f]quinoline (MeIQ), *Mutagenesis,* 3, 303, 1988.

77. **Holme, J. A., Alexander, J., Becher, G., and Trygg, B.,** Metabolism of 2-amino-3-methylimidazo-(4,5-f)quinoline (IQ) and 2-amino-3,4-dimethylimidazo(4,5-f)quinoline (MeIQ) in suspensions of isolated rat-liver cells, *Toxicol. In Vitro,* 1, 175, 1987.

78. **Weisburger, J. H., Jones, R. C., Luks, H. J., and Vavrek, T.,** Mechanisms and prevention of formation, and metabolism of food mutagens/carcinogen 2-amino-3-methylimidazo(4,5-f)quinoline (IQ), *Environ. Mol. Mutagen.,* 14 (Suppl. 15), 216, 1989.

79. **Gooderham, N. J., Watson, D., Rice, J. C., Murray, S., Taylor, G. W., and Davies, D. S.,** Metabolism of the mutagen MeIQx in vivo: metabolite screening by liquid chromatography-thermospray mass spectrometry, *Biochem. Biophys. Res. Commun.,* 148, 1377, 1987.

80. **Degawa, M., Kojima, M., Hishinuma, T., and Hashimoto, Y.,** Sex-dependent induction of hepatic enzymes for mutagenic activation of a tryptophan pyrolysate component, 3-amino-1,4-dimethyl-5H-pyrido[4,3-b]-indole, by feeding in mice, *Cancer Res.,* 45, 96, 1985.

81. **Degawa, M., Tanimura, S., Agasuma, T., and Hashimoto, Y.,** Hepatocarcinogenic heterocyclic aromatic amines that induce cytochrome P-448 isoenzymes, mainly cytochrome P-448H (P-450IA2), responsible for mutagenic activation of the carcinogens in rat liver, *Carcinogenesis,* 10, 1119, 1989.

82. **Degawa, M., Yamaya, C., and Hashimoto, Y.,** Hepatic cytochrome P-450 isozyme(s) induced by dietary carcinogenic aromatic amines preferentially in female mice of DBA/2 and other strains, *Carcinogenesis,* 9, 567, 1988.

83. **Rodriges, A. D., Ayrton, A. D., Williams, E. J., Lewis, D. W. S., Walker, R., and Ioannides, C.,** Preferential induction of P450I proteins by the food carcinogen 2-amino-3-methyl(4,5-f)quinoline, *Eur. J. Biochem.,* 181, 627, 1989.

84. **Kleman, M., Övervik, E., Lindeskog, P., and Gustafsson, J.-A.,** Studies on the induction of cytochrome P450 isoenzymes by 2-amino-1-methyl-6-phenylimidazo(4,5-b)pyridine (PhIP) and 2-amino-3,8-dimethylimidazo(4,5-f)quinoxaline (MeIQx) in various organs from male rats, *Environ. Mol. Mutagen.,* 14 (Suppl. 15), 102, 1989.

85. **Yoshimi, N., Sugie, S., Iwata, H., Mori, H., and Williams, G. M.,** Species and sex differences in genotoxity of heterocyclic amine pyrolysis and cooking products in the hepatocyte primary culture/DNA repair test using rat, mouse, and hamster hepatocytes, *Environ. Mol. Mutagen.,* 12, 53, 1988.

86. **Alexander, J., et al.,** to be published.

87. **Alexander, J., et al.,** in preparation.

Chapter 8

MECHANISMS OF FOOD-BORNE INHIBITORS OF GENOTOXICITY RELEVANT TO CANCER PREVENTION

Silvio De Flora, Patrizia Zanacchi, Alberto Izzotti, and Hikoya Hayatsu

TABLE OF CONTENTS

I. INTRODUCTION

Diet and food components play a major role in the etiopathogenesis of chronic degenerative diseases, among which are cancer and other pathological conditions associated with the occurrence of genotoxic effects in germ and somatic cells. The overall contribution of dietary and alimentary factors to cancer mortality has been estimated to fall in the range between 10 and 70%,[1] and their contribution to cancer prevalence in males and females has been estimated to be 40% and 50%[2] or 25 and 42%,[3] respectively.

Two complementary approaches are available for preventive strategies based on modulation of mutagenesis and carcinogenesis by food components. The first one involves avoidance of intake of compounds or nutritional factors which may initiate or promote cancer. The second approach aims at shifting the balance between risk and antirisk factors in favor of protective mechanisms and at fortifying the host defense machinery. Indeed, such an ambitious goal is very difficult to achieve, because quite often putative inhibitors possess multiple and pleiotropic properties resulting in the simultaneous stimulation of an array of differently oriented mechanisms.[4] Therefore, the problem is to selectively enhance protective mechanisms without affecting closely interrelated processes and without disturbing the intricate homeostasis of the organism and especially of the cellular environment. In addition, physiological mechanisms themselves often play multiple roles, and their double-edged nature renders their regulation very complex.[4,5]

These patterns explain why antirisk assessment, which basically uses the same tools exploited for risk assessment (i.e., analytical epidemiology, animal carcinogenicity tests, *in vivo* or *in vitro* short-term tests and, possibly, models based on structure-activity relationships) can be biased by the frequent occurrence of contradictory indications. A putative inhibitor which appears to trigger protective mechanisms in a given situation may become ineffective or even deleterious depending on the test system and experimental conditions used. Thus, the outcome of studies on modulators of genotoxicity or carcinogenicity can be upset by changing the mutagen or carcinogen to be challenged, doses, routes of administration, sequence of intake, target organisms and tissues, and intentional or unintentional combinations with other agents, which are particularly relevant when dealing with complex mixtures such as food.[5,6]

A matter of further uncertainty is the relationship between *in vitro* and *in vivo* results and between mutagenicity and carcinogenicity data, which can be dissociated under certain conditions. For instance, antioxidants such as certain vitamins, phenolics, and products of the Maillard reaction, which are known to protect against cancer, can produce genotoxic effects in *in vitro* test systems.[6-8] Nevertheless, *in vitro* short-term tests prove very useful for a preliminary assessment of the inhibitory properties of chemical compounds, for comparing the efficiency of structurally related molecules, and for contributing to the understanding of the mechanisms involved.[9,10]

Several reviews on the antimutagenic and/or anticarcinogenic activity of food components are available in the literature.[6,8,9,11-17] Practical public health recommendations aimed at preventing cancer by dietary means have been also tentatively addressed to the population.[18-20] However, their practical application needs further scientific evidence, and some direction for future research has been also given.[21]

In this chapter, we will deal with food-borne inhibitors of genotoxicity, either of natural origin or synthetic, such as molecules which are already used or may be proposed as food additives. Due to the strict interconnections between mutagenicity and carcinogenicity and often the complementary contribution of antimutagenesis and anticarcinogenesis studies to the identification of protective food components, relevance to cancer prevention will be also discussed. Although mutagenicity in somatic cells is the essential mechanism by which cancer is initiated, oncogene research has clearly demonstrated that genetic effects also occur

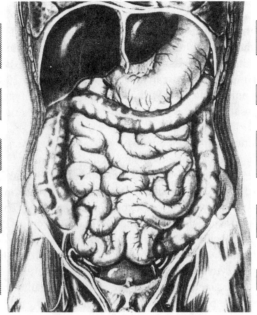

Chemical or enzymatic
deactivation of muta-
gens in the lumen of
the digestive tract

Modification of meta-
bolism during entero-
hepatic circulation

Protection of the
mucosal barrier

Inhibition of trans-
epithelial absorption

Dilution and comple-
xation of mutagens
and carcinogens

Acceleration of inte-
stinal transit and ex-
cretion

Maintenance of optimal
intragastric pH

Inhibition of nitrosation

Inhibition of metabolic
activation by gut
colonizing bacteria

Increased production of
short-chain fatty acids

Decreased fecal pH

FIGURE 1.   Mechanisms of inhibitors of mutagenesis and carcinogenesis acting extracellularly in the gastrointestinal tract.

at later stages of the carcinogenesis process. In addition, mechanisms of similar nature, e.g.,
antioxidant or antiproliferative processes, can be responsible for protective effects along
sequential steps of cancer development.

We will try to give special emphasis to the mechanisms by which food-borne inhibitors
can modulate the mutagenic and carcinogenic response. In fact, due to the already mentioned
vagaries and uncertainties of antirisk assessment, there is a stringent demand for a mechanistic
approach in this field, and investigations on the mechanisms by which food components
contribute to either increase or decrease cancer risks deserve priority.[22,23] For these reasons,
this chapter will be arranged in sections devoted to individual mechanisms of inhibition,
rather than to chemical compounds or chemical families of inhibitors. Within each mechanism, specific food-borne inhibitors will be reported. It is clear that, due to the vast literature
available, only some illustrative examples will be given. A number of inhibitors which are
known or more often suspected to work through multiple mechanisms will be cited in different
sections.

Basically, the classification proposed by De Flora and Ramel[24] will be followed. Such
a scheme involves a subdivision into several categories and subcategories, covering a variety
of mechanisms preventing mutation and cancer initiation, either in the extracellular environment or intracellularly, or modulating the promotion and progression steps.

II. INHIBITION OF MUTAGENESIS BY EXTRACELLULAR
MECHANISMS

A variety of mechanisms, which in principle can be influenced by dietary habits and
by food components, can exert protective effects in the extracellular environment of the
gastrointestinal tract. They are summarized in Figure 1.

A. INHIBITION OF THE ENDOGENOUS FORMATION OF MUTAGENS

The possibility of preventing the endogenous formation of mutagens from inactive precursors, either by chemical reaction (Section II.A.1) or by biological mechanisms (Section II.A.2), represents one of the most interesting and readily applicable approaches in cancer prevention.

1. Inhibition of the Nitrosation Reaction

Nitrite can react in the acid environment of the stomach with amines or amides, leading to the formation of *N*-nitroso compounds, one of the most extensively investigated and characterized family of mutagens and carcinogens. Of more than 300 compounds in this category, assayed in 40 animal species, 90% were found to be carcinogenic.[25] In addition, even at neutral pH, nitrosation can be catalyzed by bacteria in other sites of the organism, including the oral cavity.[26]

The nitrosation reaction can be modulated by means of catalysts or inhibitors, which generally compete with amino compounds in reacting with nitrite. Several reviews are available in the literature on this subject (e.g., References 7, 13, and 26 to 30). A number of food components have been shown to inhibit the nitrosation reaction and often to prevent mutagenicity and carcinogenicity in various experimental systems. They include vitamins, such as vitamins C (ascorbic acid) and E (α-tocoferol); phenolic compounds, such as catechol, cinnamic acid, chlorogenic acid, gallic acid, hydroquinones, phenolic acids, pyrogallol, tannic acid, tannins, thymol, and the synthetic food additives butylated hydroxyanisole (BHA) and butylated hydroxytoluene (BHT); sulfur compounds, such as bisulfite, cysteine, glutathione, *N*-acetylcysteine, methionine, sulfamic acid, and sulfur dioxide; as well as a variety of miscellaneous compounds, such as alcohols, azides, caffeine, carbohydrates, hydrazine, NADH, hydroxylamine, sorbic acid, unsaturated fatty acids, and urea.[26,31] Inhibition of nitrosation has been also observed with unfractionated food extracts and beverages, such as alcoholic beverages, betel nut extracts, coffee, tea, fruit juices, milk, radish juice, soya products, cheese, casein, pectin, and gelatin.[26,32]

The occurrence of all these natural inhibitors may explain why the mutagenicity of human gastric juice after meals is mainly dependent on preformed substances rather than on nitrosation products.[33] It should also be taken into account that the genotoxicity of *N*-nitroso compounds, either preformed or endogenously formed, can be eliminated by a variety of mechanisms (see, e.g., Reference 29 for a review) which will be discussed in the next sections. Furthermore, nitrosation of amino compounds not only can lead to formation of mutagenic/carcinogenic derivatives but, less frequently, also results in the inactivation of mutagens, such as non-IQ type pyrolysates,[34] or the conversion of direct-acting mutagens into derivatives having different genetic and metabolic properties.[35]

2. Effects on the Intestinal Microbial Flora

Another important source of endogenous formation of mutagens in the digestive tract is represented by the multitude of bacteria colonizing the intestine, which possess a variety of metabolic pathways having various physiological functions, including the biotransformation of xenobiotics. In some instances, this results in the formation of mutagenic metabolites or products, such as the aglycone methylazoxymethanol from the mycotoxin cycasin following β-D-glucosidation,[36] or fecapentenes, which are produced by anaerobic bacteria belonging to the genus *Bacteroides*.[37] On the other hand, the bacterial decarboxylation of amino acids can produce biogenic amines having anticarcinogenic activity.[38]

Although antibiotics have been shown to reduce the formation of mutagens in the rat intestine,[39] a safer and more realistic approach is to try to modify the microecological profile of the intestinal bacterial flora by favoring suppression of putrefactive organisms involved in the production of putative carcinogens. At the beginning of this century, Metchikoff ascribed the unusually long life span of Balkan peasants to the consumption of large amounts of fermented milk, but its beneficial effects on longevity are even reported in Old Testament Scriptures. A number of recent studies have pointed out that dairy products may possess antitumor effects. In particular, such a role has been ascribed to *Lactobacillus acidophilus*, which is capable of implanting in the intestinal tract, thereby lowering the activity of enzymes of bacterial source involved in the metabolic activation of mutagens and carcinogens, such as β-glucosidase, nitroreductase, azoreductase, and α-dehydroxylase. This subject has been recently reviewed.[40] It should also be noted that one of the advantages of accelerating the intestinal transit (e.g., by means of dietary fibers) is to shorten the contact of promutagens with metabolizing bacteria (see Section II.B.3).

B. MODULATION OF MUTAGEN UPTAKE AND/OR REMOVAL FROM THE INTESTINE

Once mutagens have been introduced with food or endogenously synthesized from food-borne precursor molecules, prevention of their effects can be pursued by inhibiting their uptake and/or by favoring their removal by a variety of mechanisms.

1. Protection of the Mucosal Barrier

A high-fat diet, combined with low calcium intake, is likely to produce an increased permeability of the colon epithelium and a loss in integrity of the mucosal barrier, thus increasing exposure of the stem cells of the crypt to mutagens in the fecal stream.[41] The damage of colon epithelium by fat and bile acids can be reduced by dietary calcium.[41]

A newly proposed approach is to reduce the transepithelial absorption of carcinogens by means of oral vaccines, stimulating formation of secretory immunoglobulin A (IgA) in the intestine. Such a goal was experimentally pursued in rabbits immunized with a 2-acetylaminofluorene-cholera toxin conjugate.[42] Clearly, the practical applicability of vaccines against carcinogens and toxicants remains to be established.

2. Chelation and Formation of Complexes

A further mechanism by which calcium can prevent colon carcinogenesis is the formation of insoluble salts with cholic acids and fatty acids.[43] Formation of complexes in the extra-cellular environment has been postulated as a protective mechanism also in the case of carbonyl compounds[44] and porphyrins (reviewed in Reference 9). The protection by porphyrins against mutagenesis has been extensively investigated, using food pyrolysis mutagens[45] and benzo(a)pyrene.[46] An interaction occurs between a multi-ring mutagen and porphyrin by forming, presumably, a face-to-face complex. This type of complexing is most clearly demonstrated by the selective adsorption of multi-ring mutagens on blue cotton, a cotton preparation bearing covalently linked copper-phthalocyanine trisulfonate.[47,48] The protection by hemoglobin and myoglobin against the action of food pyrolysis mutagens, observed in bacterial test systems, is possibly due to similar complexing mechanisms.[49]

Chlorophyll and chlorophyllin, belonging to the porphyrin family, may be expected to behave like other porphyrins. Indeed, chlorophyll and chlorophyllin have been reported to act as inhibitors of mutagenesis.[45,50-53] The genotoxicity of Trp-P-2, as detected in Drosophila, can be effectively suppressed by simultaneously feeding the larvae with chlorophyll and the mutagen.[54] In this case, again, there is evidence that a complex formation between the mutagen and the pigment is the cause of the protection.

Entrapping of a variety of mutagens by micelles of oleic acid and C16-24 unsaturated fatty acids may contribute to their protective effects.[9,55,56]

Various compounds, such as bioflavonoids,[57] creatinine,[58] and ergothioneine,[10] are capable of chelating metal ions, especially copper, thereby preventing the generation of reactive oxygen species. Plant tannins, which are introduced in high amounts with tea and coffee, are also well known for their ability to chelate heavy metals.[10]

It is also likely that similar effects may be also produced when the same molecules penetrate into cells.

3. Dilution and Acceleration of the Intestinal Transit

A major role in accelerating removal of mutagens and carcinogens from the intestine is played by dietary fibers, including heterogeneous carbohydrate compounds, such as cellulose, hemicellulose, and pectin, as well as noncarbohydrate compounds, such as lignin. The dietary intake of these substances of plant origin (cereals, fruit, vegetables) has been inversely related to the prevalence of colonic-rectal cancer.[59] The epidemiologic data have been supported by the results of experimental studies in animals and in *in vitro* test systems, which, though with some conflicting indications, have confirmed the inhibitory properties of fibers and have also contributed to clarifying their mechanisms of action.[13,14,60-62]

A variety of mechanisms concur to the protective activity of fibers. First of all, fibers increase the fecal bulk, both because they are resistant to enzymatic digestion in the intestinal tract and because they increase the mass of intestinal bacteria, which constitute a considerable proportion of the stool mass. This results in a dilution of genotoxins and cancer promoters, either introduced with food or produced endogenously (e.g., bile acids), and in an accelerated transit time and excretion, thereby minimizing the possibility of contact with metabolizing and target cells. Certain types of fibers also bind bile salts,[63] and adsorb food pyrolysis products and other mutagens.[64,65] Several additional mechanisms can be involved, such as lowering of colonic pH, decreased production of fecal mutagens, increase in the production of short-chain fatty acids, and modification of metabolism during enterohepatic circulation, especially of promoters (see the aforementioned reviews). The relative contribution of these mechanisms varies depending on the type of fiber, which may also help to explain some controversial results reported in the literature. For instance, cereal bran fibers mainly act by increasing the fecal bulk, whereas lignin and pectin fibers have multiple effects.[13] Also in the case of fibers, adverse effects have been reported, which may be ascribed to the unavailability of certain essential trace metals that are involved in carcinogen detoxification.[66]

C. EXTRACELLULAR DEACTIVATION OF FOOD-BORNE MUTAGENS

Molecules having a genotoxic potential, which are present in food or are locally formed from precursor compounds, can be deactivated by various mechanisms during transit through the digestive tract or during subsequent transportation to target tissues and cells (see a review in Reference 24).

1. Deactivation by Physicochemical Mechanisms

Variations in normal pH values in the different sections of the digestive tract are associated with deleterious effects, including cancer. For instance, this is the case for the oral cavity, where the alkaline environment induced by the lime in betel chewers creates a condition favoring oxidation of phenolics and formation of free radicals.[7] Similarly, enhancement of intragastric pH (e.g., in individuals with chronic gastritis) is associated with an increased incidence of gastric cancer.[67] This may be ascribed to colonization of the stomach by intestinal bacteria catalyzing nitrosation phenomena.[26]

Clearly, it is not simple to correct abnormal situations due to noxious habits, pathological conditions or even pharmacological treatments. For instance, the elevation of intragastric pH consequent to the administration of histamine H_2-receptor antagonists, which have proven quite beneficial in the medical therapy of peptic ulcer, produces a plateau of gastric juice mutagenicity during the night, following bed-time administration of these drugs.[33] Under these conditions, the administration of dietary or pharmacological protectors of the gastric mucosa may be advisable.

The acidic gastric environment also affects the stability of certain mutagens, either by protecting them from inactivation or by accelerating their deactivation.[68] Deactivation can be stimulated by both drug treatments and food ingestion. For instance, enhancement of the reduction of hexavalent chromium compounds and reversal of their mutagenicity can be obtained by stimulating gastric secretion with either food or drugs (e.g., pentagastrin).[69]

Chemicals having nucleophilic and/or antioxidant properties can favor deactivation of direct-acting mutagens in the gastrointestinal tract or during transportation in the blood, with mechanisms similar to those occurring inside cells, although in some instances with opposite effects. A typical example is provided by the interaction between N-methyl-N'-nitro-N-nitrosoguanidine (MNNG) and reduced glutathione (GSH), which is present not only in cells of the mucosa but also in enterobacteria, in millimolar ranges, and is exported extracellularly,[70] as it may occur in the lumen of the gastrointestinal tract. GSH concentrations in enterobacteria can be modulated by adding GSH itself, or its precursors (e.g., N-acetylcysteine) or depletors (e.g., diethyl maleate).[71] It is noteworthy that the products of the reaction of thiols with MNNG are harmless when they are formed outside cells, while they can methylate DNA once generated within the cell.[70-73] The protective effect of thiols is likely to occur not only in the gastrointestinal lumen but also in the blood plasma, where GSH levels of about 25 μM can be dynamically maintained by stimulating its hepatic synthesis and efflux.[74]

A prerequisite for inhibitors acting in the gastric environment is maintenance of their properties in acidic pH ranges, as demonstrated for instance in the case of nucleophilic sulfur compounds (such as sodium thiosulfate and 4-mercaptobenzene sulfonate)[75] and the thiols GSH and N-acetylcysteine, inhibiting MNNG mutagenicity at both pH 1.5 and 7.4.[71]

Extracellular spaces normally contain high concentrations of molecules having a marked antioxidant activity, such as bilirubin[76] and uric acid,[77] which is the main antioxidant molecule in saliva.[78] Although the concentrations of these catabolites can be modulated by either dietary or pharmacological means, their enhancement, which also occurs under pathological conditions, is not proposable for cancer prevention.

2. Deactivation by Enzyme-Catalyzed Mechanisms

More readily applicable is the modulation of antioxyenzymes, which also play a protective role in the lumen of the digestive tract. For instance, it was reported that saliva can inactivate mutagens, which was mainly ascribed to the presence of peroxidase.[79]

Juices prepared from various vegetables reduced in vitro the mutagenicity of tryptophan pyrolysis products,[80] with peroxidase and NADPH-oxidase activities being responsible for deactivation in cabbage,[81] broccoli,[82] and horseradish.[83]

D. INHIBITION OF PENETRATION INTO CELLS

The last intervention step which can be envisaged in the extracellular environment is prevention of penetration of mutagens into cells. A major defensive mechanism of the cell is provided by surface polysaccharides, acting as molecular sieves preventing the uptake of molecules of various molecular weights and electric charge.[10] Protective agents may act by modifying the permeability of cells to toxic compounds or by competitive cellular uptake. A well-known application of the latter mechanism is provided by the administration of iodide, which competes with the uptake by the thyroid cells of ^{131}I introduced with the diet in cases of radioactive contamination, as recently occurred on a large scale after the Chernobyl accident. Other specific examples are putrescin, which prevents the cellular uptake of the oxidative mutagen paraquat,[84] and aromatic amino acids, which compete with the transport of azaserine into bacterial cells.[85] Short-chain fatty acids, such as caproate and caprylate (C_6), inhibited the mutagenicity of N-nitrosodimethylamine in bacteria. As assessed by using the radioactive mutagen, such effect was ascribed to lack of penetration into target cells, due to the formation of complexes between the fatty acid and the alpha-hydroxylated derivative of N-nitrosodimethylamine.[86] The ability of acylglucosylsterols isolated from green fruits (Momordica charantia) to counteract the increase in the frequency of micronucleated

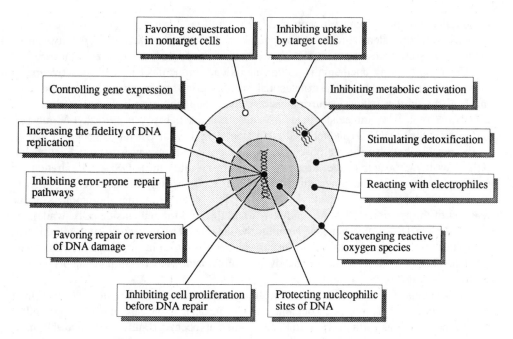

FIGURE 2.   Mechanisms of inhibitors of mutagenesis and carcinogenesis acting at the cellular level.

erythrocytes induced by mitomycin C in mice was also ascribed to membrane absorption of this inhibitor, altering the cell permeability to the mutagen.[87]

III. INHIBITION OF MUTAGENESIS BY INTRACELLULAR MECHANISMS

As summarized in Figure 2, a variety of interconnected mechanisms can be modulated within the cell in order to prevent occurrence and fixation of DNA damage and, consequently, initiation of cancer and other mutation-related diseases.

A. SEQUESTRATION OF MUTAGENS IN NONTARGET CELLS
In this domain, a first mechanism which could be exploited for preventive purposes is sequestration and detoxification of mutagens in long-living nontarget cells, such as carrying cells (e.g., erythrocytes) and sweeping cells (e.g., macrophages).[24] For example, hexavalent chromium entering the bloodstream is selectively accumulated into erythrocytes, where it is reduced, mainly by GSH, and bound to hemoglobin and other molecules.[88] Thiols, such as *N*-acetylcysteine, enhance GSH concentrations in erythrocytes following *in vivo* treatment,[89] and therefore their administration may prove useful for potentiating this protective mechanism.

It is obvious that the distinction between target and nontarget cells is linked to various factors and that the differential efficiency of intracellular defense mechanisms plays a major role in selecting target cells.[90] Accordingly, the exogenous modulation of metabolism and DNA repair processes, which will be discussed in the next sections, can contribute to convert target cells to nontarget cells or vice versa. In fact, it should be noted that certain modulators of carcinogenesis, such as ethanol and disulfiram, have been shown to shift the organotropism of carcinogens, such as nitrosamines.[25,91]

B. MODULATION OF METABOLISM AND BLOCKING OF REACTIVE MOLECULES
Food-borne inhibitors acting at this stage, collectively referred to as blocking agents by

Wattenberg,[92] have a prominent importance in cancer prevention. Although they can be classified into several subcategories (see Figure 2), most of them have multiple metabolic effects, and several modulators can influence both protective and activating mechanisms at the same time. In fact, modulation of enzyme-catalyzed biotransformations, modifying or creating functional groups in the molecule (phase I reactions), is largely connected with the effects of conjugating (phase II) reactions and with scavenging of reactive chemical species.[24]

1. Modulation of Metabolic Activation

It is difficult to selectively inhibit the activation of promutagens/procarcinogens to ultimate metabolites without influencing other metabolic stages. In principle, due to the low substrate specificity, a plethora of compounds can interfere with the microsomal monooxygenase system by a variety of mechanisms, which have been investigated more often for biochemical studies than for practical preventive applications.[93]

The influence of food components on metabolic activation has been mainly explored in liver cells. However, the local metabolism should be also taken into account. For example, it has been shown that the rat intestinal mucosa contains lipid components capable of inhibiting the metabolic activation of food pyrolysis products.[94]

A typical example of metabolic effects produced by food components, but at the same time of the problems involved, is provided by cruciferous plants. The active principles of these plants, e.g., phenols, isothiocyanates, indole derivatives, and dithiolthiones, possess multiple properties, among which are induction of both activating enzymes (such as microsomal arylhydrocarbon hydroxylase activity)[95] and detoxifying enzymes (such as cytosolic GSH S-transferase).[96] On the whole, both epidemiologic and experimental studies show, with few exceptions,[97] that the anticarcinogenic effect of these plants is prevailing. This is probably the expression of a general detoxification procedure for procarcinogens, i.e., a delicate coordination between addition of active polar groups and the subsequent conjugation or trapping by nucleophiles.[24,98] This mechanism has also been postulated in the case of indoles contained in cruciferous plants.[99] It has been calculated that even in the U.K., where the intake of cruciferous vegetables is particularly high, the *pro capite* average daily consumption of indoles contained in the glucobrassicin molecule is less than 10 mg/day, i.e., substantially lower than the amounts used in animal studies.[100] However, it should be added that dose-response studies, performed in rainbow trout exposed to aflatoxin B_1 in the diet, showed that even low doses of indole-3-carbinol inhibit carcinogen-DNA binding.[101]

Another mechanism by which food components modulating the monooxygenase system may prevent cancer is the induction of a shift in isoenzymes, leading to decreased production of ultimate mutagens. For example, the food additive BHT, under conditions inhibiting the induction of tumors in mice, produced a shift from benzo(a)pyrene 4,5-epoxidation and 9-hydroxylation to 3-hydroxylation.[102]

Fatty acids contained in milk and other foodstuffs can also inhibit metabolic activation processes. For example, inhibition of the mutagenicity of food pyrolysis mutagens, polycyclic aromatic hydrocarbons, and nitrosamines by oleic acid has been ascribed to a block of metabolic activation, due either to entrapping phenomena (see Section II.B.2) or to interactions with the enzyme proteins.[55,56] Among short-chain saturated fatty acids, laurate (C_{12}) has been shown to block the metabolic demethylation of N-nitrosodimethylamine,[86] whereas caproate (C_6) inhibits cellular uptake (see Section II.D.)

Hemin blocks the metabolic activation of Trp-P-1 and Trp-P-2, possibly because the complex between hemin and these heterocyclic amines is no longer a suitable substrate for activating enzymes.[9,45,46]

Several plant polyphenols have also been shown to interfere with metabolic activation processes.[7,9,10] For instance, in addition to other effects (see Sections III.B.2 and III.B.3), ellagic acid is toxic to proteins, including activating enzymes.[103] Gallic acid, chlorogenic acid, and caffeic acid have been reported to suppress the mutagenicity of aflatoxin B_1 by inhibiting its metabolic activation.[104] Flavonoids, such as anthraflavic acid, inhibit the metabolic activation of the food pyrolysis product IQ.[105]

Monocyclic monoterpenoids, such as α-limonene, menthol, and carveol, which are naturally occurring hydrocarbons contained in orange and other plant species, inhibit cancer initiation by 7,12-dimethylbenz(a)anthracene, probably by modifying its activation and detoxification.[106] The ability of various spice extracts to decrease the mutagenicity of amino acid pyrolysis products was ascribed to inhibition of the N-hydroxylation by eugenol, one of the main essential oils contained in these spices.[107]

Among vitamins, vitamin A has been shown to inhibit the metabolic activation of benzo(a)pyrene,[108] cyclophosphamide,[109] and o-aminoazotoluene.[110] One of the mechanisms involved may be the ready epoxidation of vitamin A, which may compete with promutagens as an alternative susbstrate for epoxide generating systems.[111]

Nicotinamide, a member of the vitamin B group, is not only an alkyl scavenger,[112] but can also interfere with the metabolic activation of diethylnitrosamine during the initial stages of rat hepatocarcinogenesis.[113]

It should be noted that not only the presence of certain food constituents but also their deficiency can impair the activation of procarcinogens, as suggested in the case of a choline-deficient diet.[114]

2. Stimulation of Detoxification, Trapping of Electrophiles, and Antioxidant Mechanisms

These mechanisms will be discussed together, because stimulation by food components of detoxifying enzymes, including antioxienzymes, is frequently associated with their ability to block reactive species, either exogenous or intracellularly generated. These inhibitors often share the properties of antioxidants, scavenging oxygen radicals, and nucleophiles, trapping positively charged electrophilic metabolites or direct-acting molecules. Such a goal can be achieved either by direct reaction or by enzymatic conjugation, which is catalyzed by enzymes involved in phase II reactions, such as uridine diphosphateglucuronyl (UDPG) transferases, sulfotransferases, and acetyltransferases. Special importance is ascribed to the family of GSH S-transferase isozymes,[115] catalyzing the conjugation of GSH with a variety of electrophiles. Stimulation of this enzyme has been proposed as a marker for predicting the efficiency of certain inhibitors.[12] In addition to other cytosolic enzymes involved in the GSH cycle (such as GSSG reductase and the selenium-containing GSH peroxidase) and other antioxienzymes (such as superoxide dismutase and catalase), an important protective role in prevention of mutation and cancer is ascribed to DT diaphorase activity.[116]

A prototype of food-borne inhibitors exerting coordinated protective effects at this stage is represented by the phenolic food additives BHA and BHT. These synthetic antioxidants displayed antimutagenic and anticarcinogenic properties in a number of studies,[117] but also showed enhancing effects under certain conditions, for example, when administered after test carcinogens.[118] They have been shown to induce multiple conjugating and detoxifying enzymes, such as UDPG transferases, GSH S-transferase, DT diaphorase, and epoxide hydrolase, without a concomitant induction of phase I monooxygenases.[98,119,120] These patterns differentiate selective inhibitors from other inducers, such as flat planar molecules, which at the same time induce detoxifying cytosolic enzymes and cytochrome PI-450, by binding a specific Ah receptor.[121]

A selective induction of multiple detoxifying enzymes is also produced by synthetic aminothiols, such as N-acetylcysteine (NAC), which is an analog and precursor of intracellular GSH. This compound is extensively prescribed as a pharmacological agent and has also been used as an additive in certain foods. For example, treatment of raw soy flour with cysteine or NAC resulted in the introduction of new half-cysteine residues into sulfur-pool legume proteins, with a corresponding improvement of nutritional quality, as measured by the protein efficiency ratio in rats.[122] NAC, which inhibited spontaneous and induced mutagenicity and prevented carcinogenicity in various rodent models when administered with the diet, has been shown to be an efficient nucleophile and antioxidant, as well as an inducer of cytosolic enzymes (see Reference 123 for a review). GSH, NAC, and α-mercaptopropionylglycine also inhibited the mutagenicity induced by the generation of reactive oxygen species.[124]

Dithiolthiones, such as oltipraz, which are components of cruciferous vegetables capable of preventing cancer in experimental animals, are nucleophiles and at the same time increase GSH levels and stimulate multiple detoxifying enzymes.[125,126] Sodium thiosulfate is also a very effective trapping agent.[127]

Other sulfhydryl compounds or their precursors are contained in garlic and onion. The medicinal properties of garlic have been known for 5 thousand years, and consumption of allium vegetables is associated with reduced risk of stomach cancer.[128] After pioneer studies showing the tumor-inhibiting ability of a sulfhydryl blocking agent related to an active principle of garlic,[129] attention has recently been focused on active components, such as diallylsulfide, allyl methyl disulfide, diallyl trisulfide, and allyl methyl trisulfide. In addition to effects at the promotion step (see Section IV), diallyl sulfide has been suspected to act by conjugating toxic metabolites, e.g., in the case of cyclophosphamide.[130]

Stimulation of GSH S-transferase activity in mouse tissues and prevention of chemically induced cancer were also produced by the oral administration of citrus fruits oils, such as orange, tangerine, lemon, and grapefruit oils, as well as by two diterpene esters, kahweol palmitate and cafestol palmitate, which have been isolated from coffee beans.[100]

Several plant polyphenols, such as ellagic acid, tannins, and flavonoids, have been shown to trap electrophilic metabolites of promutagens and to accelerate their degradation.[9] In particular, ellagic acid has been shown to mediate the benzo(a)pyrene epoxide ring opening by taking a sterically favorable position to form a covalently linked adduct.[131] Extracts of catechu (the nonmutagenic compound of betel quid) and its active principle (i.e., the flavonoid catechin, a free-radical scavenger) were found to inhibit the direct mutagenicity of smoked meat.[132]

Even food cooking can produce inhibitors. Thus, while Maillard reaction products generated in heated meat and fish are potential carcinogens,[133] those formed from carbohydrate-rich sources, such as α-carbonyl and α-hydroxycarbonyl compounds, can inhibit the mutagenicity of food pyrolysis products by binding their molecules.[44]

Among vitamins, riboflavin 5'-phosphate was reported to accelerate the decomposition of benzo(a)pyrene diol epoxide,[134] and nicotinamide reacted with several alkylating mutagens by forming N'-alkylnicotinamide.[112]

In addition to those already discussed, a variety of other vitamins and food-borne molecules can prevent mutation and cancer by antioxidant mechanisms. We refer to other reviews for a general discussion on the antioxidant systems, including antioxienzymes and scavengers of active oxygen species,[116] and the mechanisms of inhibition of free-radical processes.[135]

A quite extensive literature is available on the antioxidant activity of vitamins, which is one of the mechanisms contributing to their anticarcinogenic activity, inferred from the indications of epidemiologic studies and from a number of experimental findings. Several chemoprevention trials, already completed or now in progress, involve the individual or combined administration of vitamins in high-risk groups.[136] Antioxidant activity is well recognized for the lipid-soluble vitamins A and E. Within the vitamin A family, the provitamins α-carotene and especially β-carotene, which are widely contained in several foods and typically in yellow-green vegetables, and retinoids, including vitamin A alcohol (retinol), esters (retinyl acetate and retinyl palmitate), aldehyde (retinal) and acid (retinoids), as well as their physiological metabolites and numerous synthetic analogs, have multiple mechanisms of action, among which is an antioxidant activity exerting protective effects along all stages of carcinogenesis.[111,137-140] Vitamin E is a mixture of phenols contained in plants, especially in wheat germ and corn. Its major component, i.e., α-tocopherol, is well known as a major radical-trapping agent in lipids of biological membranes which is the main mechanism responsible for its ability to prevent mutation and cancer.[13,117,141,142] Besides its antimutagenicity and anticarcinogenicity as an inhibitor of the nitrosation reaction (see Section II.A.1), the water-soluble vitamin C (ascorbic acid), which is ingested not only as a natural component of foods and beverages but also as a food additive, is an effective antioxidant, scavenging hypochlorous acid, tyrosyl radicals, and ozone; reacting with superoxide and hydroxyl radicals; quenching singlet oxygen; and preventing peroxidation.[10,117,142]

Selenium is a trace component of the diet mainly present in flour, in amounts depending on its concentration in soil. As a component of cytosolic GSH peroxidase, selenium is an antioxidant possessing antimutagenic[143] and anticarcinogenic[11] properties. Possibly due to different cellular localization, a synergism may occur between selenium and vitamin E.[144]

Food-borne anticarcinogens of particular interest are the conjugated dienoic derivatives of linoleic acid (CLA) which, besides being generated endogenously, are ingested in large amounts with grilled beef, cheese, and related foods. At variance with linoleic acid, which is the only fatty acid enhancing carcinogenesis in experimental animals, its conjugated derivatives have been shown to inhibit mutation and cancer. Recently, these protective effects have been ascribed to the considerable efficacy of CLA as antioxidants.[145] It is noteworthy that linoleic acid derivatives, such as arachidonic acid, exhibited no anticlastogenic properties, while all the linolenic acid derivatives produced a distinct anticlastogenic effect.[146]

Further to the already mentioned sulfhydryl compounds, an inhibitor acting as a nitrite trapping agent and as an antioxidant is thioproline, which is contained in boiled vegetables.[10] Several natural antioxidants were detected and investigated, such as lignan-type antioxidants in sesame seed, catechin-type antioxidants in tea leaves, and tannin-type antioxidants in herbs.[147] Antioxidant properties are also ascribed to other food-borne mutation and/or cancer inhibitors working through multiple mechanisms, such as the bioflavonoid quercetin, trapping superoxide radicals,[148] and the potato chymotrypsin inhibitor, decreasing hydrogen peroxide formation by activated human polymorphonuclear leucocytes.[149] Fungal products such as ergothioneine, and components of mammalian and fish muscles, such as L-histidine and β-alanyl-L-histidine (L-carnosine), are quenchers of singlet oxygen (reviewed in Reference 10).

3. Protection of Nucleophilic Sites of DNA

As just discussed, in most cases the reaction between electrophilic molecules and nucleophilic sites of DNA is blocked by inhibitors trapping electrophiles. However, the possibility has been also envisaged that food-borne inhibitors may act by masking specific sites of DNA, thereby protecting them from the attack of reactive species.

This has been demonstrated to be the case for ellagic acid, a naturally occurring phenol present in plants (e.g., coffee, nuts, and grapes), which, as already discussed, also appears to inhibit metabolic activation processes (Section III.B.1) and to bind reactive epoxides (Section III.B.2). In addition, ellagic acid caused a decrease of O^6-methylguanine adducts of N-methyl-N-nitrosourea (MNU), but only in double-stranded DNA, which rules out the hypothesis of a direct scavenging of MNU electrophilic intermediates.[150] Rather, the observed patterns suggest that inhibition of MNU mutagenicity may depend on prevention of methylation at the O^6 position of guanine through an ellagic acid-duplex DNA affinity mechanism.[150]

In some way, vitamin A can also protect DNA from the attack of reactive molecules. As already discussed (Section III.B.1), this may occur through formation of harmless epoxides, such as 5,6-epoxyretinoic acid competing with mutagenic/carcinogenic epoxides in

DNA binding. In addition, retinoids are capable of stimulating the production of prostaglandin, which inhibits the binding of carcinogens to DNA.[111]

C. CONTROL OF DNA FUNCTIONS AND METABOLISM

The mechanisms so far discussed are directed to the inhibition of genotoxicity by reducing the dosimetric levels of noxious agents, either introduced with food or formed endogenously from food-borne precursors, in tissues (pharmacokinetic dose), in the cell cytoplasm (cellular dose), in the nucleus (target dose), and finally in their adduction to target molecules (molecular dose). The subsequent possibility of preventing genotoxic and related health effects by food components is at the level of the genome of the host cell and is aimed at either modulating DNA repair processes by avoiding fixation or amplification of DNA damage (Section III.C.1) or at controlling gene expression and the effects of growth factors (Section III.C.2). It is evident that control of these cellular processes, affecting all stages of carcinogenesis, is particularly delicate, a unidirectional influence on protective mechanisms being extremely difficult.

1. Modulation of DNA Repair

First of all, it should be noted that error-free repair of DNA lesions must occur before cell division. Therefore, inhibition of cell proliferation has an important protective role not only on subsequent stages of carcinogenesis but also on the initiation stage consequent to mutations in somatic cells. In fact, replicating cells are more vulnerable to the action of the initiators than nonreplicating cells,[111] and cells treated in G_2, when a whole cell cycle is available for repair, often exhibit a threshold-like dose response, with no or few mutations at low dosages.[151]

Antiproliferative effects and prolongation of G_2 phase are produced by food components, including, for example, retinoids,[138] isothiocyanates,[152] and aldehydes.[10] An opposite effect is caused by caffeine.[153] Since a high-fat, low-calcium diet is likely to lead to an increase in the proliferation of cells in the colonic crypts,[41] a reduction in fat as well as calcium supplementation would be expected to reduce cell proliferation.

A large number of compounds can inhibit both DNA replication and repair. The main mechanism by which DNA repair modulators may prevent carcinogenicity is that DNA-damaged cells are preferentially killed rather than being repaired defectively and potentially subjected to neoplastic development.[153] Kada et al.[154] distinguished three protective mechanisms involved in modulation of DNA repair: (1) increase of the fidelity of DNA replication, (2) stimulation of the repair of DNA damage, and (3) inhibition of error-prone repair systems (see Figure 2).

Examples of food-borne inhibitors affecting these mechanisms have been mainly investigated in bacteria (reviewed in Reference 155). They include trace metal compounds, such as cobaltous chloride, which increases the fidelity of DNA replication and enhances recombination repair,[156] and sodium arsenite, which inhibits the umuC gene expression and enhances error-free repair in bacteria.[157] Vannilin and cinnamaldehyde, the principal ingredients of vanilla essence and cinnamon oil, respectively, display multiple inhibitory effects, among which is a transient inhibition of growth[158] and stimulation of recA-dependent recombinational repair, which is thought to be error-free.[155] The antimutagenicity of umbelliferone and coumarin, which however induced gastric cancer in rodents,[159] was also ascribed to stimulation of DNA damage repair.[156] Tannic acid, a plant-derived hydrolysate of Chinese gallotannin, also inhibited mutagenic effects in bacteria, possibly by stimulating excision repair mechanisms.[160] Protease inhibitors, constituting a broad family of naturally occurring compounds present in the diet, are likely to suppress an error-prone repair system in bacteria.[161] One of the mechanisms by which they might prevent carcinogenesis is by inhibiting an error-prone process due to inducible proteases leading to mutation and ultimately to cancer.[162] para-Aminobenzoic acid (PABA), an inhibitor of mutagenicity present in large amounts in baker's and brewer's yeast,[163] also inhibited inducible SOS functions in bacteria.[164] When *Escherichia coli* cells were in the log phase, α-tocopherol affected bacterial DNA polymerase III.[165] Studies in 2-acetylaminofluorene-treated rats provided evidence that the dietary administration of *N*-acetylcysteine can prevent the sharp decrease in liver ADP-ribosyl transferase (ADPRT), which catalyzes the ADP ribosylation of nuclear proteins and is involved in the modulation of cellular responses to DNA damage.[166]

2. Control of Gene Expression

Control of gene expression by genetic engineering techniques is still questionable, although it is now possible to correct genetic changes associated with cancer in both *in vitro* and *in vivo* experimental systems.[4,167] Studies exploring the possibility of modulating genotoxicity by correcting wrong cellular messages both at the level of genes and gene products, forming a chain of products acting from the outer cell membrane via the cytoplasm to the nucleus, have been recently reviewed.[168]

This kind of mechanism may also be influenced by food components. However, as is the case for other cellular processes involved in mutagenesis and carcinogenesis, it should be kept in mind that modulation of these steps may represent not only an actual regulatory mechanism but also a consequence of other mechanisms, e.g., those blocking reactive molecules and preventing their effects. Several food-borne agents, like retinoids or vitamin D derivatives, have already been shown to mimic, in appropriate cell systems, the action of growth suppressor genes or actually induce their expression.[169]

Inhibition of oncogene expression has been obtained, both *in vitro* and *in vivo,* by food components. For instance, the protease inhibitor antipain reduced c-*myc* expression in proliferating cultured cells,[170] and the same effect was produced in nonproliferating cells by retinoic acid[171,172] and by a metabolite of vitamin D_3.[173] Administration of *S*-adenosyl-L-methionine (SAM) in a rat hepatocarcinogenesis model inhibited the expression of c-*Ha*-*ras* and c-*myc* in hyperplastic nodules.[174]

Particular importance is ascribed to the control of gene expression as a key mechanism of action of retinoids in the control of cell proliferation and differentiation.[175] In addition to the already reported examples of inhibition of oncogene expression, there is evidence that retinoic acid counteracts the proliferative effects of transforming growth factor-beta (TGF-beta) and platelet-derived growth factor (PDGF) in c-*myc*-transfected fibroblasts.[175] Retinoic acid can also interact with membrane-bound receptors of specific peptide growth factors in susceptible cells. For example, it caused a considerable increase in the number of available receptor sites for epidermal growth factor (EGF or TGF-alpha) in a variety of cells, possibly by stimulating the synthesis of the specific mRNA.[176]

A connection has been established between the cholesterol biosynthetic pathway and transformation by the *ras* oncogene, which may offer a novel pharmacological approach to controlling *ras*-mediated malignant transformation.[177]

IV. MODULATION OF TUMOR PROMOTION AND PROGRESSION

We will briefly discuss some aspects of modulation of tumor promotion and progression because, together with modulation of genotoxicity and cancer initiation, this subject bears considerable relevance in cancer prevention by food-borne inhibitors. While cancer initiation involves a rapid succession of events leading to irreversible DNA damage, promotion involves slow and only partially elucidated processes leading in a reversible fashion to a mass of neoplastic cells, which acquire malignant characters during progression. Promotion is evaluated to last years or decades in humans, and progression requires nearly an additional year.[136] Therefore, exposure to antipromotors and/or antiprogressors can further delay these steps, thereby decreasing the incidence of cancer. Due to their continuous administration with the diet, food-borne inhibitors appear to be ideal candidates as antipromoters and antiprogressors. On the other hand, the diet is also an important source of promotors and progressors, and therefore any suggestion to vary dietary habits must be weighed very carefully.

From a mechanistic point of view, it should be taken into account that the distinction between the different carcinogenesis steps is becoming more and more faded, with events of a similar nature occurring along all stages of cancer development. Therefore, mechanisms which have already been discussed with respect to modulation of genotoxicity are also involved and in some cases are even more important in the control of tumor promotion and progression.

This applies, first of all, to all the mechanisms discussed in Sections II and III, because, although tumor promotion and progression are generally viewed as epigenetic processes, various genotoxic effects have also been detected during the promotion step,[178] and chromosomal and ploidy changes appear to be involved in tumor progression.[179] Activation of oncogenes occurs along all steps of carcinogenesis, and even in very advanced stages of evolution into malignancy.[180] As discussed in Section III.C.2, food components can interact with oncogenes, peptide growth factors, and their receptors. An additional example is provided by protease inhibitors, which antagonize proteolytic activity (the Boc-Val-Pro-Arg-MCA hydrolyzing activity) involved in processing a growth factor associated with the malignant transformation of cells.[181] Modulation of signal transduction can be achieved by various approaches,[168,169] and food-borne molecules can also affect the unfolding of the cellular signal systems. For example, the flavonoid quercetin is an inhibitor of protein kinase C, acting on its catalytic domain.[182]

Inhibitors of cell proliferation have been already discussed (Section III.C.1). Caloric intake may exert an important influence on this mechanism, since dietary restriction leads to a reduction of cell proliferation in various mouse tissues.[183]

Diallyl sulfide is also likely to inhibit tumor promotion by reducing postnecrotic liver regeneration.[184] Food-borne inhibitors also include differentiation inducers, such as vitamin D_3 analogs, and retinoids, which exert a hormone-like control of either cell proliferation or differentiation.[24]

Oxidative stress plays a crucial role in tumor promotion[116,135,185-187] and is also involved in tumor progression.[188] Therefore, all the antioxidant vitamins and molecules discussed in Sections II.C.2, II.C.3, and III.B.2 as inhibitors of genotoxicity and cancer initiation are also expected to affect subsequent stages of carcinogenesis.

An additional inhibitory mechanism occurring at these stages is cytotoxicity to initiated and neoplastic cells, which has been shown to be produced, for example, by vitamin A, releasing hydrolytic enzymes destroying neoplastic cells,[189] as well as by methylglyoxal,[190] phenolic compounds,[191] and protease inhibitors.[192] However, it is unlikely that these components may exert a selective cytotoxic effect *in vivo* at the normal concentrations present in food.[8] On the other hand, restoration of cell membrane integrity is another important mechanism of inhibition of carcinogenesis. Several studies have evaluated the ability of retinoids to modulate the cell surface composition (glycoproteins, glycolipids, phospholipids) and functions (adhesion, spreading, ion transport, cell-to-cell communication) (reviewed in Reference 111).

Food components can even influence the efficiency of the immune response, which represents a well-recognized defensive process against cell transformation and tumor growth. For instance, lipotropes such as choline, folate, methionine, and vitamin B_{12}, display immunoregulatory effects, possibly due to their role in lipid metabolism.[193] Immune-related mechanisms may contribute to the protective effects of *n*-3 polyunsaturated fatty acids typically present in fish oil, such as eicosapentenoic acid. At variance with *n*-6 polyunsaturated fatty acids, such as linoleic acid, eicosapentenoic acid was found to suppress excessive production in colon tumors of prostaglandin E_2, which may have an integral role in cellular proliferation and differentiation, and in tumor dissemination. Prostaglandin involvement in late stages of carcinogenesis may depend on modulation of cellular and tumoral immune responses (reviewed in reference 194). The possibility that certain food components, such as the monoterpenoid d-limonene, may inhibit mammary cancer by modulating endocrine functions has been explored but ruled out.[195]

V. CONCLUSIONS

The examples reported in this chapter give an idea of the large number and variety of mechanisms accounting for the protective effects of food components on genotoxicity and cancer initiation, as well as on subsequent steps in the carcinogenesis process. It is evident that in many cases the mechanisms of action are not fully elucidated or sufficiently substantiated by well-grounded scientific evidence. It is often difficult to discriminate whether modulation of a given end-point reflects the actual regulatory mechanism or the consequence of another mechanism operating at earlier stages in the sequence of events leading to mutation and cancer.

It is rather typical that a number of food-borne inhibitors work through multiple and coordinated mechanisms, which in principle is expected to warrant a broader range of intervention levels towards mutagens and carcinogens of various chemical nature. On the other hand, the possession of pleiotropic biological properties of putative inhibitors, together with the strict interconnections occurring in a delicate balance between toxifying and detoxifying processes of the organism, renders unidirectional modulation extremely difficult. Interactions can also occur between different modulators, which are likely to affect the regulatory properties of complex mixtures like foodstuffs.

There appears to be a considerable expansion of scientific interest in this field. Future research aimed at preventing cancer and other mutation-related diseases will hopefully provide more information on the mechanisms of food-borne inhibitors, on their efficacy, and on their safe applicability.

REFERENCES

1. **Doll, R. and Peto, R.,** The causes of cancer: quantitive estimates of avoidable risks of cancer in the United States today, *J. Natl. Cancer Inst.,* 66, 1191, 1981.
2. **Wynder, E. L. and Gori, G. B.,** Contribution of the environment to cancer incidence: an epidemiologic exercise, *J. Natl. Cancer Inst.,* 58, 825, 1977.
3. **Higginson, J. and Muir, C. S.,** Environmental carcinogenesis: misconceptions and limitations to cancer control, *J. Natl. Cancer Inst.,* 63, 1291, 1979.
4. **De Flora, S.,** Mechanisms of inhibitors of genotoxicity. Relevance in preventive medicine, in *Mutation and the Environment — Part E: Environmental Genotoxicity, Risk and Modulation,* Mendelsohn, M. L. and Albertini, R. J., Eds., Wiley-Liss, New York, 1990, 307.
5. **De Flora, S.,** Editorial. Problems and prospects in antimutagenesis and anticarcinogenesis, *Mutat. Res.,* 202, 279, 1988.
6. **Aeschbacher, H. U.,** Antimutagens and anticarcinogens in the diet, *Biol. Zeantral bl.,* 108, 303, 1989.
7. **Stich, H. F. and Rosin, M. P.,** Naturally occurring phenolics as antimutagenic and anticarcinogenic agents, in *Nutritional and Toxicological Aspects of Food Safety,* Friedman, M., Ed., Plenum Press, New York, 1984, 1.
8. **Aeschbacher, H. U.,** Antimutagenic/anticarcinogenic food components, in *Carcinogens and Mutagens in the Diet,* Pariza, M. W., Ed., Alan R. Liss, New York, in press.
9. **Hayatsu, H., Arimoto, S., and Negishi, T.,** Dietary inhibitors of mutagenesis and carcinogenesis, *Mutat. Res.,* 202, 429, 1988.
10. **Hartman, P. E. and Shankel, D. M.,** Antimutagens and anticarcinogens: a survey of putative interceptor molecules, *Environ. Mol. Mutagen.,* 15, 145, 1990.

11. **Ames, B. N.,** Dietary carcinogens and anticarcinogens, *Science,* 221, 1256, 1983.

12. **Wattenberg, L. W.,** Inhibition of neoplasia by minor dietary constituents, *Cancer Res.,* 43, 2448s, 1983.

13. **Fiala, E. S., Reddy, B. S., and Weisburger, J. H.,** Naturally occurring anticarcinogenic substances in foodstuffs, *Annu. Rev. Nutr.,* 5, 295, 1985.

14. **Wargovich, M. J.,** Dietary promotors and antipromotors, in *Antimutagenesis and Anticarcinogenesis Mechanisms,* Shankel, D. M., Hartman, P. E., Kada, T., and Hollaender, A., Eds., Plenum Press, New York, 1986, 409.

15. **Reddy, B. S. and Cohen, L. A., Eds.,** *Diet, Nutrition and Cancer: A Critical Evaluation,* Vols. I and II, CRC Press, Boca Raton, FL, 1986.

16. **Hayashi, Y., Nagao, M., Sugimura, T., Takayama, S., Tomatis, L., Wattenberg, L. W., and Wogan, G. N., Eds.,** *Diet, Nutrition and Cancer,* Japan Scientific Societies Press, Tokyo, VNU Science Press, Utrecht, The Netherlands, 1986.

17. **Pariza, M. W., Aeschbucher, H.-U., Felton, J. S., and Sato, S., Eds.,** *Carcinogens and Mutagens in the Diet,* Wiley-Liss, New York, 1990.

18. **National Research Council,** *Diet, Nutrition and Cancer,* National Academy Press, Washington, D. C., 1982.

19. **American Cancer Society,** Nutrition and cancer: cause and prevention, *Ca-A Cancer J. Clin.* 34, 121, 1984.

20. **Knudsen, I. and Møller, Jensen,** Diet and cancer: regulatory perspectives in Europe, in *Carcinogens and Mutagens in the Diet,* Pariza, M. W., Aeschbucher, H.-U., Felton, J. S., and Sato, S., Eds., Wiley-Liss, New York, 1990, 285.

21. **Palmer, S. and Bakshi, K.,** Diet, nutrition and cancer: directions for research, in *Diet, Nutrition and Cancer: A Critical Evaluation,* Vol. II, Reddy, B. S. and Cohen, L. A., Eds., CRC Press, Boca Raton, FL 1986, 161.

22. **Tomatis, L.,** Diet, nutrition and cancer: concluding remarks and future perspectives, in *Diet, Nutrition and Cancer,* Hayashi, Y., Nagao, M., Sugimura, T., Takayama, S., Tomatis, L., Wattenberg, L. W., and Wogan, G. N., Eds., Japan Scientific Societies Press, Tokyo, Japan/VNU Science Press BV, Utrecht, The Netherlands, 1986, 325.

23. **De Flora, S., Ed.,** *Role and Mechanisms of Inhibitors in Prevention of Mutation and Cancer, Mutat. Res.,* Vol. 202, No. 2, 1988.

24. **De Flora, S. and Ramel, C.,** Mechanisms of inhibitors of mutagenesis and carcinogenesis. Classification and overview, *Mutat. Res.,* 202, 285, 1988.

25. **Bartsch, H., Ohshima, H., Nair, J., and Pignatelli, B.,** Modifiers of endogenous nitrosamine synthesis and metabolism, in *Antimutagenesis and Anticarcinogenesis Mechanisms,* Shankel, D. M., Hartman, P. E., Kada, T., and Hollaender, A., Eds., Plenum Press, New York, 1986, 87.

26. **Bartsch, H., Ohshima, H., and Pignatelli, B.,** Inhibitors of endogenous nitrosation. Mechanisms and implications in human cancer prevention, *Mutat. Res.,* 202, 307, 1988.

27. **Mirvish, S. S.,** Inhibition of the formation of carcinogenic *N*-nitroso compounds by ascorbic acid and other compounds, in *Cancer 1980: Achievements, Challenge, Projects,* Vol. 1, Burchenal, J. H. and Oettgen, H. P., Eds., Grune and Stratton, New York, 1981, 557.

28. **Hartman, P. E.,** Nitrates and nitrites: ingestion, pharmacodynamics, and toxicology, in *Chemical Mutagens,* Vol. 7, de Serres, F. J. and Hollaender, A., Eds., Plenum Press, New York, 1982, 211.

29. **Gichner, T. and Veleminsky, J.,** Inhibitors of *N*-nitroso compounds-induced mutagenicity, *Mutat. Res.,* 195, 21, 1988.

30. **Wakabayashi, K., Nagao, M., Ochiai, M., Fijita, Y., Tahira, T., Nakayasu, M., Ohgati, H., Takayama, S., and Sugimura, T.,** Recently identified nitrite-reactive compounds in food: occurrence and biological properties of the nitrosated products, in *Relevance of N-Nitroso Compounds to Human Cancer: Exposure and Mechanisms,* Bartsch, H., O'Neill, I. K., and Schulte-Hermann, R., Eds., IARC Sci. Publ. No. 84, International Agency for Research on Cancer, Lyon, 1987, 287.

31. **De Flora, S., Cesarone, C. F., Bennicelli, C., Camoirano, A., Serra, D., Bagnasco, M., Scovassi, A. I., Scarabelli, L., and Bertazzoni, U.,** Antigenotoxic and anticarcinogenic effects of thiols. *In vitro* inhibition of the mutagenicity of drug nitrosation products and protection of rat liver ADP-ribosyl transferase activity, in *Chemical Carcinogenesis: Models and Mechanisms,* Feo, F., Pani, P., Columbano, A., and Garcea, R., Eds., Plenum Press, New York, 1988, 75.

32. **Jongen, W. M. F., van Boekel, M. A. J. S., and van Broekhoven, L. W.,** Inhibitory effect of cheese and some food constituents on mutagenicity generated in *Vicia faba* after treatment with nitrite, *Food Chem. Toxicol.,* 25, 141, 1987.

33. **De Flora, S. Picciotto, A. Savarino, V., Bennicelli, C., Camoirano, A., Garibotto, G., and Celle, G.,** Circadian monitoring of gastric juice mutagenecity, *Mutagenesis,* 2, 115, 1987.

34. **Tsuda, M., Takahashi, Y., Nagao, M., Hirayama, T., and Sugimura, T.,** Inactivation of mutagens from pyrolysates of tryptophan and glutamic acid by nitrite in acidic solution, *Mutat. Res.,* 78, 331, 1980.

35. **De Flora, S. and De Flora, A.,** Variation of the frameshift activity of a mutagen (ICR 191) following nitrosation in human gastric juice, *Cancer Lett.,* 11, 185, 1981.

36. **Kobayashi, A. and Matsumoto, H.,** Studies on methylazoxymethanol, the aglycone of cycasin, *Arch. Biochem.,* 110, 373, 1965.

37. **Gupta, I., Suzuki, K., Bruce, W. R., Krepinsky, J. J., and Yates, P.,** A model study of fecapentenes: mutagens of bacterial origin with alkylating properties, *Science,* 225, 521, 1984.

38. **Alldrick, A. J. and Rowland, I. R.,** Counteraction of the genotoxicity of some cooked-food mutagens by biogenic amines, *Food Chem. Toxicol.,* 25, 575, 1987.

39. **Batzinger, R., Bueding, E., Reddy, B., and Weisburger, J. H.,** Formation of a mutagenic drug metabolite by intestinal microorganisms, *Cancer Res.,* 38, 608, 1978.

40. **Rao, D. R., Pulusani, S. R., and Chawan, C. B.,** Natural inhibitors of carcinogenesis: fermented milk products, in *Diet, Nutrition and Cancer: A Critical Evaluation,* Vol. II, Reddy, B. S. and Cohen, L. A., Eds., CRC Press, Boca Raton, FL, 1986, 63.

41. **Bruce, W. R., Bird, R. P., and Rafter, J. J.,** The effect of calcium on the pathogenicity of high fat diets to the colon, in *Diet, Nutrition and Cancer,* Hayashi, Y., Nagao, M., Sugimura, T., Takayama, S., Tomatis, L., Wattenberg, L. W., and Wogan, G. N., Eds., Japan Scientific Societies Press, Tokyo, Japan/ VNU Science Press BV, Utrecht, The Netherlands, 1986, 291.

42. **Silbart, L. K. and Keren, D. F.,** Reduction of intestinal absorption by carcinogen-specific secretory immunity, *Science,* 243, 1462, 1989.

43. **Newmark, H. L.,** A hypothesis for dietary components as blocking agents of chemical carcinogenesis: plant phenolics and pyrrole pigments, *Nutri. Cancer,* 6, 58, 1984.

44. **Kim S. B., Hayase, F., and Kato, H.,** Desmutagenic effect of α-dicarbonyl and α-hydroxycarbonyl compounds against mutagenic heterocyclic amines, *Mutat. Res.,* 177, 9, 1987.

45. **Arimoto, S., Ohara, Y., Namba, T., Negishi, T., and Hayatsu, H.,** Inhibition of the mutagenicity of amino acid pyrolysis products by hemin and other biological pyrrole pigments, *Biochem. Biophys, Res. Commun.,* 92, 662, 1980.

46. **Arimoto, S., Negishi, T., and Hayatsu, H.,** Inhibitory effect of hemin on the mutagenic activities of carcinogens, *Cancer Lett.,* 11, 29, 1980.

47. **Hayatsu, H., Oka, T., Wakata, A., Ohara, Y., Hayatsu, T., Kobayashi, H., and Arimoto, S.,** Adsorption of mutagens to cotton bearing covalently bound trisulfo-copper-phthalocyanine, *Mutat. Res.,* 119, 233, 1983.

48. **Hayatsu, H., Kobayashi, H., Michi-ue, A., and Arimoto, S.,** Affinity of aromatic compounds having three fused rings to copper phthalocyanine trisulfonate, *Chem. Pharm. Bull.,* 34, 944, 1986.

49. **Arimoto, S and Hayatsu, H.,** Role of hemin in the inhibition of mutagenic activity of 3-amino-l-methyl-5H-pyrido[4,3b]indole (Trp-P-2) and other aminoazaarenes, *Mutat. Res.,* 213, 217, 1989.

50. **Lai, C. N., Burler, M. A., and Matney, T. S.,** Antimutagenic activities of common vegetables and their chlorophyll contents, *Mutat. Res.,* 77, 245, 1980.

51. **Tarwell, L. and van der Hoeven, C. M.,** Antimutagenic activity of some naturally occurring compounds towards cigarette-smoke condensate and benzo(a)pyrene in Salmonella/microsome assay, *Mutat. Res.,* 152, 1, 1985.

52. **Ong, T. M., Whang, W. Z., Stewart, J., and Brockman, H. E.,** Chlorophyllin: a potent antimutagen against environmental and dietary complex mixtures, *Mutat. Res.,* 173, 111, 1986.

53. **Bronzetti, G., Della Croce, C., and Galli, A.,** Chlorophyllin as protector against X-rays, in *3rd Int. Conf. Anticarcinogenesis and Radiation Protection,* Dubrovnik, Yugoslavia, October 15 to 21, 1989.

54. **Negishi, T., Arimoto, S., Nishizaki,C., and Hayatsu, H.,** Inhibitory effect of chlorophyll on the genotoxicity of 3-amino-l-methyl-5H-pyrido[4,3b]indole (Trp-P-2), *Carcinogenesis,* 10, 145, 1989.

55. **Hayatsu, H., Inoue, K., Ohta, H., Namba, T., Togawa, K., Hayatsu, T., Makita, M., and Wataya, Y.,** Inhibition of the mutagenicity of cooked-beef basic fraction by its acidic fraction, *Mutat. Res.,* 91, 437, 1981.

56. **Hayatsu, H., Hamasaki, K., Togawa, K., Arimoto, S., and Negishi, T.,** Antimutagenic activity in extracts of human feces, in *Carcinogens and Mutagens in the Environment, Naturally Occurring Compounds,* Vol. 2, Stich, H. F., Ed., CRC Press, Boca Raton, FL 1983, 91.

57. **Brown, J. P.,** A review of the genetic effects on naturally occurring flavonoids, anthraquinones and related compounds, *Mutat. Res.,* 75, 243, 1980.

58. **Glazer, A. N.,** Fluorescence-based assay for reactive oxygen species: a protective role for creatinine, *FASEB J.,* 2, 2487, 1988.

59. **Burkitt, D. P.,** Colonic-rectal cancer: fiber and other dietary factors, *Am. J. Clin. Nutr.,* 31, 58s, 1978.

60. **Kritchewsky, D.,** Fiber, steroids, and cancer, *Cancer Res.,* 43, 2491s, 1983.

61. **Bingham, S. A.,** Non-starch polysaccharides as a protective factor in human large bowel cancer, in *Diet, Nutrition and Cancer,* Hayashi, Y., Nagao, M., Sugimura, T., Takayama, S., Tomatis, L., Wattenberg, L. W., and Wogan, G. N., Eds., Japan Scientific Societies Press, Tokyo, Japan/VNU Science Press BV, Utrecht, The Netherlands, 1986, 183.

62. **Reddy, B. S., Sharma, C., Simi, B., Engle, A., Laasko, K., Puska, P., and Korpela, R.,** Metabolic epidemiology of colon cancer: effect of dietary fiber on fecal mutagens and bile acids in healthy subjects, *Cancer Res.,* 47, 644, 1987.

63. **Kern, F., Birkner, H. J., and Ostrower, V. S.,** Binding of bile acids by dietary fiber, *Am. J. Clin. Nutr.,* 31, 175s, 1978.

64. **Barnes, W. S., Maiello, J., and Weisburger, J. H.,** *In vitro* binding of the food mutagen 2-amino-3-methylimidazole(*e*,5-*f*)quinoline to dietary fibers, *J. Nat. Cancer Inst.,* 70, 757, 1983.

65. **Kada, T., Kato, M., Aikawa, K., and Kiriyama, S.,** Adsorption of yurolysate mutagens by vegetable fibers, *Mutat. Res.,* 141, 149, 1984.

66. **Reddy, B. S.,** Diet and colon cancer: evidence from human and animal model studies, in *Diet, Nutrition and Cancer: A Critical Evaluation,* Vol. I, Reddy, B. S. and Cohen, L. A., Eds., CRC Press, Boca Raton, FL, 1986, 47.

67. **Correa, P., Haenszel, W., Cuello, C., Archer, M., and Tannenbaum, S.,** A model for gastric cancer epidemiology, *Lancet,* 2, 58, 1975.

68. **De Flora, S. and Boido, V.,** Effect of human gastric juice on the mutagenicity of chemicals, *Mutat. Res.,* 77, 307, 1980.

69. **De Flora, S., Badolati, G. S., Serra, D., Picciotto, A., Magnolia, M. R., and Savarino, V.,** Circadian reduction of chromium in the gastric environment, *Mutat. Res.,* 192, 169, 1987.

70. **Owens, R. A. and Hartman, P. E.,** Glutathione: a protective agent in *Salmonella typhimurium* and *Escherichia coli* as measured by mutagenicity and by growth delay assays, *Environ. Mutagen.,* 8, 659, 1986.

71. **Camoirano, A., Badolati, G. S., Zanacchi, P., Bagnasco, M., and De Flora, S.,** Dual role of thiols in *N*-methyl-*N*-nitro-*N*-nitrosoguanidine genotoxicity, *Exp. Oncol. (Life Sci. Adv.),* 7, 21, 1988.

72. **Margison, G. P. and O'Connor, P. J.,** Nucleic acid modification by *N*-nitroso compounds, in *Chemical Carcinogens and DNA,* Vol. I, Grover, P. L., Ed., CRC Press, Boca Raton, FL, 1979, 111.

73. **Romert, L. and Jenssen, D.,** Mechanism of *N*-acetylcysteine (NAC) and other thiols as both positive and negative modifiers of MNNG-induced mutagenicity in V79 Chinese hamster cells, *Carcinogenesis,* 8, 1531, 1987.

74. **Anderson, M. E. and Meister, A.,** Dynamic state of glutathione in blood plasma, *J. Biol. Chem.,* 255, 9530, 1980.

75. **Wattenberg, L. W., Hochalter, J. B., Prabhu, U. D. G., and Galbraith, A. R.,** Nucleophiles as anticarcinogens, in *Anticarcinogenesis and Radiation Protection,* Cerutti, P., Nygaard, O. F., and Simic, M. G., Eds., Plenum Press, New York, 1987, 233.

76. **Stocker, R., Yamamoto, Y., McDonagh, A. F., Glazer, A. N., and Ames, B. N.,** Bilirubin is an antioxidant of possible physiological importance, *Science,* 235, 1043, 1987.

77. **Ames, B. N., Cathcart, R., Schwiers, E., and Hochstein, P.,** Uric acid provides an antioxidant defense in humans against oxidant- and radical-caused aging and cancer: a hypothesis, *Proc. Natl. Acad. Sci. U.S.A.,* 78, 6858, 1981.

78. **Hochstein, P., Hatch, L., and Sevanian, A.,** Uric acid: functions and determination, *Methods Enzymol.,* 105, 162, 1984.

79. **Nishioka, H. and Nunoshiba, T.,** Role of enzymes in antimutagenesis of human saliva and serum, in *Antimutagenesis and Anticarcinogenic Mechanisms,* Shankel, D. M., Hartman, P. E., Kada, T., and Hollander, A., Eds., Plenum Press, New York, 1986, 143.

80. **Kada, T., Morita, K., and Inoue, T.,** Anti-mutagenic action of vegetable factor(s) on the mutagenic principle of tryptophan pyrolysate, *Mutat. Res.,* 53, 351, 1978.

81. **Inoue, T., Morita, K., and Kada, T.,** Purification and properties of a plant desmutagenic factor for the mutagenic principle of tryptophan pyrolysates, *Agric. Biol. Chem.,* 45, 345, 1981.

82. **Morita, K., Yamada, H., Iwamoto, S., Sotomura, M., and Suzuki, A.,** Purification and properties of desmutagenic factor from broccoli (*Brassica oleracea* var. *italica plenck*), *J. Food Safety,* 4, 139, 1982.

83. **Dolara, P., Caderni, G., and Lodovici, M.,** Inactivation of 2-amino-3-methyl-imidazo(4,5-*f*)quinoline by horse radish and intestinal peroxidase, *Arch. Toxicol.,* 7, (Suppl.) 253, 1984.

84. **Brooke-Taylor, S., Smith L. L., and Cohen, G. M.,** The accumulation of polyamines and paraquat by human peripheral lung, *Biochem. Pharmacol.,* 32, 717, 1983.

85. **Ames, G. F.,** Uptake of amino acids by *Salmonella typhimurium, Arch. Biochem. Biophys.,* 18, 1, 1964.

86. **Negishi, T. and Hayatsu, H.,** Inhibitory effect of saturated fatty acids on the mutagenicity of *N*-nitrosodimethylamine, *Mutat. Res.,* 135, 87, 1984.

87. **Guevara, A., Sylianco, C., Dayrit, F., and Finch, P.,** Acylglucosylsterols, antimutagens from *Momordica charantia,* in Abstr. 2nd Int. Conf. Mechanisms of Antimutagenesis and Anticarcinogenesis, Ohito, Japan, December 4 to 9, 1988, 58.

88. **De Flora, S. and Wetterhahn, K. E.,** Mechanisms of chromium metabolism and genotoxicity, *Life Chem. Rep.,* 7, 169, 1989.

89. **De Flora, S., Bennicelli, C., Camoirano, A., Serra, D., Romano, M., Rossi, G. A., Morelli, A., and De Flora, A.,** *In vivo* effects of *N*-acetylcysteine on glutathione metabolism and the biotransformation of carcinogenic and/or mutagenic compounds, *Carcinogenesis,* 6, 1735, 1985.

90. **De Flora, S., Bennicelli, C., Zanacchi, P., Basso, C., and Badolati, G. S.,** Metabolic deactivation of genotoxic chemicals as a phenomenon limiting the reliability of in vitro tests, in *Short-Term Tests for Genotoxicity,* Loprieno, N. and Pantarotto, C. A., Eds., Plenum Press, New York, in press.

91. **Schmähl, D., Kruger, F. W., Habs, M., and Diehl, B.,** Influence of disulfiram on the organotropy of the carcinogenic effect of dimethylnitrosamine in rats, *Z. Krebsforsch.* 85, 271, 1976.

92. **Wattenberg, L. W.,** Inhibitors of chemical carcinogens, in *Cancer: Achievement, Challenge and Prospects for the 1980's,* Burchenal, J. H. and Oettgen, H. F., Eds., Grune and Stratton, New York, 1981, 517.

93. **Netter, K. J.,** Inhibition of oxidative drug metabolism in microsomes, in *Hepatic Cytochrome P-450 Monooxygenase System,* Sckenkman, J. B. and Kupfer, D., Eds., Pergamon Press, Oxford, 1981, 741.

94. **Caderni, G., Lodovici, M., Salvadori, M., Bianchini, F., and Dolara, P.,** Inhibition of the mutagenic activity of some heterocyclic dietary carcinogens and other mutagenic/carcinogenic compounds by rat organ preparations, *Mutat. Res.,* 169, 35, 1986.

95. **Wattenberg, L. W., Loub, W. D., Lam, L. K., and Speier, J. L.,** Dietary constituents altering the responses to chemical carcinogens, *Fed. Proc. Fed. Am. Soc. Exp. Biol.,* 35, 1327, 1976.

96. **Sparnins, V. L., Venegas, P. L., and Wattenberg, L. W.,** Glutathione S-transferase activity: enhancement by compounds inhibiting chemical carcinogens and by dietary constituents, *J. Natl. Cancer Inst.,* 68, 493, 1982.

97. **Birt, D. F., Pelling, J. C., Pour, P. M., Tibbels, M. G., Schweickert, L., and Bresnick, E.,** Enhanced pancreatic and skin tumorigenesis in cabbage-fed hamsters and mice, *Carcinogenesis,* 8, 913, 1987.

98. **Ramel, C., Alekperov, U. K., Ames, B. N., Kada, T., and Wattenberg, L. W.,** Inhibitors of mutagenesis and their relevance to carcinogenesis, *Mutat. Res.* 168, 47, 1986.

99. **Jongen, W. M. F., Topp, R. J., Tiedink, H. G. M., and Brink, E. J.,** A co-cultivation system as model for *in vitro* studies of modulating effects of naturally occurring indoles in the genotoxicity of model compounds, *Toxicol. In Vitro,* 1, 105, 1987.

100. **Wattenberg, L. W., Hanley, A. B., Barany, G., Sparnins, V. L., Lam, L. K. T., and Fenwick, G. R.,** Inhibition of carcinogenesis by some minor dietary constituents, in *Diet, Nutrition and Cancer,* Hayashi, Y., Nagao, M., Sugimura, T., Takayama, S., Tomatis, L., Wattenberg, L. W., and Wogan, G. N.. Eds., Japan Scientific Societies Press, Tokyo, Japan/VNU Science Press BV, Utrecht, The Netherlands, 1986, 193.

101. **Dashwood, R. H., Arbogast, D. N., Fong, A. T., Hendricks, J. D., and Bailey, G. S.,** Mechanisms of anti-carcinogenesis by indole-3-carbinol: detailed *in vivo* DNA binding dose-response studies after dietary administration with aflatoxin B_1, *Carcinogenesis,* 9, 427, 1988.

102. **Speier, J. L., Lam, L. K. T., and Wattenberg, L. W.,** Effects of administration to mice of butylated hydroxyanisole by oral intubation on benzo(a)pyrene-induced pulmonary adenoma formation and metabolism of benzo(a)pyrene, *J. Natl. Cancer Inst.,* 60, 605, 1978.

103. **Del Tito, B. J., Jr., Mukhtar, H., and Bickers, D. R.,** Inhibition of epidermal metabolism and DNA-binding of benzo(a)pyrene by ellagic acid, *Biochem. Biophys, Res. Commun.,* 114, 388, 1983.

104. **San, R. H. C. and Chan, R. I. M.,** Inhibitory effect of phenolic compounds on aflatoxin B_1 metabolism and induced mutagenesis, *Mutat. Res.,* 177, 229, 1987.

105. **Ayrton, A. D., Ioannides, C., and Walker, R.,** Anthraflavic acid inhibits the mutagenicity of the food mutagen IQ: mechanism of action, *Mutat. Res.,* 207, 121, 1988.

106. **Gould, M. N., Elson, C. E., Maltzman, T. H., Wacker, W. D., and Crowell, P. L.,** Inhibition of carcinogenesis by monoterpenoids, in *Carcinogens and Mutagens in the Diet,* Pariza, M. W., Aeschbucher, H.-U., Felton, J. S., and Sato, S., Eds., Wiley-Liss, 1990, 255.

107. **Hirayama, T., Ogawa, S., Kumo, S., Hanasaki, Y., Yamada, T., and Fukui, S.,** Desmutagenic effect of spices on amino acid pyrolysates in *Salmonella typhimurium* mutagenicity assay, in *Carcinogens and Mutagens in the Diet,* Madison, WI, July 5 to 8, 1989.

108. **Hill, D. L. and Shih, T. W.,** Vitamin A compounds and analogs as inhibitors of mixed-function oxidases that metabolize carcinogenic polycyclic hydrocarbons and other compounds, *Cancer Res.,* 34, 564, 1974.

109. **Busk, L., Sjögren, B., and Ahlborg, U. G.,** The effect of vitamin A on cyclophosphamide induced mutagenicity both *in vitro* and *in vivo* using Ames test and the micronucleus test in mice, *Food Chem. Toxicol.,* 22, 725, 1984.

110. **Victorin, K., Busk, L., and Ahlborg, U. G.,** Retinol (vitamin A) inhibits the mutagenicity of o-aminoazotoluene activated by liver microsomes from several species in the Ames test, *Mutat. Res.,* 179, 41, 1987.

111. **Welsh, C. W., Zile, M. H., and Cullum, M. E.,** Retinoids and mammary gland tumorigenesis: a critique, in *Diet, Nutrition and Cancer: A Critical Evaluation,* Vol. II, Reddy, B. S. and Cohen, L. A., Eds., CRC Press, Boca Raton, FL 1986, 1.

112. **Hirayama, T., Ogawa, S., Watanabe, T., and Fukui, S.,** Nicotinamide as a scavenger of alkylating agents, in *Carcinogens and Mutagens in the Diet,* Madison, WI, July 5 to 8, 1989.

113. **Narurkar, L. M., Jagtiana, S. K., Kamat, J. P., Krishnamoorthy, R. R., D'Souza, S. J., and Narurkar, M. V.,** Inhibition of diethylnitrosamine-induced hepatocarcinogenesis in rats by nicotinamide, in *Carcinogens and Mutagens in the Diet,* Madison, WI, July 5 to 8, 1989.

114. **Reddy, T. V., Ramanathan, R., Shinozuka, H., and Lombardi, B.,** Effects of dietary choline deficiency on the mutagenic activation of chemical carcinogens by rat liver fractions, *Cancer Lett.,* 18, 41, 1983.

115. **Ketterer, B.,** Protective role of glutathione and glutathione transferases in mutagenesis and carcinogenesis, *Mutat. Res.,* 202, 343, 1988.

116. **Hochstein, P. and Atallah, A. S.,** The nature of oxidants and antioxidant systems in the inhibition of mutation and cancer, *Mutat. Res.* 202, 363, 1988.

117. **Shamberger, R. J.,** Chemoprevention of cancer, in *Diet, Nutrition and Cancer: A Critical Evaluation,* Vol. II, Reddy, B. S. and Cohen, L. A., Eds., CRC Press, Boca Raton, FL, 1986, 43.

118. **Malkinson, A. M.,** Multiple modulatory effects of butylated hydroxytoluene on tumorigenesis, *Cancer Invest.,* 3, 209, 1985.

119. **Lam, L. K. and Wattenberg, L. W.,** Effects of butylated hydroxyanisole on the metabolism of benzo(a)pyrene by mouse liver microsomes, *J. Natl. Cancer Inst.,* 58, 413, 1977.

120. **Wattenberg, L. W.,** Inhibitors of chemical carcinogens, *J. Environ. Pathol. Toxicol.,* 3, 35, 1980.

121. **Talalay, P. and Prochaska, H. J.,** Mechanisms of induction of NADP(P)H: quinone reductase, *Chem. Scripta,* 27A, 61, 1987.

122. **Friedman, M. and Gumbmann, M. R.,** Nutritional improvement of legume proteins through disulfide interchange, in *Nutritional and Toxicological Significance of Enzyme Inhibitors in Foods,* Friedman, M., Ed., Plenum Press, New York, 1986, 357.

123. **De Flora, S., Bennicelli, C., Serra, D., Izzotti, A., and Cesarone, C. F.,** Role of glutathione and N-acetylcysteine as inhibitors of mutagenesis and carcinogenesis, in *Absorption and Utilization of Amino Acids,* Vol. III, Friedman, M., Ed., CRC Press, Boca Raton, FL, 1989, 19.

124. **De Flora, S., Bennicelli, C., D'Agostini, F., and Camoirano, A.,** Generation of mutagenic oxygen species in bacteria and their enzymatic or chemical inhibition, *Mutat. Res.,* 214, 153, 1989.

125. **Kensler, T. W., Egner, P. A., Trush, M. A., Bueding, E., and Groopman, M. D.,** Modification of aflatoxin B$_1$ binding to DNA in vivo in rats fed phenolic antioxidants and dithiolthione, *Carcinogenesis,* 6, 759, 1985.

126. **Ansher, S. S., Dolan, P., and Bueding, E.,** Biochemical effects of dithiolthiones, *Food Chem. Toxicol.,* 24, 405, 1986.

127. **Wattenberg, L. W., Hochalter, J. B., and Galbraith, A. R.,** Inhibition of β-propiolactone-induced mutagenesis and neoplasia by sodium thiosulfate, *Cancer Res.,* 47, 4351, 1987.

128. **You, W. -C., Blot, W. J., Chang, Y. -S., Ershow, A., Yang, Z. T., An, Q., Henderson, B. E., Fraumeni, J. F., Jr., and Wang, T. -G.,** Allium vegetables and reduced risk of stomach cancer, *J. Natl. Cancer Inst.,* 81, 162, 1989.

129. **Weisberger, A. S. and Pensky, J.,** Tumor inhibition by a sulfhydryl-blocking agent related to an active principle of garlic, *Cancer Res.,* 18, 1301, 1958.

130. **Goldberg, M. T. and Josephy, P. D.,** Studies on the mechanism of action of diallyl sulfide, an inhibitor of the genotoxic effects of cylcophsophamide, *Can. J. Physiol. Pharmacol.,* 65, 467, 1987.

131. **Sayer, J. M., Yagi, H., Wood, A. W., Conney, A. H., and Jerina, D. M.,** Extremely facile reaction between the ultimate carcinogen benzo(a)pyrene-7,8-diol 9,10-epoxide and ellagic acid, *J. Am. Chem. Soc.,* 104, 5562, 1982.

132. **Nagabhushan, M. and Bhide, S. V.,** Anti-mutagenicity of catechin against environmental mutagens, *Mutagenesis,* 3, 293, 1988.

133. **Sugimura, T.,** Past, present and future of mutagens in cooked food, *Environ, Health Perspect.,* 67, 5, 1986.

134. **Wood, A. W., Sayer, J. M., Newmark, H. L., Yagi, H., Michaud, D. P., Jerina, D. M., and Conney, A. H.,** Mechanism of the inhibition of mutagenicity of benzo(a)pyrene 7,8-diol 9,10-epixode by riboflavin 5'-phosphate, *Proc. Natl. Acad. Sci. U. S. A.,* 79, 5122, 1982.

135. **Simic, M. G.,** Mechanisms of inhibition of free-radical processes in mutagenesis and carcinogenesis, *Mutat. Res.,* 202, 377, 1988.

136. **Bertram, J. S., Kolonel, L. N., and Meyskens, F. L., Jr.,** Rationale and strategies for chemoprevention of cancer in humans, *Cancer Res.,* 47, 3012, 1987.

137. **Hill, D. L. and Grubbs, C. J.,** Retinoids as chemopreventive and anticancer agents in intact animals (review), *Anticancer Res.,* 2, 111, 1982.

138. **Sporn, M. B. and Roberts, A. B.,** Role of retinoids in differentiation and carcinogenesis, *Cancer Res.,* 43, 3034, 1983.

139. **Moon, R. C. and Itri, L. M.,** Retinoids and cancer, in *The Retinoids,* Vol. 2, Sporn, M. B., Roberts, A. B., and Goodman, D. S, Eds., Academic Press, New York, 1984, 327.

140. **Willett, W.,** Vitamin A and selenium intake in relation to human cancer risk, in *Diet, Nutrition and Cancer,* Hayashi, Y., Nagao, M., Sugimura, T., Takayama, S., Tomatis, L., Wattenberg, L. W., and Wogan, G. N., Eds., Japan Scientific Societies Press, Tokyo, Japan/VNU Science Press BV, Utrecht,, The Netherlands, 1986, 237.

141. **Wattenberg, L. W.,** Inhibition of chemical carcinogens by minor dietary components, in *Molecular Interrelations of Nutrition and Cancer,* Arnott, M. S., van Eys, J., and Wang, Y. M., Eds., Raven Press, New York, 1982, 43.

142. **Chen, L. H., Boissonneault, G. A., and Glauert, H. P.,** Vitamin C, vitamin E and cancer (review), *Anticancer Res.,* 8, 739, 1988.

143. **Martin, S. E. and Schillaci, M.,** Inhibitory effects of selenium on mutagenicity, *J. Agric. Food Chem.,* 32, 426, 1984.

144. **Ip, C. and White, G.,** Mammary cancer chemoprevention by inorganic and organic selenium: single agent treatment or in combination with vitamin E and their effects on *in vitro* immune functions, *Carcinogenesis,* 8, 1763, 1987.

145. **Pariza, M. W. and Ha, L. Y.,** Fatty acids that inhibit cancer: conjugated dienoic derivatives and linoleic acid, in *Carcinogens and Mutagens in the Diet,* Pariza, M. W., Aeschbucher, H.-U., Felton, J. S., and Sato, S., Eds., Wiley-Liss, New York, 1990, 217.

146. **Renner, H. W.,** Anticlastogenic dietary factors assessed in mammalian cells, in *Antimutagenesis and Anticarcinogenesis Mechanisms,* II, Kuroda, Y., Shankel, D. M., and Waters, M. D., Eds., Plenum Press, New York, 1990, 35.

147. **Osawa, T., Namiki, M., and Kawakishi, S.,** Role of dietary antioxidants in protection against oxidative damage, in *Antimutagenesis and Anticarcinogenesis Mechanisms,* II, Kuroda, Y., Shankel, D. M., and Waters, M. D., Eds, Plenum Press, New York, 1990, 139.

148. **Ueno, I., Kohno, M., Haraikawa, K., and Hirono, I.,** Interaction between quercetin and superoxide radicals. Reduction of the quercetin mutagenicity, *J. Pharmacobiodyn.,* 7, 798, 1984.

149. **Frenkel, K., Chrzan, K., Ryan, C. A., Wiesner, R., and Troll, W.,** Chymotrypsin-specific proteasae inhibitors decrease H_2O_2 formation by activated human polymorphonuclear leucocytes, *Carcinogenesis,* 8, 1207, 1987.

150. **Dixit, R. and Gold, B.,** Inhibition of *N*-methyl-*N*-nitrosourea-induced mutagenicity and DNA methylation by ellagic acid, *Proc. Natl. Acad. Sci. U. S. A.,* 83, 8039, 1986.

151. **Jenssen, D.,** Elimination of MNU-induced mutational lesions in V79 Chinese hamster cells, *Mutat. Res.,* 106, 291, 1982.

152. **Kawazoe, Y. and Kato, M.,** Antimutagenic effect of isothiocyanates and related compounds in *Escherichia coli, Gann,* 73, 255, 1982.

153. **Boothman, D. A., Schlegel, R., and Pardee, A.,** Anticarcinogenic potential of DNA-repair modulators, *Mutat. Res.,* 202, 393, 1988.

154. **Kada, T., Inoue, T., and Namiki, N.,** Environmental desmutagens and antimutagens, in *Environmental Mutagenesis and Plant Biology,* Klekowski, E. J., Ed., Praeger, New York, 1982, 137.

155. **Kuroda, Y. and Inoue, T.,** Antimutagenesis by factors affecting DNA repair in bacteria, *Mutat. Res.,* 202, 387, 1988.

156. **Kada, T., Inoue, T., Ohta, T., and Shirasu, Y.,** Antimutagens and their mode of action, in *Antimutagenesis and Anticarcinogenesis Mechanisms,* Shankel, D. M., Hartman, P. E., Kada, T., and Hollaender, A., Eds., Plenum Press, New York, 1986, 181.

157. **Nunoshiba, T. and Nishioka, H.,** Sodium arsenite inhibits spontaneous and induced mutations in *Escherichia coli, Mutat. Res.,* 184, 99, 1987.

158. **de Silva, H. V. and Shankel, D. M.,** Effect of the antimutagen cinnamaldehyde on reversion and survival of selected Salmonella tester strains, *Mutat. Res.,* 187, 11, 1987.

159. **Wattenberg, L. W., Lam, L. K. T., and Fladmoe, A. V.,** Inhibition of chemical carcinogen-induced neoplasia by coumarins and α-angelicalactone, *Cancer Res.* 39, 1651, 1979.

160. **Shimoi, K., Nakamura, Y., Tomita, I., Hara, Y., and Kada, T.,** The pyrogallol related compounds reduce UV-induced mutation in *Escherichia coli* B/r WP2, *Mutat. Res.,* 173, 239, 1986.

161. **Meyn, M. S., Rossman, T., and Troll, W.,** A protease inhibitor blocks SOS functions in *Escherichia coli:* antipain prevents lambda repressor inactivation, ultraviolet mutagenesis and filamentous growth, *Proc. Natl. Acad. Sci. U. S. A.,* 74, 1152, 1977.

162. **Baturay, N. and Kennedy, A. R.,** Pyrene acts as a cocarcinogen with the carcinogens benzo(a)pyrene, beta-propiolactone and radiation in the induction of malignant transformation of cultured mouse fibroblasts. Soybean extract containing the Bowman-Birk inhibitor acts as an anticarcinogen, *Cell. Biol. Toxicol.,* 2, 21, 1986.

163. **Gichner, T. and Velemínsky, J.,** Mechanisms of inhibition of *N*-nitroso compounds-induced mutagenicity, *Mutat. Res.,* 202, 325, 1988.

164. **Vasilieva, S. V., Davnichenko, L. S., and Rapoport, I. A.,** Intensification of DNA repair processes by *para*-aminobenzoic acid in *Escherichia coli* K-12 (in Russian), *Genetika (Moscow),* 19, 1952, 1983.

165. **Kalinina, L. M., Tarasow, V. A., Sardarly, G. M., and Alekperov, U. K.,** The effect of α-tocopherol on the mutagenic repair pathways in *Escherichia coli* (in Russian), *Genetika (Moscow),* 17, 1644, 1981.

166. **Cesarone, C. F., Scovassi, A. I., Scarabelli, L., Izzo, R., Orunesu, M., and Bertazzoni, U.,** Depletion of adenosine diphosphate-ribosyl transferase activity in rat liver during exposure to *N*-2-acetylaminofluorene: effect of thiols, *Cancer Res.,* 48, 3581, 1988.

167. **Friedman, T.,** Progress toward human gene therapy, *Science,* 244, 1225, 1989.

168. **Ramel, C.,** Modulation of genotoxicity, in *5th Int. Conf. Environmental Mutagens,* Mendelsohn, M., Ed., Alan R. Liss, New York, in press.

169. **Weinstein, I. B.,** Strategies for inhibiting multistage carcinogenesis based on signal transduction pathways, *Mutat. Res.,* 202, 413, 1988.

170. **Chang, J. D., Billings, P. C., and Kennedy, A. R.,** C-*myc* expression is reduced in antipain-treated proliferating C3H 10T1/2 cells, *Biochem. Biophys. Res. Commun.,* 133, 830, 1985.

171. **Westin, E. H., Wong-Staal, F., Gelmann, E. P., Dalla Favera, R., Papas, T. S., Lautenberger, J. A., Eva, A., Reddy, E. P., Tronick, S. R., Aaronson, S. A., and Gallo, R. C.,** Expression of cellular homologues of retroviral *onc* genes in human hemopoietic cells, *Proc. Natl. Acad. Sci. U. S. A.,* 79, 2490, 1982.

172. **Campisi, J., Gray, H. E., Pardee, A. B., Dean, M., and Sonenshein, G. E.,** Cell-cycle control of c-*myc* but not c-*ras* expression is lost following chemical transformation, *Cell,* 36, 241, 1984.

173. **Reitsma, P. H., Rothberg, P. G., Astrin, S. M., Trial, J., Bar-Shavit, Z., Hall, A., Teitelbaum, S. L., and Kahn, A. J.,** Regulation of *myc* gene expression in HL-60 leukaemia cells by a vitamin D metabolite, *Nature,* 306, 492, 1983.

174. **Feo, F., Garcea, R., Daino, L., Pascale, R., Frassetto, S., Cozzolino, P., Vannini, M. G., Ruggiu, M. E., Simile, M. M., and Puddu, M.,** S-Adenosylmethionine antipromotion and antiprogression effect in hepatocarcinogenesis. Its associations with inhibition of gene expression, in *Chemical Carcinogenesis: Models and Mechanisms,* Feo, F., Pani, P., Columbano, A., and Garcea, R., Eds., Plenum Press, New York, 1988, 407.

175. **Sporn, M. B. and Roberts, A. B.,** Suppression of carcinogenesis by retinoids: interactions with peptide growth factors and their receptors as a key mechanism, in *Diet, Nutrition and Cancer,* Hayashi, Y., Nagao, M., Sugimura, T., Takayama, S., Tomatis, L., Wattenberg, L. W., and Wogan, G. N., Eds., Japan Scientific Societies Press, Tokyo, Japan/VNU Science Press BV, Utrecht, The Netherlands, 1986, 149.

176. **Jetten, A. M.,** Retinoids and their modulation of cell growth, in *Growth and Maturation Factors,* Vol. 3, Guroff, G., Ed., John Wiley & Sons, New York, 1985, 252.

177. **Schafer, W. R., Kim, R., Sterne, T., Thorner, J., Kim, S. -H., and Rine, J.,** Genetic and pharmacological suppression of oncogenic mutations in *ras* genes of yeast and humans, *Science,* 245, 379, 1989.

178. **Ramel, C.,** Deployment of short-term assays for the detection of carcinogens: genetic and molecular considerations, *Mutat. Res.,* 168, 327, 1986.

179. **Klein, G. and Klein, E.,** Oncogene activation and tumor progression, *Carcinogenesis,* 5, 429, 1984.

180. **Bishop, J. M.,** The molecular genetics of cancer, *Science,* 235, 305, 1987.

181. **Kennedy, A. R. and Billings, P. C.,** Anticarcinogenic actions of protease inhibitors, in *Anticarcinogenesis and Radiation Protection,* Cerutti, P., Nygaard, O. F., and Simic, M. G., Eds., Plenum Press, New York, 1987, 285.

182. **Gschwendt, M., Horn, F., Kittstein, W., Furstenberger, G., Besemfelder, L., and Marks, F.,** Calcium and phospholipid dependent protein kinase activity in mouse epidermis cytosol. Stimulation by complete and incomplete tumor promotors and inhibition by various compounds, *Biochem. Biophys. Res. Commun.,* 124, 63, 1984.

183. **Lok, E., Nera, E. A., Iverson, F., Scott, F., So, Y., and Clayson, D. B.,** Dietary restriction, cell proliferation and carcinogenesis: a preliminary study, *Cancer Lett.,* 38, 249, 1988.

184. **Hayes, M. A., Rushmore, T. H., and Goldberg, M. T.,** Inhibition of hepatocarcinogenic responses to 1,2-dimethylhydrazine by diallyl sulfide, a component of garlic oil, *Carcinogenesis,* 8, 1155, 1987.

185. **Cerutti, P. A.,** Prooxidant states and tumor production, *Science,* 227, 375, 1985.

186. **Ames, B. N.,** Carcinogens and anticarcinogens, in *Antimutagenesis and Anticarcinogenesis Mechanisms,* Shankel, D. M., Hartman, P. E., Kada, T., and Hollaender, A., Eds., Plenum Press, New York, 1986, 7.

187. **Pryor, W. A.,** Cancer and free radicals, in *Antimutagenesis and Anticarcinogenesis Mechanisms,* Shankel, D. M., Hartman, P. E., Kada, T., and Hollaender, A., Eds., Plenum Press, New York, 1986, 45.

188. **Rotstein, J. B. and Slaga, T. J.,** Anticarcinogenesis mechanisms, as evaluated in the multistage mouse skin model, *Mutat. Res.,* 202, 421, 1988.

189. **Shamberger, R. J.,** Inhibitory effect of vitamin A on carcinogenesis, *J. Natl. Cancer Inst.,* 47, 667, 1971.

190. **Hibasami, H., Tsukada, T., Maekawa, S., and Nakashima, K.,** Antitumor effect of methylglyoxal (butylamidino hydrazone), a new inhibitor of S-adenosylmethionine decarboxylase, against human erythroid leukemia K562 cells, *Cancer Lett.,* 30, 17, 1986.

191. **Grunderberger, D., Banerjee, R., Eisinger, K., Oltz, E. M., Efros, L., Caldwell, M., Estevez, V., and Nakanishi, K.,** Preferential cytotoxicity on tumor cells by caffeic acid phenethyl ester isolated from propolis, *Experientia,* 44, 230, 1988.

192. **Billings, P. C., St Clair, W., Rayan, C. A., and Kennedy, A. R.,** Inhibition of radiation-induced transformation of C3H/10T1/2 by chymotrypsin inhibitor 1 from potatoes, *Carcinogenesis,* 8, 809, 1987.

193. **Newberne, P. M. and Rogers, A. E.,** The role of nutrients in cancer causation, in *Diet, Nutrition and Cancer,* Hayashi, Y., Nagao, M., Sugimura, T., Takayama, S., Tomatis, L., Wattenberg, L. W., and Wogan, G. N., Eds., Japan Scientific Societies Press, Tokyo, Japan/VNU Science Press BV, Utrecht, The Netherlands, 1986, 205.

194. **Minoura, T., Takata, T., Sakaguchi, M., Takada, H., Yamamura, M., Hioki, K., and Yamamoto, M.,** Effect of dietary eicosapentaenoic acid on azoxymethane-induced colon carcinogenesis in rats, *Cancer Res.,* 48, 4790, 1988.

195. **Elson, C. E., Maltzman, T. H., Boston, J. L., Tanner, M. A., and Gould, M. N.,** Anti-carcinogenic activity of *d*-limonene during the initiation and promotion/progression stages of DMBA-induced rat mammary carcinogenesis, *Carcinogenesis,* 9, 331, 1988.

Chapter 9

MUTAGENIC AND ANTIMUTAGENIC COMPOUNDS IN BEVERAGES

Hans-Ulrich Aeschbacher

TABLE OF CONTENTS

I. INTRODUCTION

Traditional beverages such as coffee, tea, wine, and beer have been prepared and consumed by mankind for centuries. Recently, however, certain compounds, occurring naturally as components or formed from them during processing such as roasting and fermentation, have been shown to have possible adverse biological effects. Of special concern was the possible existence of compounds in the final beverage which may cause DNA damage. Some classes of compounds induced mutagenic effects in a variety of *in vitro* mutagenicity test systems ranging from microorganisms to mammalian cells in culture. However in the majority of cases the relatively strong mutagenic effect observed *in vitro* could not be confirmed under *in vivo* conditions. In fact, many even showed antimutagenic and anticarcinogenic effects, especially in the case of constituents with antioxidant properties.

II. COFFEE

A. COFFEE AS A WHOLE

Coffee was shown to cause a relatively weak mutagenic effect per se in the Ames bacterial mutagenicity test and also in some *in vitro* mammalian test systems. A number of studies were carried out with different blends and brands both with home brew and instant coffee.[1-14]

Coffee needed to be roasted to cause a mutagenic effect[5,9,12] and hence heat reaction products, some of which are also present in coffee aroma, are suspected of being involved. This was confirmed by the observation that coffee aroma caused mutagenic damage in cultured human lymphocytes[8] and by the fact that aliphatic carbonyl compounds occurring in coffee aroma were positive in the Ames test.[15]

Home brew and instant coffees generally were shown to cause comparable mutagenic activity when expressed on the same basis.[1,3,12,14] No distinct mutagenic effects were observed between various coffee brands[1,5,14] or coffee blends such as different *Arabica* and *Robusta* varieties.[5,9] Caffeine containing and decaffeinated coffee gave comparable results in various *in vitro* test systems,[1,3,5,8,13] hence suggesting that caffeine was not involved in the mutagenicity observed. It has in fact repeatedly been suggested that the interaction of various coffee components is responsible for the direct mutagenicity of coffee.[13,16,17]

Although the mode of preparation of coffee and the test methodology (e.g., direct plating, preincubation) did not significantly alter the results,[1,3] a considerable variation was observed between different laboratories. In fact, no effect was observed in two studies,[4,7] whereas a relatively weak effect was obtained in the majority of studies[1,2,3,5,11,14] with a higher activity in some others.[9,10] In particular, a relatively strong effect was observed with very sensitive bacterial tester strains.[12,13] To obtain a mutagenic effect, very high doses of coffee were usually required ranging between 10 to 60 mg per plate for the sensitive Ames tester strains,[17] whereas for the Ara Salmonella tester strain lower concentrations were sufficient.[13] Despite this, there is general agreement that coffee causes very specific effects at the molecular level. Coffee caused base-pair substitutions but only when the bacterial tester strains contained a plasmid[17] which increased the sensitivity of mutagen detection by enhancing error-prone DNA repair.[18] Another *Salmonella typhymurium* tester strain which affects forward mutations at any of several specific genes in the Ara operon[19] responded even more readily to coffee.[12,13] It has, however, been speculated that effects induced by coffee might be efficiently repaired and that major mutagens in coffee might not have DNA cross-linking activity.[20] This is confirmed by another study using repair-competent *E. coli* which did not respond to coffee.[5] Furthermore, coffee caused only a relatively low mutagenic activity in mammalian cells[6] and in human lymphocytes.[8] Again, relatively high doses of coffee were required to obtain an effect in test systems which were optimized for maximal sensitivity.[17]

It is important to remember that all of the above *in vitro* test systems which gave mutagenic responses were devoid of an optimally functioning "detoxifying enzyme system". When mammalian liver microsomal enzymes were added either to bacterial or mammalian cells in culture, mutagenic activity was decreased.[17]

Evidence exists that active detoxifying enzymes are involved in this deactivation,[3,10] such as glutathione and catalases.[10,13,21] These observations support the hypothesis that reactive coffee compounds can be easily handled *in vivo* by the "detoxifying" enzyme system and so avoid DNA damage. This is confirmed with *Drosophila* which possess a metabolizing system comparable to that of mammals. In *Drosophila* no distinct genotoxic effect, such as point mutations, chromosome aberrations, or nondisjunction, were observed either with instant or home brew coffee.[22] Also, nonmutagens were extracted from the urine of human coffee drinkers, as shown by the negative effect of the extract in the Ames test.[23] Although in another study extracted urine of coffee drinkers was claimed to cause chromosome damage in Chinese hamster cells,[24] these results are difficult to interpret as different coffee fractions gave conflicting results without a meaningful explanation. Yet another study with humans which implicated coffee in chromatid exchanges is also difficult to interpret and the authors disqualify their own work by stating that the design of the study was focused on smokers and hence was inadequate to study the influence of coffee.[25] No induction of sister chromatid exchanges or chromosomal aberrations was observed when rodents were given per os high doses of coffee.[26,27] Hence, it can be concluded that coffee as a whole is readily detoxified by mammalian enzyme systems as demonstrated by negative *in vivo* mutagenicity data. This is also in line with the absence of a carcinogenic activity of coffee observed in rodents.[28,29]

On the other hand, there is recent and accumulating evidence that coffee or coffee components have antimutagenic or anticarcinogenic properties towards environmental compounds. An example is the inhibition by coffee of carcinogenic nitrosamine formation,[30] whereas coffee had no effect on preformed nitrosamine as shown in a micronucleus test.[27] Furthermore, coffee has been shown to inhibit some mutagenic compounds.[31,32] Mutagenicity of coffee is in turn inhibited by sulfite.[33]

B. COFFEE COMPONENTS

1. Methylxanthines

Of the methylxanthines, caffeine is predominantly present in green and roasted coffee and its content varies according to the type of coffee.[34] Based on dry matter, caffeine content of roasted coffee varies between 2.8 and 4.6% in instant coffee and from 1 to 2.5% in roasted coffee beans.[34]

Although caffeine is mutagenic and in particular clastogenic in a variety of *in vitro* test systems,[35] the mutagenicity of coffee in these test systems was not attributed to caffeine.[1,3,5,8,13] In fact, caffeine was not mutagenic in the respective bacterial test systems when tested in its pure form,[17,36,37] nor was it present in sufficient amounts in the coffee doses used in human lymphocyte cultures to exert a chromosome breaking effect.[8] Hence, caffeine at the concentrations ingested in coffee does not seem to present a mutagenic risk to man.[35,38]

2. Phenolic Acids

Phenolic acids, in particular chlorogenic acid and its derivatives (cinnamic, caffeic, ferulic acids) are present in soluble coffee powder at relatively high levels between 3 and 12%.[39]

There are some reports that chlorogenic and caffeic acid cause mutagenic activity in *in vitro* systems[13,40,42] or enhance mutagenicity.[43] There are, on the other hand, an increasing number of reports which show that chlorogenic (caffeic, ferulic) acids have relatively strong antimutagenic properties *in vitro*,[44,45] and *in vivo*[46] and that they can block the formation of mutagenic/carcinogenic nitroso compounds.[47] Green coffee beans[48] rich in phenolic acids and pure chlorogenic acid[49] inhibited chemically induced carcinogenicity.

3. Aliphatic Carbonyl Compounds

As already mentioned, suspect mutagenic compounds are mainly those formed during roasting. Systematic evaluation of such substances[11,13-15,50,51] revealed that aliphatic carbonyl compounds were the main mutagens when tested as pure compounds. Methylglyoxal was analytically detected in coffee at concentrations between 25 and 500 ppb,[14,15,52] glyoxal at 10 to 60 ppb,[14,15,] and diacetyl at 20 to 45 ppb.[14,15]

The mutagenic activity observed in bacterial test systems gave the following ranking in decreasing order of potential: methylglyoxal, glyoxal, diacetyl, maltol.[11,13,15,50] Methylglyoxal was also reported to induce mutations in cultured mammalian cells.[53] While the presence of microsomal enzymes or catalases did not markedly deactivate the mutagenic activity of aromatic carbonyl compounds (methylglyoxal),[14,15,21] it was shown that specific microsomal enzyme fractions[10,21] reduced the mutagenicity of methylglyoxal. Methylglyoxal-induced mutagenicity was also reduced when mammalian cells were co-cultured with freshly isolated rat hepatocytes.[53] Although methylglyoxal was determined as the major factor in *in vitro* mutagenicity and its contribution to coffee mutagenicity was previously estimated at about 50%, several observations led to a revision of this estimation.[17] It was finally concluded that methylglyoxal contributed very little to the direct mutagenicity of coffee[13,21] and therefore aliphatic carbonyl compounds do not seem to be of great importance in coffee mutagenicity.

As a matter of fact, carbonyl compounds and other heat reaction products such as melanoidins which are present in roasted coffee were shown to have antimutagenic properties.[54,55]

4. Hydrogen Peroxide

Hydrogen peroxide (H_2O_2) content is difficult to determine by analytical means in roasted coffee, as reflected by the wide variation in results. H_2O_2 content was on average about 1 to 10 ppb when calculated on the basis of a normally percolated coffee solution containing ca. 1% dry matter.[13-16] This H_2O_2 concentration has been reported to considerably increase with time after fresh percolation,[13,16] although another group observed no increase using two different analytical methods.[15] In coffee samples in which H_2O_2 content was observed to increase with time after percolation, no increase in mutagenicity was observed.[10,15] Furthermore, a large part of coffee mutagenicity was attributed to H_2O_2 in the Ara Salmonella test,[13] whereas H_2O_2 itself causes no mutagenic effect in the standard Ames Strains.[16] The role of hydrogen peroxide in the mutagenic activity of coffee is therefore not yet clear, although it is possible that it causes mutagenicity by interaction with other coffee compounds, e.g., methylglyoxal.[16]

III. TEA

A. TEA AS A WHOLE

Black and green tea when freshly infused or extracted caused a relatively weak direct mutagenic effect and mainly affected "base-pair" bacterial strains,[1,4] This effect was potentiated when test systems with increased sensitivity were used. A relatively strong effect was especially reported for the Ara forward mutation assay.[13,56] The presence of a plasmid and error-prone repair system, which increases sensitivity, was required in bacterial tester strains to obtain a positive effect with tea. This suggests that mutagenic damage induced by tea might be repaired in mammals. Since a preincubation procedure was required to obtain clear-cut mutagenic effects, this also suggests the involvement of heat-labile chemical compounds in tea mutagenicity.

The mutagenic effect of tea was also abolished either in the presence of microsomal enzymes or catalases.[1,56] When, however, tea or tea components were hydrolyzed in the presence of glucuronidases or fecalases a relatively strong "frameshift" mutagenicity was observed upon metabolic activation.[1,57,58] This suggests that mutagenic principles in tea are made available in the lower digestive tract through cleavage by gut enzymes.

On the other hand, there are recent reports which show that black and green tea infusions or tea extracts exert an antimutagenic effect by inhibiting nitroso compounds *in vitro*[59-61] and *in vivo*.[61] It was also reported that different food-borne mutagens were suppressed by tea or tea factors *in vitro* and *in vivo*.[62-64] The major inhibitory factors might be attributed to the phenolic compounds and possibly to some extent to caffeine.[64]

B. TEA COMPONENTS
1. Flavanols

Polyphenolic flavanols (flavan-3-ols) and glycosides contribute up to 30% of the dry matter of fresh tea flush.[65] Catechin derivatives account for the major part of the phenolic compounds present in tea whereas glycosides such as kampferol, quercetin, and myricetin only occur in small amounts in tea leaves.[65]

In fact, the total mutagenic activity of tea was attributed to the fraction containing the polyphenols which were identified as gallocatechines, pyrogallol, kampferol, quercetin, and myricetin. Hence the mutagenic activity of tea is mainly attributed to quercetin, myricetin, and kampferol, which are among the most mutagenic flavanols.[66] It should be noted, however, that for one of the strongest flavanol mutagens, namely quercetin, no convincing evidence of carcinogenicity exists although it has been thoroughly investigated.[66] Most recent experiment even provided evidence that quercetin exerts antimutagenic/anticarcinogenic properties.[45,67] It is most likely that the two other structurally related flavanols, kampferol and myricetin, have properties similar to quercetin. Furthermore flavan-3-ols, which were generally not shown to be mutagenic,[68,69] exert antimutagenic activity in *in vitro* test systems.[61, 69-71] This inhibitory effect was mainly attributed to the involvement of catechin derivatives in monooxygenase bioactivation.[71]

2. Methylxanthines

Methylxanthines account for about 4% of the dry matter content of fresh black tea flush. The majority of the methylxanthine content is contributed by caffeine, with theobromine below 0.5% and theophylline only occurring in traces.[65]

As seen with coffee, the methylxanthines do not account for the *in vitro* mutagenicity observed with tea.[57] The methylxanthines caffeine and theobromine are generally mutagenic in *in vitro* test systems[35,71] (see also Section II.B.1), although they are negative in standard bacterial tests. Furthermore, the methylxanthines (caffeine) are not considered as presenting either a mutagenic[35] or a carcinogenic[38] risk for man.

3. Hydrogen Peroxide

As for coffee, hydrogen peroxide (H_2O_2) has been reported in tea (at about 170 μM)[56] and was claimed to play an essential role in tea mutagenicity.[56] However, the authors had considerable difficulty in interpreting the results on the involvement of H_2O_2 and its role is again far from clear.

IV. WINE

A. WINE AS A WHOLE

Wine was found to exert mutagenic effects on both "base-pair" and "frameshift" Ames tester strains but required activation by microsomal enzyme system and fecalases. It was, in principle, only red wine which caused a mutagenic effect.[4,58,73,76] but the urine of red wine drinkers was negative when tested *in vitro* upon extraction.[74] White wine only exerted slight mutagenicity.[76] Some of the variations have been attributed to different grape species, wine processing, vinification, and aging factors.[75] The fact that mutagenicity was predominantly in red wine suggests that plant phenolics are mainly responsible for this effect. This was confirmed in one study[73] where rutin was identified by HPLC/MS as the major factor in a mutagenic wine fraction.

B. WINE COMPONENTS

1. Alcohols

Of the alcohols, ethanol is present in white and red wine in large amounts, with levels of about 10 to 14% v/v but ranging up to about 25% v/v for dessert wines.[77] The methanol content is much lower and varies widely, between 30 to 100 mg/l for white and 70 to 160 mg/l for rose and red wine. Higher alcohols (e.g., propanol, butanol, hexanol) are also present but in relatively low concentrations and between levels of 250 to 600 mg/l.[77]

Ethanol caused no significant effect in bacterial tests and hence probably does not contribute to the mutagenicity of whole wine in the Ames test. Ethanol was also evaluated in numerous other *in vitro* tests and was shown to induce sister chromatid exchanges and chromosone aberrations in mammalian cells including those of alcoholics.[78] This effect was, however, only relatively weak and a considerable number of *in vitro* mutagenicity test systems did not respond to alcohol.[78] The mutagenic effect of ethanol might be mainly attributed to acetaldehyde because sister chromatid exchanges were induced by aldehyde hydrogenase in human lymphocytes.[78]

2. Acetaldehyde

Acetaldehyde content varies considerably and in table wine ranges from about 25 to 100 mg/l while sherry contains up to ca. 200 mg/l.[77] As a matter of fact, acetaldehyde was observed to be mutagenic in various *in vitro* test systems.[79]

3. Phenolic Compounds

Total phenol content was estimated at about 300 mg/l in white and about 1800 mg/l in red wines with maxima of 1300 and 3800 mg/l, respectively.[77] Numerous phenolic compounds such as anthocyanins (cyanidin, delphinidin, malvidin, petunidin), flavane-3-ols [(+)-catechin, (−)-epicatechin, (+)-gallocatechin, (−)-epigallocatechin], flavanols (kampferol, quercetin, quercitrin, myricetin, rutin), traces of flavones, as well as phenolic acids (e.g., ferulic, *p*-coumaric, caffeic, sinapic, cholorogenic, quinic, and gallic acids, etc. or their esters) were also reported to occur in wine.[77] Proanthocyanidins were reported at concentrations ranging from about 460 to 740 mg/l[73] and hence are major contributors to the total phenolics. Flavanols may be mainly responsible for the effect in wine since flavanols are relatively strong bacterial mutagens.[66,68] Although some of the other plant phenolics were also shown to be mutagenic *in vitro*, there is no convincing evidence of mutagenic/carcinogenic property in mammals for any of them. There is, however, extensive literature which shows that the majority of these plant phenolics including flavanols present in wine have antimutagenic or even anticarcinogenic properties[66,80] (see also Section III.B.1).

4. Biogenic Amines

The biogenic amine content is relatively low but shows great variation in different wines and values range between 0.2 and about 5.5 mg/l for histamine and likewise for tyramine.[77] There is some limited evidence that biogenic amines (histamine) are mutagenic *in vitro*, but they were also shown to be antimutagens.[82]

5. Ethyl Carbamate

Ehtyl carbamate (urethane) contamination was identified in fermented food and especially in alcoholic beverages such as wine, where average levels ranging from about 1 to 20 μg/l were determined.[83] With the exception of spirits, the main occurrence of ethyl carbamate may presumably be the result of the reaction of ethanol with urea or other N-carbamyl compounds. A number of studies have been published concerning the mutagenicity of ethyl carbamate which caused an effect in a wide range of organisms but mainly in eukaryotic cell systems.[84,85] This is explained by an insufficient oxidation of ethyl carbamate to vinyl carbamate by the endogenous microsome enzyme system used in the bacterial test systems. Ethyl carbamate was also shown to have carcinogenic properties in rodents.[85]

V. BEER

Very little data is available on the mutagenicity of beer, but the results so far from extracted commercial beer were generally negative in *in vitro* tests.[4,72] The major components of beer with possible biological activity are the same as in wine, i.e., alcohol and phenolic compounds. However, these compounds are generally present at lower concentrations. Therefore acetaldehyde and ethyl carbamate contamination would account for even lower levels in beer than in wine.

Levels of 150 to 250 mg/l of phenolic compounds have been reported,[87,88] with two thirds coming from barley and one third from hops. These phenols are phenolic acids and esters (*p*-hydroxybenzoic, caffeic, vanillic, and ferulic acids) flavan-3-ols (catechin), procyanidins, flavan-3,4-diols, and flavones (kampferol, myricetin).

Contamination with low levels of nitrosamines has been observed.[89] Although nitrosamines are generally strong mutagens,[90] it was concluded that the mutagenicity of alcoholic beverages cannot be attributed to nitrosamines.[91] However, it is possible that the alcohols and nitrosamines react to yield reactive compounds which might damage the DNA.[92]

VI. CONCLUSION

In traditional beverages such as coffee, tea, wine, and beer, mainly the plant phenolics and to some extent heat reaction products, methylxanthines, or biogenic amines may be implicated in the *in vitro* mutagenicity observed with these beverages. There is evidence, however, that these classes of compounds are not generally mutagenic *in vivo* and in particular are not convincingly carcinogenic in mammals. On the other hand, evidence is increasing to suggest that phenolics, heat reaction compounds, and to some extent also the methylxanthines and biogenic amines may protect man from mutagenic or carcinogenic damage induced by other environmental compounds. Ingestion of high amounts of wine or beer leads to significant intake of ethanol or acetaldehyde, respectively, as well as ethyl carbamate. These components are suspected of having a carcinogenic potential and hence might represent a risk to man. Nitrosamine contamination in beer might also affect human health and should therefore be kept at a low level.

ACKNOWLEDGEMENTS

The author thanks Drs. H. P. Würzner, D. Magnolato, and R. Liardon for their advice and Ms. R. Acheson-Shalom for editorial help.

REFERENCES

1. **Nagao, M., Takahjashi, Y., Yamanaka, H., and Sugimura, T.,** Mutagens in coffee and tea, *Mutat. Res.,* 68, 101, 1979.
2. **Aeschbacher, H. U. and Würzner, H. P.,** An evaluation of instant and regular coffee in the Ames mutagenicity test, *Toxicol. Lett.,* 5, 139, 1980.
3. **Aeschbacher, H. U., Chappuis, C., and Würzner, H. P.,** Mutagenicity testing of coffee: a study of problems encountered with the Ames Salmonella test system, *Food Cosmet. Toxicol.* 18, 605, 1980.
4. **Stoltz, D. R., Stavric, B., Krewsky, D., Klassen, R., Bendall, R., and Junkins, B.,** Mutagenicity screening of foods. I. Results with beverages, *Environ. Mutagen.,* 4, 477, 1982.
5. **Kosugi, A., Nagao, M., Suwa, Y., Wakabayashi, K., and Sugimura, T.,** Roasting coffee beans produces compounds that induce prophage λ in *E. coli* and are mutagenic in *E. coli* and *S. typhimurium, Mutat. Res.,* 116, 179, 1983.
6. **Nakasato, F., Nakayasu, M., Fujita, Y., Nagao, M., Terada, M., and Sugimura, T.,** Mutagenicity of instant coffee on cultured hamster lung cells, *Mutat. Res.,* 141, 109, 1984.
7. **Blair, C. A. and Shibamoto, T.,** Ames mutagenicity tests of overheated brewed coffee, *Food Chem. Toxicol.,* 22, 971, 1984.
8. **Aeschbacher, H. U., Ruch, E., Meier, H., Würzner, H. P., and Munoz-Box, R.,** Instant and brewed coffees in the *in vitro* human lymphocyte mutagenicity test, *Food Chem. Toxicol.,* 23, 747, 1985.
9. **Albertini, S., Friederich, U., Schlatter, Ch., and Würgler, F. E.,** The influence of roasting procedure on the formation of mutagenic compounds in coffee, *Food Chem. Toxicol.,* 23, 593, 1985.
10. **Friederich, U., Hann, D., Albertini, S., Schlatter, Ch., and Würgler, F. E.,** Mutagenicity studies on coffee. The influence of various factors on the mutagenic activity in the Salmonella/mammalian microsome assay, *Mutat. Res.,* 156, 39, 1985.
11. **Nagao, M., Fujita, Y., Wakabayashi, K., Nukaya, H., Kosuge, T., and Sugimura, T.,** Mutagens in coffee and other beverages, *Environ. Health Perspect.,* 67, 89, 1986.
12. **Dorado, G., Barbancho, M., and Pueyo, C.,** Coffee is highly mutagenic in the L-arabinose resistant test in *Salmonella typhimurium, Environ. Mutagen.,* 9, 251, 1987.
13. **Ariza, R. R., Dorado, G., Barbancho, M., and Pueyo, C.,** Study of the causes of direct-acting mutagenicity in coffee and tea using the Ara test in *Salmonella typhimurium, Mutat. Res.,* 201, 89, 1988.
14. **Shane, B. S., Troxclair, A. M., McMillan, D. J., and Henry C. B.,** Comparative mutagenicity of nine brands of coffee to *Salmonella typhimurium* TA 100, TA 102 and TA 104, *Environ. Mol. Mutagen.,* 11, 195, 1988.
15. **Aeschbacher, H. U., Wolleb, U., Löliger, J., Spadone, I. C., and Liardon, R.,** Contribution of coffee aroma constituents to the mutagenicity of coffee, *Food Chem. Toxicol.,* 27, 227, 1989.
16. **Fujita, Y., Wakabayashi, K., Nagao, M., and Sugimura, T.,** Implication of hydrogen peroxide in the mutagenicity of coffee, *Mutat. Res.,* 144, 227, 1985b.
17. **Aeschbacher, H. U.,** Mutagenicity of coffee, in *Coffee Physiology,* Clarke, R. J., and Macrae, R., Eds., Elsevier Applied Science, London, 1988, 195.
18. **Claxton, L. D., Allen, J., Aulettte, A., Mortelmans, K., Nestmann, E., and Zeiger, E.,** Guide for the *Salmonella typhimurium*/mammalian microsome tests for bacterial mutagenicity, *Mutat. Res.,* 189, 83, 1987.
19. **Ruiz-Rubine, M., Alejandre-Duran, E., and Pueyo, C.,** Oxidative mutagens specific for A. T. base pair induced mutations to L-arabinose resistance in *Salmonella typhimurium, Mutat. Res.,* 147, 153, 1985.
20. **Nagao, M., Suwa, Y., Yoshizumi, H., and Sugimura, T.,** Mutagens in coffee, in *Coffee and Health,* Banbury Report, Vol. 17, MacMahon, B., and Sugimura, T., Eds., New York, 1984, 69.
21. **Fujita, Y., Wakabayashi, K., Nagao, M., and Sugimura, T.,** Characteristics of major mutagenicity of instant coffee, *Mutat. Res.,* 142, 145, 1985.
22. **Graf, U. and Würgler, F. E.,** Investigation of coffee in Drosophila genotoxicity tests, *Food Chem. Toxicol.,* 24, 835, 1986.
23. **Aeschbacher, H. U. and Chappuis, C.,** Non-mutagenicity of urine from coffee drinkers compared with that from cigarette smokers, *Mutat. Res.,* 89, 161, 1981.
24. **Dunn, B. P. and Curtin, J. R.,** Clastogenic agents in the urine of coffee drinkers and cigarette smokers, *Mutat. Res.,* 147, 179, 1985.
25. **Reidy, J. A., Annest, J. L., Chen, A. T. L., and Welty, T. K.,** Increased sister chromatid exchange associated with smoking and coffee consumption, *Environ. Mol. Mutagen.,* 12, 311, 1988.
26. **Aeschbacher, H. U., Meier, H., Ruch, E., and Würzner, H. P.,** Investigation of coffee in sister chromatid exchange and micronucleus test *in vivo, Food Chem. Toxicol.,* 22, 803, 1984.
27. **Shimuzu, M. and Yano, E.,** Mutagenicity of instant coffee and its interaction with dimethylnitrosamine in the micronucleus test, *Mutat. Res.,* 189, 307, 1987.
28. **Würzner, H. P., Lindström, E., Vuataz, L., and Luginbühl, H.,** A 2-year feeding study of instant coffees in rats. II. Incidence and types of neoplasms, *Food Cosmet. Toxicol.,* 15, 289, 1977.

29. **Stalder, R., Luginbühl, H., Bexter, A., and Würzner, H. P.,** Preliminary findings of a carcinogen bioassay of coffee in mice, in *Coffee and Health,* Banbury Report, Vol. 17, MacMahon, B. and Sugimura, T., Eds., New York, 1984, 79.

30. **Nishikawa, A., Tanaka, T., and Mori, H.,** An inhibitory effect of coffee on nitrosamine hepatocarcinogenesis with aminopyrine and sodium nitrite in rats, *J. Nutr. Growth Cancer,* 3, 161, 1986.

31. **Yamaguchi, T. and Iki, M.,** Inhibitory effect of coffee extract against some mutagens, *Agric. Biol. Chem.,* 50, 2983, 1986.

32. **Obana, H., Nakamura, S., and Tanaka, R.,** Suppressive effects of coffee on the SOS response induced by UV and chemical mutagens, *Mutat. Res.,* 175, 47, 1986.

33. **Suwa, Y., Nagao, M., Kosugi, A., and Sugimura, T.,** Sulfate suppresses the mutagenic property of coffee, *Mutat. Res.,* 102, 383, 1982.

34. **Macrea, R.,** Nitrogenous components in *Coffee Chemistry,* Vol. 1, Clarke, R. J., and Macrae, R., Eds., Elsevier Applied Science, London, 1985, 115.

35. **Timson, J.,** Caffeine, *Mutat. Res.,* 47, 1, 1977.

36. **Kier, L. E., Brusick, D. J., Aulette, A. E., Von Halle, E. S., Brown, M. M., Simmon, V. F., Dunkel, V., McCann, J., Mortelmans, K., Prival, M., Rao, T. K., and Ray, V.,** The *Salmonella typhimurium/* mammalian microsomal assay. A report of the U. S. Environmental Protection Agency Gene-Tox Program, *Mutat. Res.,* 168, 69, 1986.

37. **Quillardet, P., de Bellecombe, C., and Hofnung, M.,** The SOS Chromotest, a colorimetric bacterial assay for genotoxins validation: study with 83 compounds, *Mutat. Res.,* 147, 79, 1985.

38. **Grice, H. C. and Murray, T. K.,** Caffeine, a perspective of current concerns, *Nutr. Today,* 22, 36, 1987.

39. **Clifford. M. N.,** Chlorogenic acids, in *Coffee Chemistry,* Vol. 1, Clarke, R. J., and Macrea, R., Eds., Elsevier Applied Science, London, 1985, 153.

40. **Stich, H. F., Rosin, M. P.. Wu, C. H., and Powrie, W. D.,** A comparative genotoxicity study of chlorogenic acid (3-O-caffeoylquinic acid), *Mutat. Res.,* 90, 201, 1981.

41. **Hanham, A. F., Dunn, B. P., and Stich, H. F.,** Clastogenic activity of caffeic acid and its relationship to hydrogen peroxide generated during autooxidation, *Mutat. Res.,* 116, 333, 1983.

42. **Fung, V. A., Cameron, T. P., Hughes, T. J., Kirby, P. E., and Dunkel, V. C.,** Mutagenic activity of some coffee flavor ingredients, *Mutat. Res.,* 204, 219, 1988.

43. **Kaul, B. L. and Tandon, V.,** Modification of the mutagenic activity of propane sulfone by some phenolic antioxidants, *Mutat. Res.,* 89, 57, 1981.

44. **San, R. H. C. and Chan, R. I. M.,** Inhibitory effect of phenolic compounds on aflatoxin B_1 metabolism and induced mutagenesis, *Mutat. Res.,* 177, 229, 1987.

45. **Alldrick, A. J., Flynn, J., and Rowland, I. R.,** Effects of plant-derived flavonoids and polyphenolic acids on the activity of mutagens from cooked food, *Mutat. Res.,* 163, 225, 1986.

46. **Raj, A. S., Heddle, J. A., Newmark, H. L., and Katz, M.,** Caffeic acid as an inhibitor of DMBA-induced chromosomal breakage in mice assessed by bone-marrow micronucleus test, *Mutat. Res.,* 124, 247, 1983.

47. **Kuenzing, W., Chau, J., Norkus, E., Holowaschenko, H., Newmark, H., Mergens, W., and Conney, A. H.,** Caffeic and ferulic acid as blockers of nitrosamine formation, *Carcinogenesis,* 5, 309, 1984.

48. **Wattenberg, L. W.,** Inhibition of neoplasia by minor dietary constituents, *Cancer Res.,* 43, 2240, 1983.

49. **Mori, H., Tanaka, T., Shima, H., Kuniyasu, T., and Takahashi, M.,** Inhibitory effect of chlorogenic acid on methylazoxymethanol acetate-induced carcinogenesis in large intestine and liver of hamsters, *Cancer Lett.,* 30, 49, 1986.

50. **Bjeldanes, L.F. and Chew, H.,** Mutagenicity of 1,2 dicarbonyl compounds: maltol, kojic acid, diacetyl and related substances, *Mutat. Res.,* 67, 367, 1979

51. **Kasai, H., Kumeno, K., Yamaizumi, Z., Nishimura, S., Nagao, M., Fujita, Y., Sugimura, T., Nukaya, H., and Kosuge, T.,** Mutagenicity of methylglyoxal in coffee, *Gann,* 73, 681, 1982.

52. **Hayashi, T. and Shibamoto, T.,** Analysis of methylglyoxal in foods and beverages, *J. Agric. Food Chem.,* 33, 1090, 1985.

53. **Cajelli, E., Canonero, R., Martelli, A., and Brambilla, G.,** Methylglyoxal-induced mutation to 6-thioguanine resistance in V79 cells, *Mutat. Res.,* 190, 47, 1987.

54. **Kim, S. B., Hayase, F., and Kato, H.,** Desmutagenic effect of α-dicarbonyl and α-hydroxy carbonyl compounds against mutagenic heterocyclic amines, *Mutat. Res.,* 177, 9, 1987.

55. **Kato, H., Kim, S. B., Hayase, F., and Van Chuyen, N.,** Desmutagenicity of melanoidines against mutagenic pyrolysates, *Agric. Biol. Chem.,* 49, 3093, 1985.

56. **Alejandre-Duran, E., Alonso-Morago, A., and Pueyo, C.,** Implication of active oxygen species in the direct-acting mutagenicity of tea, *Mutat. Res.,* 188, 251, 1987.

57. **Uyeta, M., Taue, T., and Mazaki, M.,** Mutagenicity of tea infusions, *Mutat. Res.,* 88, 233, 1981.

58. **Tamura, G., Gold, C., Ferro-Suzzi, A., and Ames, B. N.,** Fecalase: a model for activation of dietary glycosides to mutagens by intestinal flora, *Proc. Natl. Acad. Sci. U. S. A.,* 77, 4961, 1980.

59. **Stich, H. F., Chan, K. L., and Rosin, M. P.,** Inhibitory effects of phenolics, teas and saliva on the formation of mutagenic nitrosation products of salted fish, *Int. J. Cancer,* 30, 719, 1982a.

60. **Stich, H. F., Rosin, M. P., and Bryson, L.,** Inhibition of mutagenicity of a model nitrosation reaction by naturally occurring phenolics, coffee and tea, *Mutat. Res.,* 95, 119, 1982b.

61. **Jain, A. K., Shimoi, K., Nakamura, Y., Kada, T., Hara, Y., and Tomita, I.,** Crude tea extracts decrease the mutagenic activity of N-methyl-N'-nitro-N-nitrosoguanidine *in vitro* and in intragastric tract of rats, *Mutat. Res.,* 210, 1, 1989.

62. **Joner, P. E. and Dommarsnes, K.,** Effects of herbal and ordinary teas on the mutagenicity of benzo(a)pyrene in the Ames test, *Acta Agric. Scand.,* 33, 53, 1983.

63. **Kada, T., Kaneko, K., Matsuzaki, S., Matsuzaki, T., and Hara, Y.,** Detection and chemical identification of natural bio-antimutagens. A case of the green tea factor, *Mutat. Res.,* 150, 127, 1985.

64. **Ito, Y., Ohnishi, S., and Fujie, K.,** Chromosome aberrations induced by aflatoxin B_1, in rat bone marrow cells *in vivo* and their suppression by green tea, *Mutat. Res.,* 222, 253, 1989.

65. **Spiller, G. A.,** Composition of tea, in *The Methylxanthine Beverages and Foods,* Vol. 158, Spiller, G. A., Ed. Alan R. Liss, New York, 1984, 29.

66. **Mac Gregor, J. T.,** Genetic and carcinogenic effects of plant flavonoids: an overview, *Adv. Exp. Med. Biol.,* 177, 497, 1984.

67. **Verma, A. K., Johnson, J. A., Gould, M. N., and Tanner, M. A.,** Inhibition of 7,12-dimethyl-benz(a)anthracene and N-nitromethylurea-induced rat mammary cancer by dietary flavonol quercetin, *Cancer Res.,* 48, 5754, 1988.

68. **Brown, J. P.,** A review of the genetic effects of naturally occurring flavonoids, anthraquinones and related compounds. *Mutat. Res.,* 75, 243, 1980.

69. **Nagabhushan, M. and Bhide, S. V.,** Anti-mutagenicity of catechin against environmental mutagens, *Mutagenesis,* 3, 293, 1988.

70. **Shimoi, K., Nakamura, Y., Tomita, I., Hara, Y., and Kada, T.,** The pyrogallol related compounds reduce UV-induced mutations in *E. coli* B/r WP2, *Mutat. Res.,* 173, 239, 1986.

71. **Steele, C. M., Lallies, M., and Ionnides, C.,** Inhibition of the mutagenicity of aromatic amines by the plant flavonoid (+)-catechin, *Cancer Res.,* 45, 3573, 1985.

72. **Brusick, D. J., Jagannath, D. R., Myhr, B., Sernau, R., and Chappel, C. I.,** Genetic toxicology evaluation of commercial beers. II. Mutagenic activity of commercial beer products in *Salmonella typhimurium* strains TA 98, TA 100 and TA 102, *Mutat. Res.,* 206, 41, 1988.

73. **Yu, C. L., Swaminathan, B., Butler, L. G., and Pratt, D. E.,** Isolation and identification of rutin as the major mutagen of wine, *Mutat. Res.,* 170, 103, 1986.

74. **Sousa, J., Nath, J., and Ong, T.,** Dietary factors affecting the urinary mutagenicity assay system. II. The absence of mutagenic activity in human urine following consumption of red wine or grape juice, *Mutat. Res.,* 156, 171, 1985.

75. **Subden, R. E., Krizus, A., and Rancourt, D.,** Mutagen content of table wines made from various grape species and hybrid cultivars, *Food Chem. Toxicol.,* 22, 309, 1984.

76. **Bull, P., Yanez, L., and Nervi, F.,** Mutagenic substances in red and white wine in Chile, a high risk area for gastric cancer, *Mutat. Res.,* 187, 113, 1987.

77. **Amerine, M. A. and Ough, C.S.,** *Methods for Analysis of Musts and Wines,* John Wiley & Sons, New York, 1980, 79.

78. **Obe, G. and Anderson, D.,** Genetic effects of ethanol, *Mutat. Res.,* 186, 177, 1987.

79. **Dellarco, V. L.,** A mutagenicity assessment of acetaldehyde, *Mutat. Res.,* 195, 1, 1988.

80. **Hayatsu, H., Arimoto, S., and Negishi, T.,** Dietary inhibitors of mutagenesis and carcinogenesis, *Mutat. Res.,* 202, 429, 1988.

81. **Sokorova, S. V. and Kerkis, I.,** Mutagenic effect of histamine and bradykinin in a culture of human and mammalian cells, *Dokl. Akad. SSSR,* 232, 478, 1977.

82. **Alldrick, A. J. and Rowland, I. R.,** Counteraction of the genotoxicity of some cooked-food mutagens by biogenic amines, *Food Chem. Toxicol.,* 25, 575, 1987.

83. **Dennis, J. J., Howarth, N., Key, P. E., Pointer, M., and Massey, R. C.,** Investigation of ethyl carbamate levels in some fermented foods and alcoholic beveragess, *Food Addit. Contamin.,* 3, 383, 1989.

84. **Field, K. J. and Long, C. M.,** Hazards of urethane (ethyl carbamate): a review of the literature, *Lab. Anim.,* 22, 255, 1988.

85. **Dahl, G. A., Miller, E. C., and Miller, J. A.,** Comparative carcinogenicities and mutagenicities of vinyl carbamate, ethyl carbamate and ethyl N-hydroxycarbamate, *Cancer Res.,* 40, 1194, 1980.

86. **Brusick, D., Myher, B., Galloway, S., Rudell, J., Jagannath, D. R., and Tarka, S.,** Genotoxicity of theobromine in a series of short-term assays, *Mutat. Res.,* 169, 105, 1986.

87. **Narziss, L.,** *Abriss der Brauerei,* Enke Verlag, Stuttgart, 1972, 264.

88. **Moceck, M.,** A method for estimating total polyphenols in beer, *J. Inst. Brew.,* 79, 165, 1973.

89. **Castegnaro, M.,** International N-nitrosamine check sample programme: report on the performance in the 1st study dedicated to the determination of N-nitrosamines in beer and malt, *Food Addit. Contamin.,* 5, 283, 1988.

90. **Guttenplan, J. B.,** N-Nitrosamines: bacterial mutagenesis and *in vitro* metabolism, *Mutat. Res.*, 186, 81, 1987.
91. **Tuyns, A. J. and Gricuite, L.L.,** Carcinogenic substances in alcoholic beverages, in *Advances in Tumor Prevention, Detection and Characterization*, Vol. 5, Davis, W., et al. Eds., Excerpta Medica, Amsterdam 1980, 130.
92. **Valin, N., Haybron, D., Groves, L., and Mower, H. F.,** The nitrosation of alcohol-induced metabolites produces mutagenic substances, *Mutat. Res.*, 158, 159, 1985.

Chapter 10

PREVENTION OF MUTAGEN FORMATION

Chapter 10.1

PREVENTION OF HETEROCYCLIC AMINE FORMATION IN RELATION TO CARCINOGENESIS

John H. Weisburger

TABLE OF CONTENTS

I. INTRODUCTION

Beginning successes have been achieved in unravelling the complex causes of important types of human cancer world-wide. Also, much has been learned about the mechanisms of carcinogenesis. It is now fairly well established that cancer is a general term applicable to many distinct diseases, with different causative, enhancing, promoting, and also inhibiting elements. Yet, at the fundamental level, all cancers are characterized by one singular property, namely a specific loss of growth control elements present in normal cells.

The mechanisms of carcinogenesis indicate that the introduction of cancer involves a stepwise sequence of distinct events. It begins with the transformation of normal cells by carcinogens that can be chemical, radiation, or viruses, affecting specific codons in DNA and representing a somatic mutation.[1,2] Growth and development of cells with such a modified DNA, followed by additional alterations of the genetic element, perhaps as a result of lower fidelity of the polymerases synthesizing DNA in such cells, eventually leads to cell types with the characteristic gene structure and phenotypic expression typical of neoplastic cells. Clearly, control of cell duplication rates will also play an important role in the expression of neoplasia.[3] This sequence of early events represents the genotoxic component of the overall process.[1,2,4-8]

The subsequent steps in the development of neoplasia relate to the growth control of the neoplastic cells.[9,10] These phenomena, involving promoting and enhancing as well as inhibiting elements, are characterized usually by the fact that these aspects are a function of the amounts and potency of the promoting stimulus.[11,12] This part of the process is highly dose dependent.[2] Also, up to a certain point in the development of neoplasms, this sequence is reversible. Thus, removal of the promoting stimulus or lowering its concentration effectively inhibits or delays the production of clinical invasive metastatic cancer.

Later steps are called progression and metastasis.[13] The underlying mechanisms are as yet not well known, and in any case are outside the scope of this discussion.

Based on these mechanistic considerations, it is important to analyze environmental elements associated with each type of cancer for the presence, quantities, and potencies of genotoxic compounds and the amounts and potencies of promoting factors. Clearly, therefore, prevention of each type of cancer can be approached through lower amounts of genotoxic components or promoting elements. Alternatively, it is possible to visualize the introduction of inhibiting elements for each of these two major steps (Table 1).

II. ETIOLOGIC FACTORS FOR NUMERICALLY IMPORTANT TYPES OF CANCER

A. TOBACCO

It is now known that one set of cancer types, namely cancer of the oral cavity, esophagus, lung, pancreas, kidney, bladder, and cervix, is linked to the use of tobacco products, through chewing, snuff-dipping, and especially through smoking.[14-18] Tobacco and tobacco smoke contain genotoxic carcinogens, but a major effect leading to cancer depends on promoting and enhancing elements. For that reason, smoking cessation leads to a progressively lower risk of cancer development. In the U.S. and the Western world generally, about 30% of all cancers are associated with tobacco use (Table 2).

B. NUTRITION

The field of nutrition in cancer induction is receiving increasing attention (Table 2). In the Orient, cancer of the stomach and esophagus has been traced to the customary intake of salted and pickled foods and a relatively low traditional intake of fruits and vegetables.[19-23] The associated carcinogens have been not fully identified. For cancer of the esoph-

TABLE 1
Types of Agents Associated with the Etiology of Human Cancer

Problems or questions to be resolved for each type of cancer.
I.   Nature of genotoxic carcinogens or mixtures — can be chemical, viral, or radiation
II. Nature of any nongenotoxic promoting or enhancing stimulus — can be chemical or viral
III. Amount, duration of exposure, and potency for each kind of agent
IV. Possibility of inhibition of the action of agents under I or II

TABLE 2
Postulated Causes of Main Human Cancers

Type	% total cases
Occupational cancer	1—5
varied organs	
Cryptogenic cancers	10—15
Lymphomas, Leukemias, Sarcomas, (Cervix?), (Virus?)	
Lifestyle cancers	
1. Tobacco-related	31
Lung, pancreas, bladder, kidney, cervix(?)	
2. Diet-related	6
Nitrate-nitrite, salt, low vitamins A, C. E (yellow-green vegetables), stomach, esophagus	
Mycotoxins, hepatitis B antigen, liver	
3. High-fat, low-fiber, low-calcium, broiled, or fried foods	32
Large bowel, pancreas, prostate, breast, uterus	
Multifactorial	
1. Tobacco and alcohol	2
Oral cavity, esophagus	
2. Tobacco-asbestos, tobacco-mining, tobacco-uranium-radium	2
Lung, respiratory tract	
Iatrogenic — radiation, drugs	1
Different kinds of cancer, dependent on type of agent	
Unspecified sites and actions	7

Note: Estimated cancer deaths calculated from American Cancer Society, 1988 Cancer Facts and Figures

agus, they appear to be specific nitrosamines. For cancer of the stomach, they may be nitroso-indoles or diazophenols.[25-27] Salt itself damages the gastric mucosa and exerts a cocarcinogenic effect. However, there appear to be few promoting elements, accounting for the fact that individuals growing up in areas with a customary high intake of such foods, when migrating to a region with a lower risk, retain their propensity for cancer development because the genotoxic character of the carcinogens and the lack of promoting influences renders the disease development basically irreversible.

On the other hand, in the Western world, and particularly in North America, the main cancers associated with nutrition are those in the postmenopausal breast, distal colon, pancreas, prostate, ovary and endometrium.[19,20,22,28] Well-documented studies indicate that a key element in the development of all of these cancer types, accounting for about 32% of all cancer deaths in the U. S., is the amount of total fat and fat calories consumed.[29] In a high-risk region like the U. S., fat accounts for about 40% of calories.[30-32] In contrast, in a low-risk region, fat is only 15 to 20% of calories. The mechanism whereby the type and amount of fat translates to risk for the specific cancers mentioned has been partially delineated.[33-36] It is reasonably certain that fat and fat calories operate through promoting processes. For example, the type and amount of fat controls the production of bile acids by the liver.[37,38] These bile acids are further metabolized by the intestinal bacterial flora to secondary bile acids, that together with certain fatty acids exert a promoting effect in relation to colon cancer. Therefore, decreasing the concentration of bile acids by reducing the total fat intake to about 20% of calories, or increasing dietary intake of fibers, giving higher stool bulk and diluting bile acids, or raising calcium intake to neutralize bile acids and fatty acids and otherwise controlling intestinal cell duplication rates, provide tools for intestinal cancer reduction by lowering events associated with promotion.[21,29,33,39-43] Since this phenomenon is essentially reversible, one recommended approach to the prevention of mortality from these kinds of cancer is to modify dietary traditions, including less fat and more fiber, vegetable, and fruit intakes.

The question as to the nature of the genotoxic carcinogens for the important types of nutritionally linked cancers was open until fairly recently. At the Princess Takamatsu conference in 1976, Sugimura and associates[45] mentioned for the first time the new finding of powerful mutagenic activity at the surface of fried fish or meat in *Salmonella typhimurium* TA98 and requiring metabolic activation with a liver S9 fraction. Also available then were early data giving the chemical structure of pyrolysates of amino acids like tryptophan. These structures were rather similar to that of 2',3-dimethyl-4-aminobiphenyl. We had utilized the latter compound to induce cancer in the colon and mammary gland of rats and discovered that this chemical could induce cancer in the prostate.[46] The latter finding has been extended by the group of Ito,[47] who also noted recently an effect on the pancreas.[48] Thus, we formulated the hypothesis that these newly discovered mutagens might be the genotoxic carcinogens for the main nutritionally linked cancers in the Western world.[49] Bioassays conducted in Japan and in this laboratory now show that certain of the heterocyclic aromatic amines (HAA) found in fried or broiled meat or fish do induce cancer in the colon, pancreas, and mammary gland, as well as in certain other tissues.[26,50-54] The amounts present are fairly small, as detailed elsewhere in this monograph. However, exposure occurs regularly from childhood onward, and as discussed above, in the Western world the diet includes amounts of fat that exert powerful promoting effects at specific target organs like breast, colon, pancreas, and prostate.[29-38] Thus, one other way of preventing these types of cancer is to lower the amount of these genotoxic carcinogens in the human environment, which essentially means to lower their formation during cooking.

III. MECHANISMS OF FORMATION

The groups of Matsushima[55] and especially of Jägerstad[56] have demonstrated that the formation of the HAA involves Maillard type reactions. The important element is the fact that creatine and creatinine are key reactants. Mutagens are formed when certain amino acids and simple carbohydrates are refluxed or cooked.[57] However, these mutagens are distinct from the HAAs. The role of creatinine is to provide the essential methylaminoimidazo part of the HAA molecule.[56] We have taken advantage of this specific role of creatinine to design mechanistically and practically useful means of preventing the formation of HAAs during cooking of meats or fish or in laboratory models.

IV. PREVENTION OF HAA FORMATION

Early studies in this field showed clearly that the formation of HAAs is controlled in part by the surface temperature of the protein-containing food being cooked and the length of time high temperature is maintained.[58]

It was found that one way of lowering the surface temperature is to mix ground meat with soy proteins or other components that retain water. Thus, during cooking to a satisfactory gustatory property, such a soyburger reaches a lower maximum temperature and therefore decreases production of HAAs.[59]

Another approach, developed in an arbitrary fashion without knowledge at that time of the underlying mechanisms, was the use of antioxidants like BHA that exerted an appreciable inhibiting effect on the formation of these mutagens.[59] It is probably on the basis of current knowledge that antioxidants interfere with the development of Maillard products.

The reactive intermediates produced during Maillard reactions that interact with creatinine appear to be aldehydes or ketones. A mutagen that is IQ-like, namely 2-amino-5-ethylidene-1-methylimidazol-4-one, is formed by the reaction of acetaldehyde with creatinine in the Jägerstad model system, namely refluxing in an ethylene glycol-water mixture at 150°C. Threonine refluxed with creatinine yields the same type of IQ-like products.[60]

Chemicals with NH groups like proline or tryptophan and related indoles have the unique property of reacting with aldehydes and ketones in conventional aldol or Schiff's base type of condensation[61] (Figure 1). It was demonstrated that the formation of HAAs in Jägerstad model systems or during the realistic frying or broiling of meats could be inhibited and indeed completely blocked through such mechanisms[61,62] (Tables 3 and 4). It was found that mixtures of proline and tryptophan form a unique, practical combination that, when applied in a meat sauce to meat or fish prior to cooking, can inhibit the formation of HAAs[61-63] (Table 3).

Therefore, a number of procedures are available to reduce the formation of HAAs in the human food chain. Inasmuch as these chemicals may well be the genotoxic carcinogens for important types of cancer not only in the Western world[64] but also increasingly in the Orient as Western dietary patterns are adopted,[22,28,65] it would seem that the procedures described would be useful tools for the eventual prevention of these types of cancer (See also Chapter 5.1).

V. CONCLUSION

Multidisciplinary research has demonstrated that a sizable fraction of the existing cancer burden is associated with nutritional traditions. In the U.S. and the Western world, and beginning elsewhere, one set of such traditions has been altered because of technological developments. Until about 100 years ago, an important form of food preservation was through salting, pickling, or smoking of meats, fish, and certain vegetables. This tradition and process is now clearly related to the development of cancer of the stomach and esophagus and also hypertension and stroke. As refrigeration, particularly the home refrigerator, was introduced, the processes of food preservation changed. There is now no need to pickle or salt. As a result, stomach cancer as well as stroke assumed progressively lower incidence and mortality in the U.S. at first, and now also in other areas.[21,66]

On the other hand, mortality from cancer of the colon, breast, prostate, pancreas, ovary, and endometrium has been virtually unchanged in the U.S. for at least 60 years, since epidemiological studies have been conducted.[19,20] The underlying reason is that the nutritional traditions, namely the intake of broiled or fried foods and the type and amount of total fat, have not changed in that period.[52] Based on current knowledge, changes in nutritional habits are needed in order to meaningfully lower the risk of the nutritionally linked cancers. One recommendation, namely to lower the total fat intake from 40% to about 20% of calories, would serve to reduce promoting processes.[65] Since these actions are rapidly reversible, it can be predicted that the effects of a lower fat intake by Western populations would have an almost immediate effect in the developmental aspects of cancer production. This was already demonstrated for individuals who stop smoking cigarettes, among whom there is a progressively lower mortality from lung cancer, because the pressure of enhancing elements is eliminated.[14-17]

Another approach detailed in this report is to avoid the formation of HAAs. A number of specific procedures are at hand. One would be to replace the widely eaten fried hamburgers with soyburgers, a great advantage since such equally tasty foods have not only virtually

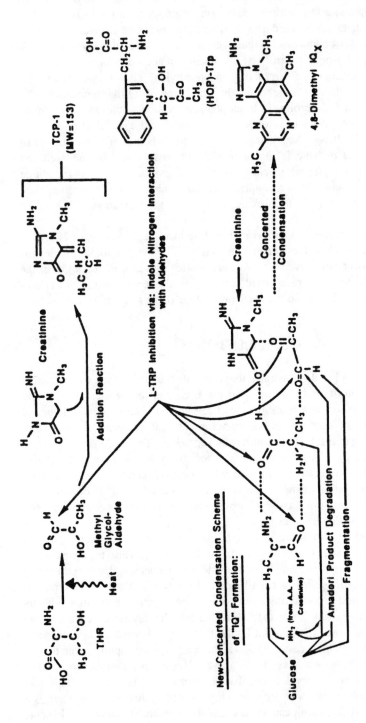

FIGURE 1.   Proposed mechanisms of heterocyclic amine formation and inhibition by L-tryptophan (L-TRP), other indoles, or L-proline. Top shows theonine yielding reactive intermediates that, in the presence of creatinine, yield aminoalkylimidazole-4-one mutagens. Bottom shows general Maillard-type reactions producing intermediates, that in a concerted fashion react with creatinine to form 2-amino-3-methylimidazo[4,5-f]quinolines or quinoxalines (IQ, IQx, PhIP, and analogues). All of these reactions are inhibited by L-TRP or proline.

TABLE 3

Dose-Dependent Inhibition of IQ-Type Mutagenicity in *Salmonella* *typhimurium* TA98 + S9 by L-Tryptophan in Fried and Broiled Lean Ground Beef

Sample	Broiled beef model		% inhibition	Fried beef model	% inhibition
	Avg. rev. col/plate			Avg. rev. col/ pl.	
	Study 1	Study 2			
Plain patty control	515 ± 15[a]			—	
SS patty control[b]	450 ± 28[a] (*p* > 0.05)	280 ± 28[a]		360 ± 21[a]	
50 mg L-trp/side (0.69 mg/cm²)	—	270 ± 13[a] (*p* > 0.05)	6	—	
60 mg L-trp/side (0.83 mg/cm²)	—	—		280 ± 20[a] (*p* < 0.05)	33
75 mg L-trp/side (1.04 mg/cm²)	—	120 ± 22[a] (*p* < 0.05)	100[c]	150 ± 8[a] (*p* < 0.05)	88
100 mg L-trp/side (1.38 mg/cm²)	—	110 ± 5 (*p* < 0.05)	100[c]	—	
150 mg L-trp/side (2.07 mg/cm²)	115 ± 7[a] (*p* < 0.05)	—	99	—	
DMSO control	55	60		60	
IQ (5 ng/plate)	635[a]	1265[a]		1025[a]	

Note: To convert values to avg. rev. col./g meat wet weight, multiply table values by 10 and divide by 50 g.

[a] Significant level of mutagenicity (2 times DMSO control values) (mean ± standard error) from triplicate Ames plates). *p* values refer to comparison with control patty basic extract.

[b] SS, steak sauce. In the broiled beef model, SS, with or without L-trp, was applied (2.5 mℓ/side or 0.83 ml/cm²) to both patty surfaces prior to broiling side 1. In the fried beef model, SS, with or without L-trp, was applied (2.5 ml/side or 0.83 ml/cm²) to side 1 prior to frying and to side 2 immediately while side 1 was cooking.

[c] 100% Inhibition in these studies is a decrease in mutagenic yield to or below the level of significance at 2 times background.

no HAAs, but also contain less total fat.[59] Another means is to apply to meat or fish before frying a sauce containing mixtures of tryptophan and proline, which specifically would block the formation of HAAs.[67]

Therefore, world-wide research has provided a number of theoretically and mechanistically sound and practically feasible means of prevention of important types of cancer in man (Table 5).

ACKNOWLEDGMENTS

Research described in this chapter was supported by USPHS Grant CA-24217 from the National Cancer Institute and BRSG Grant RR-05775 from NIH.

Mr. Matthew Leish and Ms. Julie Howard provided excellent editorial services.

TABLE 4
Percent Inhibition by L-Proline in the Glucose + Glycine + Creatinine Model

Ingredients (mM)	TA98 + S9 Avg.rev.col./plate			Percent inhibition		
	Experiment 1	Experiment 2	Experiment 3	Experiment 1	Experiment 2	Experiment 3
Control (gluc + gly + cr)	1590[a]	1665[a]	590[a,b]	—	—	—
Pro (7)	1460[a]	—	—	8	—	—
Pro (35)	1330[a]	—	—	16	—	—
Trp (7)	1015[a]	—	—	35	—	—
Pro (7) + Trp (7)	820[a]	—	—	50	—	—
Pro (35) + Trp (7)	460[a]	245[a]	—	75	90	—
Pro (35) + Trp (14)	—	—	160[a]	—	—	80
Pro (35) + Trp (35)	—	720[a]	160[a]	—	60	80
Pro (70)	—	740[a]	—	—	55	—
Trp (70)	—	1045[a]	—	—	40	—
Pro (105)	—	—	—	—	—	—
Pro (140)	—	—	—	—	—	—
DEG neg control	60	45	60	—	—	—
5 ng IQ-plate positive control	915[a]	950[a]	830[a]	—	—	—

[a] Significant level of mutagenicity where significance is two times DEG negative control value. Reflux model used contains gluc (35 mM) + gly (70 mM) + cr (70 mM) in diethylene glycol (DEG): 5% distilled water, 150°C for 2 hr.

[b] Controls vary from experiment to experiment as a function of indicator organism, age and other factors; experimental values are thus relative to respective study controls.

TABLE 5
Long Term Goals for Chronic Disease Prevention

Action	Benefit
Control smoking — less harmful cigarette	Coronary heart disease; cancers of the lung, kidney, bladder, pancreas
Lower total fat intake	Coronary heart disease; cancers of the colon, breast, prostate, ovary, endometrium
Lower salt Na^+ intake Balance $K^+ + Ca^{2+}/Na^+$ ratio	Hypertension, stroke, cardiovascular disease
Increase natural fiber, Ca^{2++} (?)	Colorectal cancer
Avoid pickled, smoked, highly salted foods	Cancer of the stomach, esophagus, nasopharyngx, liver(?); hypertension, stroke
Avoid mycotoxins, senecio alkaloids, bracken fern	Cancer of the liver, stomach, bladder(?)
Increase and balance micronutrients, vitamins, minerals	Cardiovascular disease, several types of cancer
Lower intake of fried foods	Cancers of colon, breast, pancreas(?)
Practice sexual hygiene	Cancer of the cervix, penis
Have regular moderate exercise	Coronary heart disease, cancer of the colon, breast
Maintain proper weight, avoid obesity	Coronary heart disease, cancer of the endometrium, kidneys

REFERENCES

1. **Williams, G. M.,** Methods for evaluating chemical genotoxicity, *Annu. Rev. Pharmacol. Toxicol.,* 29, 189, 1989.
2. **Williams, G. M. and Weisburger, J. H.,** Chemical carcinogens, in *Casarett and Doull's Toxicology, The Basic Science of Poisons,* Klaassen, C. D., Amdur, M. O., and Doull, Eds., Macmillan, New York, 1986, 99.
3. **Knight, G. B., Gudas, J. M., and Pardee, A. B.,** Coordinate control of S phase onset and thymidine kinase expression, *Jpn. J. Cancer Res.,* 80, 493, 1989.
4. **Pienta, K. J., Partin, A. W., and Coffey, D. S.,** Cancer as a disease of DNA organization and dynamic cell structure, *Cancer Res.,* 49, 2525, 1989.
5. **Weinberg, R. A.,** Oncogenes, antioncogenes, and the molecular bases of multistep carcinogenesis, *Cancer Res.,* 49, 3713, 1989.
6. **Barbacid, M.,** Human oncogenes, in *Important Advances in Oncology 1986,* DeVita, V. T., Jr., Hellman, S., and Rosenberg, S. A., Eds., J. B. Lippincott, Philadelphia, 1986, 3.
7. **Aaronson, S. A., Bishop, J. M., Sugimura, T., et al., Eds.,** *Oncogenes and Cancer,* Japan Scientific Societies Press, Tokyo, 1987.
8. **Rosenkranz, H. S., Ed.,** Strategies for the deployment of batteries of short-term tests, *Mutat. Res.,* 205, 1, 1988.
9. **Lutz, W. K. and Maier, P.,** Genotoxic and epigenetic chemical carcinogenesis: one process, different mechanisms, *T.I.P.S.,* 9, 322, 1988.
10. **Clayson, D. B.,** Can a mechanistic rationale be provided for non-genotoxic carcinogens identified in rodent bioassays?, *Mutat. Res.,* 221, 53, 1989.
11. **Yamasaki, H., Enomoto, T., and Martel, N.,** Intercellular communication, cell differentiation and tumour promotion, *IARC Sci. Publ.,* 56, 217, 1984.
12. **Milman, H. A. and Elmore, E., Eds.,** *Biochemical Mechanisms and Regulation of Intercellular Communication,* Princeton Scientific Publishing, Princeton, NJ, 1987.
13. **Pitot, H. C.,** Progression: the terminal stage in carcinogenesis, *Jpn. J. Cancer Res.,* 80, 599, 1989.
14. **Hoffmann, D. and Hecht, S. S., Eds.,** *Mechanisms in Tobacco Carcinogenesis,* Banbury Report 23, Cold Spring Harbor Laboratory, Cold Spring Harbor, NY, 1986.
15. *IARC Monographs on the Evaluation of the Carcinogenic Risk of Chemicals to Humans, Tobacco Smoking,* Vol. 38, International Agency for Research on Cancer, Lyon, France, 1986.
16. **Sorsa, M. and Löfroth, G., Eds.,** Environmental tobacco smoke and passive smoking (special issue), *Mutat. Res.,* 222, 1989.

17. **Wynder, E. L.,** Tobacco and health: a review of the history and suggestions for public health policy, *Publ. Health Rep.,* 103, 8, 1988.

18. **Bannasch, P., Ed.,** *Cancer Risks: Strategies for Elimination,* Springer-Verlag, Berlin, 1986.

19. **Schottenfeld, D. and Fraumeni, J. F., Jr.,** *Cancer Epidemiology and Prevention,* W. B. Saunders, Philadelphia, 1982.

20. **Parkin, D. M., Laara, E., and Muir, C. S.,** Estimates of the world wide frequency of sixteen major cancers in 1980, *Int. J. Cancer,* 41, 184, 1988.

21. **Joossens, J. V., Hill, M. J., and Geboers, J., Eds.,** *Diet and Human Carcinogenesis,* Excerpta Medica, Amsterdam, 1985.

22. **Wynder, E. L. and Hiyama, T.,** Comparative epidemiology of cancer in the United States and Japan: preventive implications, *Gann Monogr. Cancer Res.,* 33, 183, 1987.

23. **Farber, E., Kawachi, T., Nagayo, T., Sugano, H., Sugimura, T., and Weisburger, J. H., Eds.,** *Pathophysiology of Carcinogenesis in Digestive Organs,* University of Tokyo Press, Tokyo, 1977.

24. **Bartsch, H., O'Neil, I. K., and Schulte-Hermann, R., Eds.,** *The Relevance of N-Nitroso Compounds to Human Cancer: Exposures and Mechanisms,* IARC Sci Publ. 84, International Agency for Research on Cancer, Lyon, France, 1986.

25. **Ohshima, H., Friesen, M., Malaveille, C., Brouet, I., Hautefeuille, A., and Bartsch, H.,** Formation of direct-acting genotoxic substances in nitrosated smoked fish and meat products: identification of simple phenolic precursors and phenyldiazonium ions as reactive products, *Food Chem. Toxicol.,* 27, 193, 1989.

26. **Weisburger, J. H.,** Application of the mechanisms of nutritional carcinogenesis to the prevention of cancer, in *Diet, Nutrition, and Cancer,* Hayashi, Y., Nagao, M., Sugimura, S., Takayama, L., Tomatis, L., Wattenberg, L. W., and Wogan, G. N., Eds., Japan Scientific Societies Press, Tokyo, and VNU Science Press, Utrecht, 1986, 11.

27. **Tannenbaum, S. R.,** Diet and exposure to N-nitroso compounds, in *Diet, Nutrition, and Cancer,* Hayashi, Y., Nagao, M., Sugimura, T., Takayama, S., Tomatis, L., Wattenberg, L. W., and Wogan, G. N., Eds., Japan Scientific Societies Press, Tokyo, and VNU Science Press, Utrecht, 1986, 67.

28. **Tominaga, S.,** Cancer incidence in Japanese in Japan, Hawaii, and Western United States, *Natl. Cancer Inst. Monogr.,* 69, 83, 1985.

29. American Health Foundation, Proceedings of a workshop on new developments on dietary fat and fiber in carcinogenesis (optimal types and amounts of fat or fiber), *Prev. Med.,* 16, 449, 1987.

30. **Cohen, L. A.,** Diet and cancer, *Sci. Am.,* 257(5), 42, 1987.

31. **Popkin, B. M., Haines, P. S., and Reidy, K. C.,** Food consumption trends of US women: patterns and determinants between 1977 and 1985, *Am. J. Clin. Nutr.,* 49, 1307, 1989.

32. Committee on Diet and Health, Food and Nutrition Board, *Diet and Health: Implications for Reducing Chronic Disease Risk,* National Academy Press, Washington, D.C., 1989.

33. **Reddy, B. S. and Cohen, L. A., Eds.,** *Diet, Nutrition, and Cancer: A Critical Evaluation,* Vol. I, CRC Press, Boca Raton, FL, 1986.

34. **Carroll, K. K., Braden, M. S., Bell, J. A., and Kalamegham, R.,** Fat and Cancer, *Cancer,* 58, 1818, 1986.

35. **Woods, M. N., Gorbach, S. L., Longcope, C., Goldin, B. R., Dwyer, J. T., and Morrill-LaBrode, A.,** Low-fat, high-fiber diet and serum estrone sulfate in premenopausal women, *Am. J. Clin. Nutr.,* 49, 1179, 1989.

36. **Silverman, J., Powers, J., Stromberg, P., Pultz, J. A., and Kent, S.,** Effects on C3H mouse mammary cancer of changing from a high fat to a low fat diet before, at, or after puberty, *Cancer Res.,* 49, 3857, 1989.

37. **Reddy, B., Engle, A., Katsifis, S., Simi, B., Bartram, H. P., Perrino, P., and Mahan, C.,** Biochemical epidemiology of colon cancer: effect of types of dietary fiber on fecal mutagens, acid, and neutral sterols in healthy subjects, *Cancer Res.,* 49, 4629, 1989.

38. **Sato, Y., Furihata, C., and Matsushima, T.,** Effects of high fat diet on fecal contents of bile acids in rats, *Jpn. J. Cancer Res. (Gann),* 78, 1198, 1987.

39. **Hayatsu, H., Arimoto, S., and Negishi, T.,** Dietary inhibitors of mutagenesis and carcinogenesis, *Mutat. Res.,* 202, 429, 1988.

40. **Kawaura, A., Tanida, N., Sawada, K., Oda, M., and Shimoyama, T.,** Supplemental administration of 1-hydroxyvitamin D_3 inhibits promotion by intrarectal instillation of lithocholic acid in *N*-nitrosourea-induced colonic tumorigenesis in rats, *Carcinogenesis,* 10, 647, 1989.

41. **Jacobs, L. R.,** Role of dietary factors in cell replication and colon cancer, *Am. J. Clin. Nutr.,* 48, 775, 1988.

42. **Lipkin, M.,** Biomarkers of increased susceptibility to gastrointestinal cancer: new application to studies of cancer prevention in human subjects, *Cancer Res.,* 48, 235, 1988.

43. **Slattery, M. L., Sorenson, A. W., and Ford, M. H.,** Dietary calcium intake as a mitigating factor in colon cancer, *Am. J. Epidemiol.,* 128, 504, 1988.

44. **Moon, T. E. and Micozzi, M. S., Eds.,** *Nutrition and Cancer Prevention,* Marcel Dekker, New York, 1989.

45. **Nagao, M., Honda, Y., Seino, Y., Yahagi, T., and Sugimura, T.,** Mutagenicities of smoke condensates and the charred surface of fish and meat, *Cancer Lett.,* 2, 221, 1977.

46. **Katayama, S., Fiala, E., Reddy, B. S., Rivenson, A., Silverman, J., Williams, G. M., and Weisburger, J. H.,** Prostate adenocarcinoma in rats: induction by 3,2'-dimethyl-4-aminobiphenyl, *J. Natl. Cancer Inst.,* 68, 867, 1982.

47. **Tada, M., Aoki, H., Kojima, M., Morita, T., Shirai, T., Yamada, H., and Ito, N.,** Preparation and characterization of antibodies against 3,2'-dimethyl-4-aminobiphenyl-modified DNA, *Carcinogenesis,* 10, 1397, 1989.

48. **Shirai, T., Nakamura, A., Wada, S., and Ito, N.,** Pancreatic acinar cell tumors in rats induced by 3,2'-dimethyl-4-aminobiphenyl, *Carcinogenesis,* 10, 1127, 1989.

49. **Weisburger, J. H.,** Current views on mechanisms concerned with the etiology of cancers in the digestive tract, in *Pathophysiology of Carcinogenesis in Digestive Organs,* Farber, E., Kawachi, T., Nagayo, T., Sugano, H., Sugimura, T., and Weisburger, J. H., Eds., University of Tokyo Press, Tokyo, 1977.

50. **Sugimura, T.,** New environmental carcinogens in daily life, *Trends Pharmacol. Sci.,* 9, 205, 1989.

51. **Tanaka, T., Barnes, W. S., Williams, G. M., and Weisburger, J. H.,** Multipotential carcinogenicity of the fried food mutagen 2-amino-3-methylimidazo[4,5-f]quinoline in rats, *Jpn. J. Cancer. Res. (Gann),* 76, 570, 1985.

52. **Hatch, F. T., Knize, M. G., Healy, S. K., Slezak, T., and Felton, J. S.,** Cooked-food mutagen reference list and index, in *Environmental and Molecular Mutagenesis,* Vol. 12, Suppl. 14, Alan R. Liss, New York, 1988.

53. **Hayashi, Y., Nagao, M., Sugimura, T., Takayama, S., Tomatis, L., Wattenberg, L. W., and Wogan, G. N., Eds.,** *Diet, Nutrition and Cancer,* Japan Scientific Societies Press, Tokyo, and VNU Science Press, Utrecht, 1986.

54. **Knudsen, I., Ed.,** *Genetic Toxicology of the Diet,* Alan R. Liss, New York, 1986.

55. **Furihata, C. and Matsushima, T.,** Mutagens and carcinogens in foods, *Annu. Rev. Nutr.,* 6, 67, 1986.

56. **Reutersward, A. L., Skog, K., and Jägerstad, M.,** Effects of creatine and creatinine content on the mutagenic activity of meat extracts, bouillons and gravies from different sources, *Food Chem. Toxicol.,* 25, 747, 1987.

57. **Shibamoto, T.,** Genotoxicity testing of Maillard reaction products, in *The Maillard Reaction in Aging, Diabetes, and Nutrition,* Baynes, J. W. and Monnier, V. M., Eds., Alan R. Liss, New York, 1989.

58. **Spingarn, N., Kasai, H., Vuolo, L., Nishimura, S., Yamaizumi, Z., Sugimura, T., Matsushima, T., and Weisburger, J. H.,** Formation of mutagens in cooked foods. III. Isolation of a potent mutagen from beef, *Cancer Lett.,* 9, 177, 1980.

59. **Wang, Y. Y., Vuolo, L. L., Spingarn, N. E., and Weisburger, J. H.,** Formation of mutagens in cooked foods., V., The mutagen reducing effect of soy protein concentrates and antioxidants during frying of beef, *Cancer Lett.,* 16, 179, 1982.

60. **Jones, R. C. and Weisburger, J. H.,** Characterization of aminoalkylimidazol-4-one mutagens from liquid-reflux models, *Mutat. Res.,* 222, 43, 1989.

61. **Jones, R. C. and Weisburger, J. H.,** Inhibition of aminoimidazoquinoxaline-type and aminoimidazol-4-one-type mutagen formation in liquid reflux models by L-tryptophan and other selected indoles, *Jpn. J. Cancer Res. (Gann),* 79, 222, 1988.

62. **Jones, R. C. and Weisburger, J. H.,** L-Tryptophan inhibits formation of mutagens during cooking of meat and in laboratory models, *Mutat. Res.,* 206, 343, 1988.

63. **Jones, R. C. and Weisburger, J. H.,** Inhibition of aminoimidazoquinoxaline-type and aminoimidazol-4-one type mutagen formation in liquid-reflux models by the amino acids L-proline and/or L-tryptophan, *Environ. Mol. Mutagen.,* 11, 509, 1988.

64. **Schiffman, M. H.,** Epidemiology of fecal mutagenicity, *Epidemiol. Rev.,* 8, 92, 1986.

65. **Rose, D. P., Boyar, A. P., and Wynder, E. L.,** International comparisons of mortality rates for cancer of the breast, ovary, prostate, and colon, and per capita food consumption, *Cancer,* 58, 2363, 1986.

66. **Howson, C. P., Hiyama, T., and Wynder, E. L.,** The decline in gastric cancer: epidemiology of an unplanned triumph, *Epidemiol. Rev.,* 8, 1, 1986.

67. **Weisburger, J. H. and Jones, R. C.,** Nutritional toxicology: on the mechanisms of inhibition of formation of potent carcinogens during cooking, in *The Maillard Reaction in Aging, Diabetes, and Nutrition,* Alan R. Liss, New York, 1989.

Chapter 10.2

PREVENTION OF NITROSAMINE FORMATION

Kiyomi Kikugawa and Tetsuta Kato

TABLE OF CONTENTS

I. INTRODUCTION

Man may be exposed to a hazard of carcinogenic nitrosamines and *N*-nitroso compounds on two accounts: one by ingesting the compounds preformed in food and the other by synthesizing the compounds in the stomach from nitrite and *N*-nitrosatable compounds which may be ingested as natural constituents of food, drugs, and food additives. Synthesis of the compounds from nitrite and amines or amides has been demonstrated *in vitro* with simulated gastric conditions and *in vivo* in animals.

Mechanistic studies of *N*-nitrosation in aqueous solutions are complicated, since nitrite or nitrous acid (pKa 3.4) generates various effective nitrosating agents, i.e., nitrous anhydride (N_2O_3), nitrous acidium ion ($ON-OH_2^+$), and nitrosonium ion (NO^+). Nitrosamines are produced from nitrite and secondary amines under mildly acidic conditions at pH 1.5 to 5.[1] Nonprotonated forms of nitrous acid and amines are concerned in the nitrosating reactions, and the nitrosating species is nitrous anhydride. Alkylureas and carbamates are more rapidly nitrosated in the more acidic pH region.[2] The main nitrosating agent for the amides is probably nitrous acidium ion (Figure 1).

The mechanism of *N*-nitrosopyrrolidine (NPYR) formation during frying bacon has been studied, and it was suggested that nitrous acid is converted into nitrous anhydride, which in turn undergoes dissociation at high temperatures to NO· and NO_2· radicals.[3] The NO_2· radical abstracts the amino proton from proline to give a radical which combines with the NO· radical to form *N*-nitrosoproline. Decarboxylation of *N*-nitrosoproline may give NPYR (Figure 2).

Many chemicals that prevent or regulate the formation of nitrosamines or *N*-nitroso compounds are known. The present authors will briefly review prevention of the formation of nitrosamines and *N*-nitroso compounds.

II. INORGANIC AND ORGANIC SALTS

It has been demonstrated that several anions exert an accelerating effect on the nitrosation of secondary amines in strongly acidic media.[4-7] Thiocyanate, a normal constituent of human saliva, has a pronounced catalytic effect on the reaction. The effect of the anion appears to result from the formation of nitrosyl thiocyanate (NOSCN). Chloride and bromide promoted the reactions by formation of nitrosyl halides (NOCl and NOBr). On the other hand, the absence of the catalytic effect of these anions on the nitrosation of methylurea has been demonstrated.[8]

Mirvish et al.[9] reported that the rate of nitrosamine formation was reduced at pH 3.4 by sodium chloride, which was attributed to activity effects. The effect of a salt on the reaction rate is suggested to be due to at least two different reasons.[10] Increasing ionic strength will favor the stability of nitrite, thereby reducing the concentration of nitrous acid and thus active nitrous anhydride. Increased ionic strength will also tend to favor protonated amines rather than free amines which are involved in the nitrosation reaction. Whereas sodium chloride might therefore activate nitrosamine formation in strongly acidic media, the same ion would be expected to produce a moderately inhibiting effect on the nitrosation under mildly acidic or neutral conditions.

Nitrous acid-nitrite equilibrium in the presence and absence of various salts has been investigated spectrophotometrically (Table 1).[11] Plots of absorbance at the absorption maximum of nitrous acid vs. pH value show that the dissociation constant of the nitrous acid-nitrite equilibrium is decreased by addition of inorganic and organic salts. Thus, the conversion of nitrite to nitrous acid at the pH values around the pKa value is suppressed by addition of these salts, and nitrosamine formation is prevented.

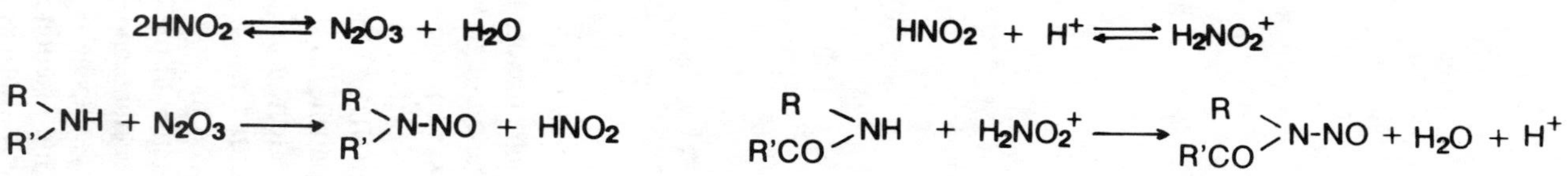

$$2HNO_2 \rightleftharpoons N_2O_3 + H_2O$$

$$\begin{array}{c} R \\ R' \end{array}\!\!>\!NH + N_2O_3 \longrightarrow \begin{array}{c} R \\ R' \end{array}\!\!>\!N\text{-}NO + HNO_2$$

$$\text{Rate} = k\,[RR'NH][HNO_2]^2$$

$$HNO_2 + H^+ \rightleftharpoons H_2NO_2^+$$

$$\begin{array}{c} R \\ R'CO \end{array}\!\!>\!NH + H_2NO_2^+ \longrightarrow \begin{array}{c} R \\ R'CO \end{array}\!\!>\!N\text{-}NO + H_2O + H^+$$

$$\text{Rate} = k\,[RNHCOR'][HNO_2][H^+]$$

FIGURE 1. Formation of nitrosamines and nitrosamides under acidic conditions.

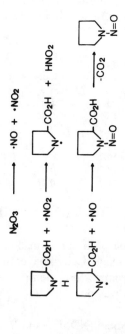

$$N_2O_3 \longrightarrow \,\cdot NO + \cdot NO_2$$

FIGURE 2. Formation of NPYR during bacon frying.

TABLE 1
**Effect of Salts on pKa of Nitrite and Formation
of Nitrosodimethylamine (NDMA) from
Dimethylamine and Nitrite**

Solution	Δ pK$_a$[a]	Formation of NDMA
3.4 *M* Sodium chloride	−0.75	Inhibited
0.6 *M* Sodium succinate	−0.30	Inhibited
1.0 *M* Sodium citrate	−0.45	Inhibited
0.5 *M* Sodium tartrate	−0.85	Inhibited

[a] Δ pKa = pKa in test solution — pKa (3.30) in control.

TABLE 2
**Effects of Alcoholic Drinks on NDMA Formation
from 0.2 *M* Nitrite and 0.051 *M* Dimethylamine in
a 3.5-hr Incubation at 37°C**

pH	Alcoholic drink	Ethanol (%)	NDMA formation	
			Ratio	% of pH-specific control value
3.0	None	0	1.00	100
	Whiskey	16.7	0.41	41
	Sake	13.3	0.11	11
5.0	None	0	0.11	100
	Whiskey	16.7	0.27	245
	Sake	13.3	0.23	209
	Wine	11.7	0.21	191

III. ORGANIC SOLVENTS AND LIPIDS

It has been reported that during bacon frying formation of NPYR occurs preferentially in the fat and that this effect is due to the nonpolar lipid of the adipose tissue which provides the environment conductive to nitrosamine formation.[12,13] An enhanced nitrosation of dihexyl-, dibutyl-, and dipropylamines was observed in a heterogeneous model system containing a *n*-decane phase.[14]

Formation of *N*-nitrosodimethylamine (NDMA) and *N*-nitrosodiethylamine (NDEA) in organic solvents such as chloroform, benzene, ethyl acetate, *n*-hexane, acetonitrile, and dioxane was much faster than that in the control aqueous buffer at pH 3.[15] The effect of the organic solvents may be due to their ability to suppress ionization of nitrous acid. The use of pH-controlled aqueous systems containing acetonitrile and dioxane accelerated the formation of the nitrosamines by increasing the concentration of nitrous acid. Formation of the nitrosamines in an aqueous buffer was accelerated or inhibited by addition of acetone and ethanol depending upon the pH conditions. Alcoholic drinks such as whiskey, Japanese sake, and wine inhibited NDMA and NDEA formation at pH 3 but enhanced the reaction at pH 5 (Table 2).[16] The inhibition at pH 3 may be due to the transformation of nitrite into inactive nitrite ester of ethanol, whereas the enhancement at pH 5 may be due to the solvent effects.[15]

The formation of NDMA in chloroform was suppressed by the presence of unsaturated fatty acid esters and chicken adipose tissues.[17] Unsaturated fatty acids are effective scavengers of the nitrosating agent. They were converted into peroxide species by reaction with nitrous acid. It was shown also that a lipid hydroperoxide inhibited NDMA formation in an aqueous

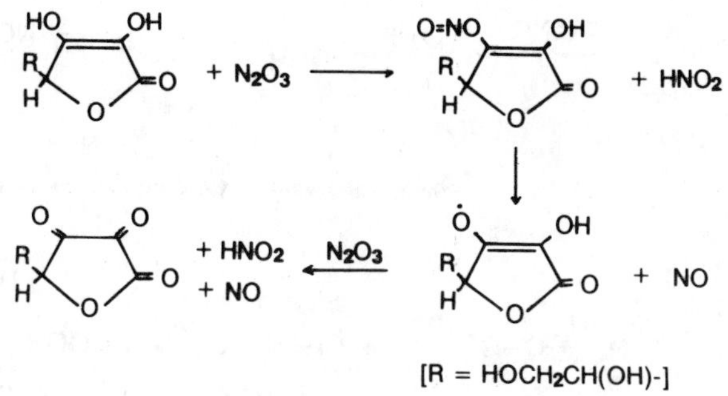

$$[R = HOCH_2CH(OH)-]$$

FIGURE 3. Reaction of nitrosating agents with ascorbic acid.

reaction mixture at pH 3 to 5.[18] The inhibition may be due to the loss of nitrite by conversion into nitrate and the formation of adducts containing nitrogen atoms. Hence, unsaturated fatty acids and their hydroperoxides may be effective inhibitors of nitrosamine formation.

Cow's milk[19,20] and soya products (soya-bean curd, soya milk, and soya-bean paste)[21] inhibited nitrosamine formation in a heterogeneous system. The inhibition may be due to the unsaturated fatty acids[20,21] or the proteins[19] in the products.

IV. ASCORBIC ACID AND RELATED COMPOUNDS

Mirvish et al.[22] demonstrated that the formation of N-nitroso compounds by the reaction between nitrite and oxytetracycline, morpholine, piperazine, methylaniline, methylurea, and dimethylamine was blocked by ascorbic acid. Fan and Tannenbaum[19] obtained similar results. Shortly thereafter it was reported that induction of lung adenomas in mice by piperazine and nitrite was inhibited when the food contained sodium ascorbate.[23] Rat liver damage was much reduced by the co-administration of ascorbate with the mixture of nitrite and aminopyrine.[24] Examination of several potential inhibitors of nitrosamine formation in human gastric juice concluded that ascorbic acid is the most suitable inhibitor because of its activity at pH values occurring in the stomach and because of its lack of toxicity.[25,26]

The reaction of ascorbic acid with nitrite is a general reaction of reductones and has characteristics of a typical oxidation-reduction process and gives quantitative formation of nitric oxide and dehydroascorbic acid according to the stoichiometric equation:

$$C_6H_8O_6 + N_2O_3 \longrightarrow C_6H_6O_6 + 2NO + 2H_2O$$

A mechanism for this reaction is shown in Figure 3.[27] Between pH 1.5 and 5, the nitrosation of secondary amines in the presence of ascorbic acid can be described by two competitive reactions. N-Methylaniline is nitrosated so rapidly that blocking of its nitrosation by ascorbate is only partial, and morpholine and piperazine, on the other hand, react less rapidly and their nitrosations can be effectively blocked.[22]

Ascorbic acid and its derivatives have been widely studied as inhibitors of the N-nitrosation reactions in bacon and frankfurters.[3,28-34] Fiddler et al.[28] used sodium ascorbate and erythorbate to inhibit NDMA formation in frankfurters. Walters et al.[29] demonstrated that the incorporation of up to 300 ppm of ascorbate eliminated NPYR formation in bacon frying. Above this level, however, increased amounts of the nitrosamine were detected. Similarly, Sen et al.[30] reported that treatment of bacon with 1000 ppm of ascorbate generally reduced the formation, although in some cases an enhancement was observed. Mottram and Patterson[31] examined the effect of ascorbate on the nitrosation in a two-phase system comprising aqueous and nonpolar solvent. In this simulated fat system it was found that after 2 hr ascorbate markedly increased nitrosamine formation as compared with the ascorbate-free controls. Similar enhancement was observed in a heterogeneous model system.[32]

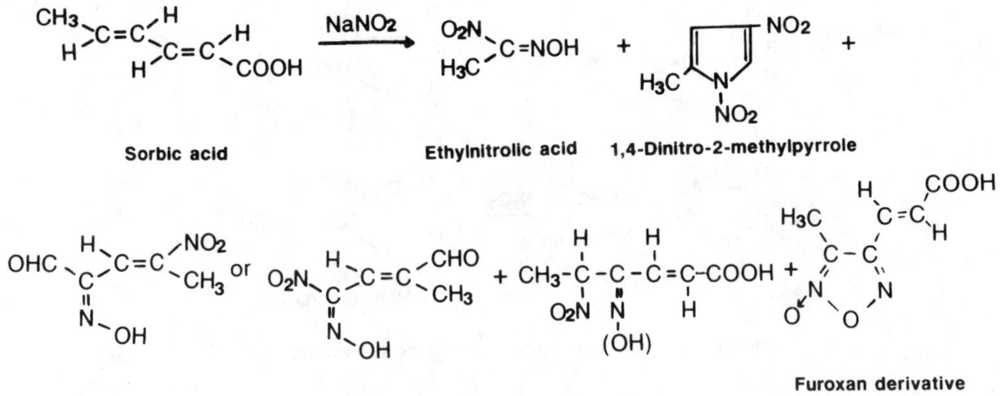

FIGURE 4.   Reaction of sorbic acid with nitrite.

On the basis of the mechanisms of formation of NPYR in bacon, it was postulated that a good nitrosamine blocking agent would, inter alia, satisfy the following requirements: (1) serve as a good trap of nitroxide radical, (2) be fat soluble, (3) be nonsteam-volatile, and (4) be stable up to the maximum frying temperature of about 174°C.[3] A large number of experiments involving the use of ascorbyl palmitate incorporated on bacon slices showed that it is far more effective than sodium ascorbate or erythorbate in reducing nitrosamine formation in bacon.[30,31,33] Long-chain acetals of ascorbic and erythorbic acids have been found to be excellent blocking agents, and they retain their efficacy in bacon even after 35 days at 3°C.[34]

V. SORBIC ACID

Sorbic acid, a food preservative, reacts rapidly with nitrite in acidic media. It inhibited the *in vitro* formation of NDMA.[35] The extent of the inhibition by sorbic acid was approximately the same as that by ascorbic acid. The inhibitory effect of the acid on the nitrosamine formation in a heterogeneous model system was described.[32] The reaction of sorbic acid with nitrite has been extensively investigated by Namiki et al.[36-39] The reaction afforded at least ten compounds, among which ethylnitrolic acid, 1,4-dinitro-2-methylpyrrole, and a furoxan derivative and its precursors were obtained (Figure 4). It should be noted that ethylnitrolic acid and 1,4-dinitro-2-methylpyrrole are mutagenic.[39] Sorbic acid inhibits the nitrosamine formation but produces other mutagens.

VI. TOCOPHEROL AND OTHER LIPID ANTIOXIDANTS

Phenolic lipid antioxidants, i.e., α-tocopherol and propyl gallate (Figure 5), are effective in the inhibition of nitrosamine formation in a heterogeneous system at pH 3.5.[40] The inhibition of formation of NPYR and NDMA in fried bacon by the use of α-tocopherol has been demonstrated.[29,41] The inhibitory effect was also observed in an oil-aqueous-protein system.[42] It has been shown that treatment just prior to frying of normal nitrite-cured bacon with 1000 ppm of propyl gallate markedly reduces the formation of NPYR during cooking.[30] Antioxidants ethoxyquin, BHA, and BHT also inhibited NPYR formation in a model system for bacon frying.[13] Ethoxyquin and *p*-alkoxyanilines have been found to be excellent inhibitors of nitrosamine formation in bacon, the latter being more efficient than the former.[43,44]

Sesamol, a natural antioxidant in sesame oil, consumed nitrite to produce nitrososesamol readily and thus inhibited or accelerated NDMA formation in an aqueous system at pH 3 and 5.[45] While nitrososesamol catalytically promoted nitrosamine formation, the presence

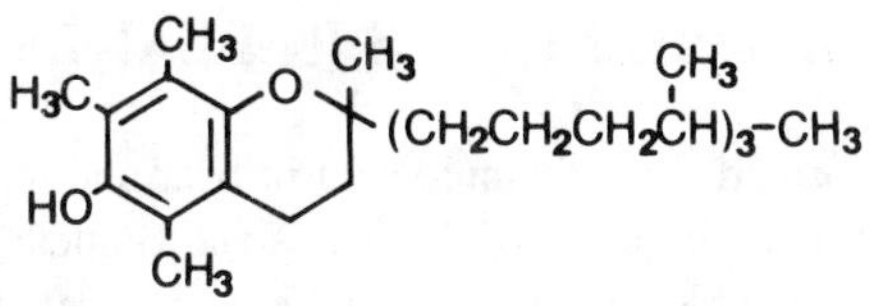

FIGURE 5. Antioxidants.

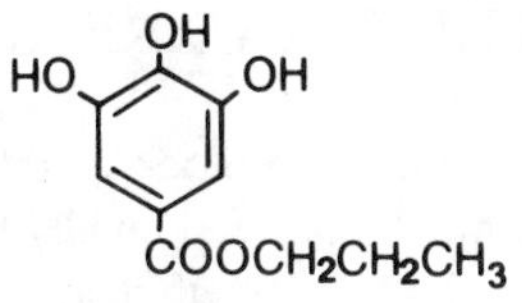

of a large amount of sesamol effectively consumed nitrite and thus reduced the available nitrite for the nitrosation. BHA prevented NDMA formation in a heterogeneous emulsion system.[46] The reaction between BHA and nitrite was shown to yield compounds including a nitro compound as a major product. α-Tocopherol, propyl gallate, and BHA were found to be inhibitory on the formation of the nitrosamide from the reaction of methylurea and nitrite at pH 2.5, due to the possible competition for available nitrite.[47] α-Tocopherol, BHA, BHT, and propyl gallate inhibit the hepatotoxicity that occurs after feeding sodium nitrite and secondary amines in rats.[48,49]

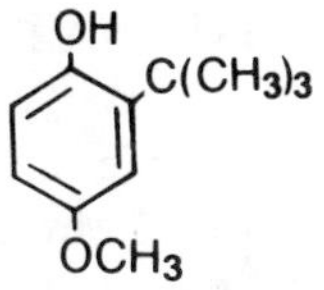

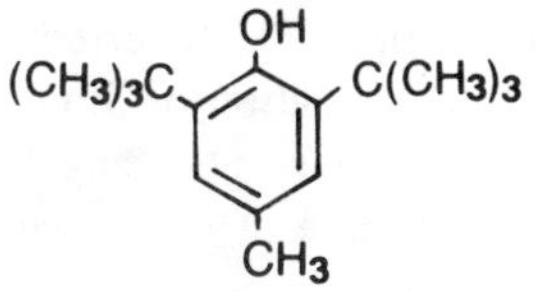

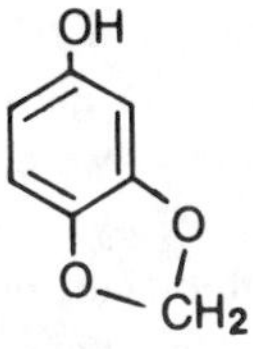

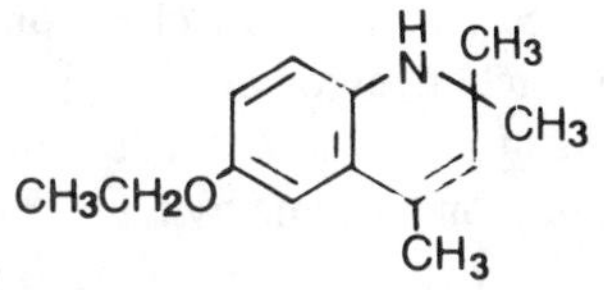

VII. VEGETABLE JUICES

Bogovski et al.[50] found that apple juice and black tea extract show an inhibitory effect on nitrosamine formation. In contrast, Japanese green tea extracts show an enhancing effect on nitrosamine formation.[51]

The present authors observed that nitrite concentration was decreased by juices from carrot, cucumber, potato, Japanese radish (Raphanus sativus), and apple, during incubation at pH 3 and 37°C for 1 hr.[52] Japanese radish juice effectively decreased nitrite at the acidic range. Nitrite-consuming activity in the juice was due to dialyzable, negatively charged, and unstable substances showing characteristic ultraviolet absorption maxima at 270 and 332 nm (pH 12). The major nitrite-reacting substance(s) was assumed to be phenolics. The nitrite-reacting substances were transformed into other compounds showing absorption maxima at 288 nm (pH 12) by reaction with nitrite. Japanese radish juice effectively inhibited NDMA and NDEA formation.

In contrast to the above observation, different results were obtained. Formation of NDMA was decreased by the addition of juices of spinach, green pepper, komatsuna, and orange, while it was increased by addition of juices of cucumber, plum, and apple.[53] The effectiveness of vegetable or fruit juices varied over a wide range from sample to sample, suggesting that the producing area, the harvesting season, and other factors influenced the inhibitory effects on nitrosamine formation. Ascorbic acid content in the juices had no apparent relation to the effects.

Juice from the Chinese wild plum (Actinidia sinensis) inhibited nitrosation reactions by efficient scavenging of nitrite.[54] This is partly due to a high concentration of ascorbic acid and partly to a nitrite scavenger, 3-hydroxy-2-pyranone.

VIII. NATURALLY OCCURRING POLYPHENOLICS IN PLANTS

Tannic acid was considered to block NDMA formation, and tannic acid in beer and tea explained why added nitrite disappeared rapidly from these beverages.[25,50] Challis and Bartlett[55] observed that readily oxidized 4-methylcatechol and chlorogenic acid, which give rapid formation of the corresponding quinones with evolution of nitric oxide, were very powerful catalysts for *N*-nitrosopiperidine formation at pH 3 to 4. They suggested that other readily oxidizable phenolics may be expected to act similarly. Gallic acid, a constituent of tannin, also showed a stimulatory effect on the formation of NDEA at around pH 4.[56] The effects of pyrocatechol, pyrogallol, 4-methylcatechol, and gallic acid on the nitrosation of dimethylamine at pH 3 showed that these polyphenolics were inhibitory rather than enhancing.[57] The nitrosation of dimethylamine, diethylamine, pyrrolidine, and piperidine at pH 3 was inhibited by pyrocatechol, pyrogallol, and gallic acid, but enhanced by catechin.[51]

Walker et al.[58] investigated the effects of natural polyphenolics such as ferulic acid, caffeic acid, chlorogenic acid, kaempferol, quercetin, fisetin, catechin, naringenin, and naringin on the nitrosation of diethylamine at pH 3 to 4 and found that they were stimulatory. They found that catechin was very stimulatory and suggested that this effect may be due to the formation of a dinitroso derivative that catalyzed the nitrosation. Pignatelli et al.[59] demonstrated the effects of resorcinol, catechin, *p*-nitrosophenol, phenol, and chlorogenic acid on *N*-nitrosoproline formation both *in vitro* and *in vivo*. The catalytic effect observed *in vivo* decreased in the same order as those observed *in vitro*: resorcinol > *p*-nitrosophenol > catechin > phenol > guaiacol; chlorogenic acid acted as an inhibitor both *in vitro* and *in vivo*. The role of catechol, 4-hydroxychavicol, eugenol, and methyleugenol on the nitrosation of pyrrolidine, piperidine, and morpholine at pH 3.5 was investigated.[60] It was observed that 4-hydroxychavicol and catechol were excellent inhibitors of the nitrosation reaction, while eugenol was less effective.

Tannic acid, gallic acid, and chlorogenic acid blocked the formation of mutagenic activity from the combination of methylurea and nitrite in an Ames bacterial mutagenesis system by inhibiting the nitrosation of methylurea.[61] Gallic acid, pyrogallol, and resorcinol were found to be inhibitory against nitrosamide formation from methylurea at pH 2.5.[47]

In conclusion, naturally occurring polyphenolics have the inhibitory or stimulatory effects on nitrosamine formation depending on their structures, reaction conditions including pH values, and basicity of the nitrosatable compounds.

IX. MONOPHENOLICS

Challis[62] showed that nitrosation of phenol was faster than that of dimethylamine. Knowles et al.[63] made the observation that nitrite interacted with a wide variety of smoke phenols that may occur in bacon during processing and frying. Vanillin, thymol, and hydroquinone were inhibitory against nitrosamine formation at pH 3.5.[40] However, the effect of *p*-nitrosocresol on nitrosamine formation from pyrrolidine at pH 5 was found to be stimulatory.[64] Nitrosation of pyrrolidine at 120°C was enhanced by *p*-cresol at pH 5 and inhibited at pH 3.[65] The catalytic effect of *p*-nitrosophenol is suggested as illustrated in Figure 6.[66,67] *Ortho*-nitrosation or nitration of monophenols by nitrite has been demonstrated.[45,46]

Virk and Issenberg[68] demonstrated that phenol enhanced NPYR formation at pH 3 and did not affect *N*-nitrosomorpholine formation, but 2,6-dimethoxyphenol (syringol) blocked the formation of both the nitrosamines. The results indicate that basicity of the amines and hence the rate of nitrosamine formation may alter the rate-modifying effect of the phenolics. Syringol deased the amount of *N*-nitrosomorpholine in both stomach and blood, while phenol had no effect, when rats were gavaged with morpholine and nitrite.[69]

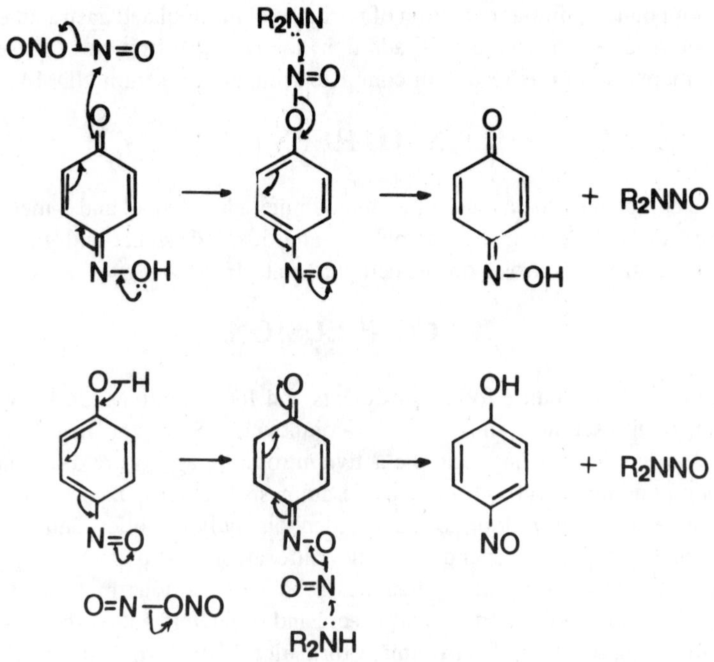

FIGURE 6. Possible mechanisms for nitrosophenol-catalyzed *N*-nitrosation.

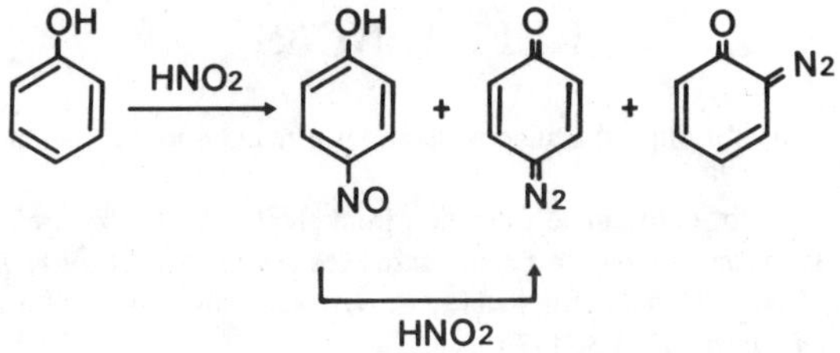

FIGURE 7. Reaction of phenol with nitrite to produce *p*-nitrosophenol, *p*-diazoquinone, and *o*-diazoquinone.

The present authors have found that mutagenic diazoquinones were produced by the reaction of phenol and nitrite.[70] *p*-Diazoquinone was found to be produced from *p*-nitrosophenol by reaction with an excess of nitrite. *p*-Nitrosophenol may serve as not only a catalyst for nitrosation but also as a precursor of the mutagenic diazoquinone (Figure 7). Monophenolic compounds such as tyramine (a component of food),[71] bamethan,[72] and etilefrin[73] (a drug orally administered) undergo diazotization at the *ortho* position of the phenol ring under mildly acidic conditions. Diazo compounds thus produced were highly mutagenic in an Ames bacterial system without metabolic activation. Although these monophenolics may consume nitrite and thus prevent nitrosamine formation, they are converted into other highly mutagenic compounds.

The present authors studied the loss of nitrite and inhibition of nitrosamine formation caused by naturally occurring *p*-coumaric acid and ferulic acid.[74] These compounds markedly reduced nitrite levels and inhibited nitrosamine formation at pH 3 to 5. The effects are attributed to the chemical reactions of the phenolics with nitrite. From the reaction of *p*-coumaric acid and ferulic acid with nitrite, many complicated products were isolated. Most of the lost nitrite may be incorporated into the olefinic groups in the substituents. Some of the products may be furoxan derivatives derived from dehydration of the nitroso-nitro or oxime-nitro compounds. Similar reduction of nitrite level in simulated gastric juice by caffeic acid and ferulic acid was observed.[75] In addition, the coadministration of these acids with nitrite and aminopyrine to rats results in complete reduction in serum NDMA levels.

X. OTHERS

Sulfhydryl compounds such as cysteine, methionine, glutathione, and 2-mercaptoethanol were inhibitory to NDMA formation at pH 3.5 and 5.5.[40] Cysteine and glutathione were also inhibitory against the nitrosation of methylurea at pH 2.5.[47]

XI. CONCLUSION

Many compounds including food constituents and food additives are known to inhibit the formation of nitrosamines and *N*-nitroso compounds. Since the mechanisms for the formation of *N*-nitroso compounds and the active nitrosating agents are different depending on the reaction conditions and the nitrosatable amines or amides, the inhibitory effect of each compound is not uniform. Inorganic salts, alcohols, polyphenolics, and monophenolics are inhibitory and stimulatory depending on the nitrosatable compounds and reaction conditions. Sorbic acid and certain monophenolics produce hazardous mutagenic compounds by inhibiting nitrosamine formation. Ascorbic acid and some lipid antioxidants may be better blocking agents for prevention of nitrosamine formation both *in vitro* and *in vivo*.

REFERENCES

1. **Mirvish, S. S.,** Kinetics of dimethylamine nitrosation in relation to nitrosamine carcinogenesis, *J. Natl. Cancer Inst.,* 44, 633, 1970.
2. **Mirvish, S. S.,** Kinetics of nitrosamide formation from alkylureas, *N*-alkylurethans, and alkylguanidines: possible implications for the etiology of human gastric cancer, *J. Natl. Cancer Inst.,* 46, 1183, 1971.
3. **Bharucha, K. R., Cross, C. K., and Rubin, L. J.,** Mechanism of *N*-nitrosopyrrolidine formation in bacon, *J. Agric Food Chem.,* 27, 63, 1979.
4. **Boyland, E., Nice, E., and Williams, K.,** Catalysis of nitrosation by thiocyanate from saliva, *Food Cosmet. Toxicol.,* 9, 639, 1971.
5. **Boyland, E. and Walker, S. A.,** Catalysis of the reaction of aminopyrine and nitrite by thiocyanate, *Artzneim. Forsch.,* 24, 1181, 1974.
6. **Boyland, E. and Walker, S. A.,** Effect of thiocyanate on nitrosation of amines, *Nature (London),* 248, 601, 1974.
7. **Fan, T. Y. and Tannenbaum, S. R.,** Factors influencing the rate of formation of nitrosomorpholine from monpholine and nitrite: acceleration by thiocyanate and other anions, *J. Agric. Food Chem.,* 21, 237, 1973.
8. **Hallett, G. and Williams, D. L. H.,** The absence of nucleophilic catalysis in the nitrosation of amides. Kinetics and mechanism of the nitrosation of methylurea and the reverse reaction, *J. Chem. Soc. Perkin II,* 1372, 1980.
9. **Mirvish, S. S., Sams, J., Fan, T. Y., and Tannenbaum, S. R.,** Kinetics of nitrosation of the amino acids proline, hydroxyproline, and sarcosine, *J. Natl. Cancer Inst.,* 51, 1833, 1973.
10. **Hildrum, K. I., Williams, J. L., and Scanlan, R. A.,** Effect of sodium chloride concentration on the nitrosation of proline at different pH levels, *J. Agric. Food Chem.,* 23, 439, 1975.
11. **Iitsuka, M., Kato, T., and Kikugawa, K.,** Inhibition of nitrosamine formation by inorganic and organic salts, *Chem. Pharm. Bull.,* 34, 3485, 1986.
12. **Mottram, D. S., Patterson, R. L. S., Edwards, R. A., and Gough, T. A.,** The preferential formation of volatile N-nitrosamines in the fat of fried bacon, *J. Sci. Food Agric.,* 28, 1025, 1977.
13. **Coleman, M. H.,** A model system for the formation of N-nitrosopyrrolidine in grilled or fried bacon, *J. Food Technol.,* 13, 55, 1978.
14. **Massey, R. C., Crews, C., Davies, R., and McWeeny, D. J.,** The enhanced N-nitrosation of lipid soluble amines in a heterogeneous model system, *J. Sci. Food Agric.,* 30, 211, 1979.
15. **Iitsuka, M., Hoshino, T., Kato, T., and Kikugawa, K.,** Formation of nitrosamines in organic solvents and aqueous systems containing organic solvents, *Chem. Pharm. Bull.,* 33, 2516, 1985.

16. **Kurechi, T., Kikugawa, K., and Kato, T.,** Effect of alcohols on nitrosamine formation, *Food Cosmet. Toxicol.,* 18, 591, 1980.

17. **Kato, T. and Kikugawa, K.,** Effect of organic solvents and unsaturated fatty acids on nitrosamine formation, *Food Chem. Toxicol.,* 22, 419, 1984.

18. **Kikugawa, K., Kato, T., Konoe, Y., and Sawamura, A.,** Effects of methyl linoleate hydroperoxide and hydrogen peroxide on *N*-nitrosation of dimethylamine, *Food Chem. Toxicol.,* 23, 339, 1985.

19. **Fan, T. Y. and Tannenbaum, S. R.,** Natural inhibitors of nitrosation reactions: the concept of available nitrite, *J. Food Sci.,* 38, 1067, 1973.

20. **Kurechi, T. and Kikugawa, K.,** Nitrite-lipid reaction in aqueous system: inhibitory effects on N-nitrosamine formation, *J. Food Sci.,* 44, 1263, 1979.

21. **Kurechi, T., Kikugawa, K., Fukuda, S., and Hasunuma, M.,** Inhibition of N-nitrosamine formation by soya products, *Food Cosmet. Toxicol.,* 19, 425, 1981.

22. **Mirvish, S. S., Wallcave, L., Eagen, M., and Shubik, P.,** Ascorbate-nitrite reaction: possible means of blocking the formation of carcinogenic N-nitroso compounds, *Science,* 177, 65, 1972.

23. **Mirvish, S. S., Cardesa, A., Wallcave, L., and Shubik, P.,** Induction of mouse lung adenomas by amines or ureas plus nitrite and by *N*-nitroso compounds: effect of ascorbate, gallic acid, thiocyanate, and caffein, *J. Natl. Cancer Inst.,* 55, 633, 1975.

24. **Kamm, J. J., Dashman, T., Conney, A. H., and Burns, J. J.,** Protective effect of ascorbic acid on hepatotoxicity caused by sodium nitrite plus aminopyrine, *Proc. Natl. Acad. Sci. U.S.A.,* 70, 747, 1973.

25. **Mirvish, S. S.,** Formation of *N*-nitroso compunds: chemistry, kinetics, and *in vivo* occurrence, *Toxicol. Appl. Pharmacol.,* 31, 325, 1975.

26. **Mirvish, S. S.,** Blocking the formation of N-nitroso compounds with ascorbic acid *in vitro* and *in vivo,* *Ann. N.Y. Acad. Sci.,* 258, 175, 1975.

27. **Bunton, C. A., Dahn, H., and Loewe, L.,** Oxidation of ascorbic acid and similar reductones by nitrous acid, *Nature (London),* 183, 163, 1959.

28. **Fiddler, W., Pensabene, J. W., Piotrowski, E. G., Doerr, R. C., and Wasserman, A.E.,** Use of sodium ascorbate or erythorbate to inhibit formation of N-nitrosodimethylamine in frankfurters, *J. Food Sci.,* 38, 1084, 1974.

29. **Walters, C. L., Edwards, M. W., Elsey, T. S., and Martin, M.,** The effect of antioxidants on the production of volatile nitrosamines during the frying of bacon, *Z. Lebensm. Unter. Forsch.,* 162, 377, 1976.

30. **Sen, N. P., Donaldson, B., Seaman, S., Iyengar, J. R., and Miles, W. F.,** Inhibition of nitrosamine formation in fried bacon by propyl gallate and L-ascorbyl palmitate, *J. Agric. Food Chem.,* 24, 397, 1976.

31. **Mottram, D. S. and Patterson, R. L. S.,** The effect of ascorbate reductants on N-nitrosamine formation in a model system resembling bacon fat, *J. Sci. Food Agric.,* 28, 352, 1977.

32. **Massey, R. C., Forsythe, L., and McWeeny, D. J.,** The effects of ascorbic acid and sorbic acid on *N*-nitrosamine formation in a heterogeneous model system, *J. Sci. Food Agric.,* 33, 294, 1982.

33. **Pensabene, J. W., Fiddler, W., Feinberg, J., and Wasserman, A. E.,** Evaluation of ascorbyl monoesters for the inhibition of nitrosopyrrolidine formation in a model system, *J. Food Sci.,* 41, 199, 1976.

34. **Bharucha, K. R., Cross, C. K., and Rubin, L. J.,** Long-chain acetals of ascorbic and erythorbic acids as antinitrosamine agents for bacon, *J. Agric. Food Chem.,* 28, 1274, 1980.

35. **Tanaka, K., Chung, K. C., Hayatsu, H., and Kada, T.,** Inhibition of nitrosamine formation in vitro by sorbic acid, *Food Cosmet. Toxicol.,* 16, 209, 1978.

36. **Namiki, M. and Kada, T.,** Formation of ethylnitrolic acid by the reaction of sorbic acid with sodium nitrite, *Agric Biol. Chem.,* 39, 1335, 1975.

37. **Kito, Y. and Namiki, M.,** A new N-nitropyrrole: 1,4-Dinitro-2-methylpyrrole, formed by the reaction of sorbic acid with sodium nitrite, *Tetrahedron,* 34, 505, 1978.

38. **Osawa, T., Kito, Y., and Namiki, M.,** A new furoxan derivative and its precursors formed by the reaction of sorbic acid with sodium nitrite, *Tetrahedron Lett.,* 1979, 4399.

39. **Namiki, M., Osawa, T., Ishibashi, H., Namiki, K., and Tsuji, K.,** Chemical aspects of mutagen formation by sorbic acid-sodium nitrite reaction, *J. Agric. Food Chem.,* 29, 407, 1981.

40. **Gray, J. I. and Dugan, L. R., Jr.,** Inhibition of N-nitrosamine formation in model food systems, *J. Food Sci.,* 40, 981, 1975.

41. **Fiddler, W., Pensabene, J. W., Piotrowski, E. G., Phillips, J. G., Keating, J., Mergens, W. J., and Newmark, H. L.,** Inhibition of formation of volatile nitrosamines in fried bacon by the use of cure-solubilized α-tocopherol, *J. Agric. Food Chem.,* 26, 653, 1987.

42. **Pensabene, J. W., Fiddler, W., Mergens, W., and Wasserman, A. E.,** Effect of α-tocopherol formulations on the inhibition of nitrosopyrrolidine formation in model systems, *J. Food Sci.,* 43, 801, 1978.

43. **Bharucha, K. R., Cross, C. K., and Rubin, L. J.,** Ethoxyquin, dihydroethoxyquin, and analogues as antinitrosamine agents for bacon, *J. Agric. Food Chem.,* 33, 834, 1985.

44. **Bharucha, K. R., Cross, C. K., and Rubin, L. J.,** *p*-Alkoxyanilines as antinitrosamine agents for bacon, *J. Agric. Food Chem.,* 34, 814, 1986.

45. **Kurechi, T., Kikugawa, K., and Kato, T.,** C-Nitrosation of sesamol and its effects on N-nitrosamine formation *in vitro, Chem. Pharm. Bull.,* 27, 2442, 1979.

46. **Kurechi, T., Kikugawa, K., and Kato, T.,** The butylated hydroxyanisole nitrite reaction: effects on N-nitrosodimethylamine formation in model systems, *Chem. Pharm. Bull.,* 28, 1314, 1980.

47. **Yamamoto, M., Yamada, T., Yoshihira, K., Tanimura, A., and Tomita, I.,** Effects of food components and additives on the formation of nitrosamides, *Food Addit. Contam.,* 5, 289, 1988.

48. **Astill, B. D. and Mulligan, L. T.,** Phenolic antioxidants and the inhibition of hepatotoxicity from *N*-dimethylnitrosamine formed *in situ* in the rat stomach, *Food Cosmet. Toxicol.,* 15, 167, 1977.

49. **Kamm, J. J., Dashman, T., Newmark, H., and Mergens, W. J.,** Inhibition of amine-nitrite hepatotoxicity by α-tocopherol, *Toxicol. Appl. Pharmacol.,* 41, 575, 1977.

50. **Bogovski, P., Castegnaro, M., Pignatelli, B., and Walker, E. A.,** The inhibiting effect of tannins on the formation of nitrosamines, in *N-Nitroso compounds, Analysis and formation.,* Bogovski, P., Preussmann, R., and Walker, E. A., Eds., IARC Scientific Publ. No. 3, International Agency for Research on Cancer, Lyon, France, 1972, 127.

51. **Nakamura, M. and Kawabata, T.,** Effect of Japanese green tea on nitrosamine formation in vitro, *J. Food Sci.,* 46, 306, 1981.

52. **Kurechi, T., Kikugawa, K., and Fukuda, S.,** Nitrite-reacting substances in Japanese radish juice and their inhibition of nitrosamine formation, *J. Agric. Food Chem.,* 28, 1265, 1980.

53. **Sato, K., Yamada, T., Yoshihira, K., and Tanimura, A.,** The effects of vegetable or fruit juices on nitrosodiethylamine formation, *Syokuhin Eiseigaku Zasshi,* 27, 619, 1986.

54. **Normington, K. W., Baker, I., Molina, M., Wishnok, J. S., Tannenbaum, S. R., and Puju, S.,** Characterization of a nitrite scavenger, 3-hydroxy-2-pyranone, from Chinese wild plum juice, *J. Agric. Food Chem.,* 34, 215, 1986.

55. **Challis, B. C. and Bartlett, C. D.,** Possible carcinogenic effects of coffee constituents, *Nature (London),* 254, 532, 1975.

56. **Walker, E. A. Pignatelli, B., and Castegnaro, M.,** Effect of gallic acid on nitrosamine formation, *Nature, (London),* 258, 176, 1975.

57. **Yamada, T., Yamamoto, M., and Tanimura, A.,** Studies on the formation of nitrosamines. VII. The effects of some polyphenols on nitrosation of diethylamine, *Syokuhin Eiseigaku Zasshi,* 19, 224, 1978.

58. **Walker, E. A., Pignatelli, B., and Friesen, M.,** The role of phenols in catalysis of nitrosamine formation, *J. Sci. Food Agric.,* 33, 81, 1982.

59. **Pignatelli, B., Bereziat, J. C., Descotes, G., and Bartsch, H.,** Catalysis of nitrosation *in vitro* and *in vivo* in rats by catechin and resorcinol and inhibition by chlorogenic acid, *Carcinogenesis,* 3, 1045, 1982.

60. **Shenoy, N. R. and Choughuley, A. S. U.,** Effect of certain plant phenolics on nitrosamine formation, *J. Agric. Food Chem.,* 37, 721, 1989.

61. **Stich, H. F., Rosin, M. P., and Bryson, L.,** Inhibition of mutagenicity of a model nitrosation reaction by naturally occurring phenolics, coffee and tea, *Mutat. Res.,* 95, 119, 1982.

62. **Challis, B. C.,** Rapid nitrosation of phenols and its implications for health hazards from dietary nitrites, *Nature (London),* 244, 466, 1973.

63. **Knowles, M. E., Gilbert, J., and McWeeny, D. J.,** Nitrosation of phenols in smoked bacon, *Nature (London),* 249, 672, 1974.

64. **Davis, R. and McWeeny, D. J.,** Catalytic effect of nitrosophenols on *N*-nitrosamine formation, *Nature (London),* 266, 657, 1977.

65. **Davies, R., Massey, R. C., and McWeeny, D. J.,** A study of the rates of the competitive nitrosations of pyrrolidine, p-cresol and L-cysteine hydrochloride, *J. Sci. Food Agric.,* 29, 62, 1978.

66. **Walker, E. A., Pignatelli, B., and Castegnaro, M.,** Catalytic effect of *p*-nitrosophenol on the nitrosation of diethylamine, *J. Agric. Food Chem.,* 27, 393, 1979.

67. **Davies, R., Massey, R. C., and McWeeny, D. J.,** The catalysis of the *N*-nitrosation of secondary amines by nitrosophenols, *Food Chem.,* 6, 115, 1980.

68. **Virk, M. S. and Issenberg, P.,** Nitrosation of phenol and 2,6-dimethoxyphenol and its effect on nitrosamine formation, *J. Agric. Food Chem.,* 33, 1082, 1985.

69. **Virk, M. S. and Issenberg, P.,** Effects of phenol and 2,6-dimethoxyphenol (syringol) on in vivo formation of *N*-nitrosomorpholine in rats, *Carcinogenesis,* 7, 867, 1986.

70. **Kikugawa, K. and Kato, T.,** Formation of a mutagenic diazoquinone by interaction of phenol with nitrite, *Food Chem. Toxicol.,* 26, 209, 1988.

71. **Ochiai, M., Wakabayashi, K., Nagao, M., and Sugimura, T.,** Tyramine is a major mutagen precursor in soy sauce, being convertible to a mutagen by nitrite, *Gann,* 75, 1, 1984.

72. **Kikugawa, K., Kato, T., and Takeda, Y.,** Formation of a highly mutagenic diazo compound from the bamethan-nitrite reaction, *Mutat. Res.,* 177, 35, 1987.

73. **Kikugawa, K., Kato, T., and Takeda, Y.,** Formation of a direct mutagen, diazo-*N*-nitrosoetilefrin, by interaction of etilefrin with nitrite, *Chem. Pharm. Bull.,* 37, 1600, 1989.

74. **Kikugawa, K., Hakamada, T., Hasunuma, M., and Kurechi, T.,** Reaction of p-hydroxycinnamic acid derivatives with nitrite and its relevance to nitrosamine formation, *J. Agric. Food Chem.,* 31, 780, 1983.
75. **Kuenzig, W., Chau, J., Norkus, E., Holowaschenko, H., Newmark, H., Mergens, W., and Conney, A. H.,** Caffeic and ferulic acid as blockers of nitrosamine formation. *Carcinogenesis,* 5, 309, 1984.

Chapter 11

CARCINOGENICITY OF FOOD MUTAGENS AND RISK ASSESSMENT

Chapter 11.1

SHORT REVIEW OF THE CARCINOGENICITIES OF MUTAGENS IN FOOD PYROLYSATES

Hiroko Ohgaki, Shozo Takayama and Takashi Sugimura

TABLE OF CONTENTS

I. INTRODUCTION

Human cancer is generally considered to be related to environmental factors.[1,2] Dietary factors, including carcinogens in the diet, seem to be very important in the development of cancers.[1-3] Charred parts of broiled fish and beefsteak were first demonstrated in our laboratory to be mutagenic to *Salmonella typhimurium* TA98 and TA100 with S9 mix.[4,5] Subsequently, a series of heterocyclic amines were isolated from pyrolysates of amino acids and proteins and were shown to be highly mutagenic.[6-9] These heterocyclic amines have also been shown to be carcinogenic in mice and rats. This chapter summarizes available data on the carcinogenicities of these heterocyclic amines.

II. CARCINOGENICITIES OF HETEROCYCLIC AMINES

A. ORAL ADMINISTRATION

Nine heterocyclic amines, namely, Trp-P-1, Trp-P-2, Glu-P-1, Glu-P-2, MeAαC, AαC, IQ, MeIQ, and MeIQx, have so far been shown to be carcinogenic. Carcinogenicity experiments have been carried out mainly in (BALB/cAnN × DBA/2N)F$_1$ (CDF$_1$) mice and F344 rats, which were given free access to diet (CE-2, CLEA, Japan) containing these heterocyclic amines from 6 to 8 weeks of age. The heterocyclic amines were added to the diet at levels of 0.01 to 0.08%, because in preliminary 4-week experiments these levels were found to cause a 5 to 10% decrease in body weight compared with that of controls. The periods of experiments are shown in Tables 1 and 2.

1. Experiments in CDF$_1$ Mice

The carcinogenicities of heterocyclic amines in CDF$_1$ mice are summarized in Table 1.[10-14] Liver tumors, identified histologically as hepatocellular carcinomas and hepatocellular adenomas, developed in all groups treated with heterocyclic amines. Multiple liver tumors developed in most animals (Figure 1). In all groups females were more susceptible to hepatocarcinogenesis than males. This difference is considered to be at least partly due to the fact that induction of P-450 by heterocyclic amines themselves is higher in females than in males.[15] Forestomach tumors developed at high incidences in mice given IQ or MeIQ (Figure 2). Histologically, these tumors were identified as squamous cell carcinomas or papillomas. It is interesting that about 40% of the squamous cell carcinomas that developed in mice given 0.04% MeIQ metastasized to the liver.[11] Blood vessel tumors were found in mice given Glu-P-1, Glu-P-2, MeAαC, or AαC. The major site of induction of blood vessel tumors was the interscapular brown adipose tissue (Figure 3), but tumors were also observed in the abdominal cavity, pleural cavity, and axilla. These tumors were identified histologically as hemangioendothelial sarcomas or hemangioendotheliomas. Lung tumors (adenocarcinomas and adenomas) were observed at higher incidences in mice fed IQ or MeIQx than in control mice. Lymphomas and leukemias were also observed at higher incidences in mice fed MeIQx than in control mice.

2. Experiments in Rats
a. Experiments in F344 Rats

Results of carcinogenicity experiments in F344 rats are summarized in Table 2.[16-20] High incidences of liver tumors were found in all experimental groups except the MeIQ group. Most of these tumors were hepatocellular carcinomas. Unlike in CDF$_1$ mice, male F344 rats tended to be more susceptible than females to hepatocarcinogenesis.

Adenocarcinomas and adenomas were induced in the small and large intestines of rats fed Glu-P-1, Glu-P-2, IQ, or MeIQ. The major sites of induction of tumors were the terminal ileum and the colon, 2 to 12 cm above the ano-rectal junction. Most intestinal tumors were multiple (Figure 4).

TABLE 1
Carcinogenicities of Heterocyclic Amines in CDF₁ Mice

Heterocyclic amine	Sex	Effective no. of mice	Concentration in diet (%)	No. of mice with tumors					Experimental period (weeks)
				Liver	Forestomach	Blood vessels	Lung	Hematopoietic system	
Trp-P-1	M	24	0.02	5 (21)	0	0	3 (13)	1 (4)	89
	F	26	0.02	16 (62)[a]	0	0	1 (4)	0	89
Trp-P-2	M	25	0.02	4 (16)	0	0	11 (44)	1 (4)	89
	F	24	0.02	22 (92)[a]	0	0	4 (17)	3 (13)	89
Glu-P-1	M	34	0.05	4 (12)	0	30 (88)[a]	6 (18)	1 (3)	57
	F	38	0.05	37 (97)[a]	0	31 (82)[a]	4 (11)	0	68
Glu-P-2	M	37	0.05	10 (27)[a]	1 (3)	27 (73)[a]	8 (22)	1 (3)	84
	F	36	0.05	36 (100)[a]	0	20 (56)[a]	1 (3)	1 (3)	83
MeAαC	M	37	0.08	21 (57)[a]	0	35 (95)[a]	5 (14)	0	73
	F	33	0.08	28 (85)[a]	0	28 (85)[a]	2 (6)	2 (6)	84
AαC	M	38	0.08	15 (39)[a]	0	20 (53)[a]	4 (11)	4 (11)	98
	F	34	0.08	33 (97)[a]	0	6 (18)	4 (12)	4 (12)	98
IQ	M	39	0.03	16 (41)[a]	16 (41)[a]	2 (5)	27 (69)[a]	8 (21)	96
	F	36	0.03	27 (75)[a]	11 (31)[a]	3 (8)	15 (42)[a]	9 (25)	96
MeIQ	M	38	0.04	7 (18)	35 (92)[a]		12 (39)	7 (18)	91
	F	38	0.04	27 (71)[a]	34 (89)[a]	1(3)	7 (16)	5 (13)	91
	M	38	0.01	11 (29)	7 (18)[a]	0	6 (16)	5 (13)	91
	F	36	0.01	4 (11)[a]	19 (53)[a]	0	12 (33)	14 (39)	91
MeIQx	M	37	0.06	16 (43)[a]	0	0	16 (43)	11 (29)[a]	84
	F	35	0.06	32 (91)[a]	0	2 (6)	15 (43)	10 (28)	84
Control[b]	M	29	—	4 (14)	0	0	6 (21)	2 (7)	91
	F	40	—	0	0	1 (3)	5 (13)	13 (33)	91

Note: Values in parentheses are percentage incidences.

[a] <0.05 vs. respective control group.
[b] Control for experiment on MeIQ.

TABLE 2
Carcinogenicities of Heterocyclic Amines in F344 Rats

Heterocyclic amine	Sex	Effective no. rats	Concentration in diet (%)	No. of rats with tumors								Experimental period (weeks)
				Liver	Large intestine	Small intestine	Oral cavity	Zymbal gland	Skin	Clitoral gland	Mammary gland	
Trp-P-1	M	40	0.015	30 (75)[a]	2	1 (3)	0	0	0	0	0	52
	F	40	0.02	37 (93)[a]	0	1 (3)	0	0	0	0	0	52
Glu-P-1	M	42	0.05	35 (83)[a]	26 (62)[a]	19 (45)[a]	3 (7)	18 (43)[a]	5 (12)[a]	—	0	64
	F	42	0.05	24 (57)[a]	10 (24)[a]	7 (17)[a]	2 (5)	18 (43)[a]	0	5 (12)[a]	0	67
Glu-P-2	M	42	0.05	11 (26)[a]	14 (35)[a]	6 (14)[a]	3 (7)	1 (2)	3 (7)	—	0	104
	F	42	0.05	2 (5)	8 (19)[a]	8 (19)[a]	2 (5)	7 (17)[a]	2 (5)	11 (26)[a]	0	104
IQ	M	40	0.03	27 (68)[a]	25 (63)[a]	12 (30)[a]	3 (7)	36 (90)[a]	17 (43)[a]	—	0	55
	F	40	0.03	18 (45)[a]	9 (23)[a]	1 (3)	1 (3)	27 (68)[a]	3 (8)	20 (50)[a]	0	72
MeIQ	M	20	0.03	1 (5)	7 (35)[a]	3 (15)	7 (35)[a]	19 (95)[a]	10 (50)[a]	—	0	40
	F	20	0.03	0	5 (25)[a]	2 (10)	7 (35)[a]	17 (85)[a]	1 (5)	0	5 (25)[a]	40
MeIQx	M	20	0.04	20 (100)[a]	2 (10)	2 (10)	0	15 (75)[a]	7 (35)[a]	12 (63)[a]	2 (10)	61
	F	19	0.04	10 (53)[a]	2 (11)	0	0	10 (53)[a]	1 (5)	—	2 (11)	61
Control[b]	M	50	—	2 (4)	0	0	0	0	0	—	0	104
	F	50	—	0	0	0	0	0	0	—	0	104

Note: Values in parentheses are percentage incidences.

[a] $p < 0.05$ *vs.* respective control group.
[b] Control for experiments on Glu-P-1 and Glu-P-2.

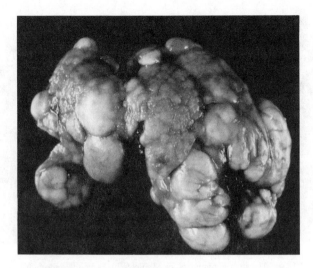

FIGURE 1. Macroscopic appearance of hepatocellular carcinoma in a female mouse fed MeIQx.

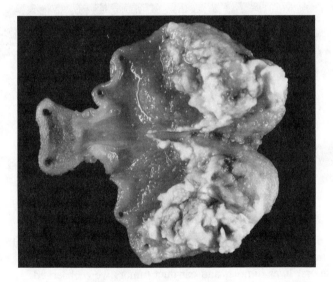

FIGURE 2. Macroscopic appearance of squamous cell carcinoma of the forestomach in a male mouse fed MeIQ.

Tumors were also induced in the Zymbal gland, skin, and clitoral glands by Glu-P-1, Glu-P-2, IQ, MeIQ, and MeIQx. These tumors were squamous cell carcinomas or sebaceous squamous cell carcinomas. Tumors in the oral cavity were found in the MeIQ group. These tumors developed in the lip and were identified histologically as squamous cell carcinomas or sebaceous squamous cell carcinomas. Mammary adenocarcinomas were also found in females fed MeIQ.

b. Experiments in Other Strains of Rats

Six neoplastic nodules and a hemangioendothelial sarcoma of the liver were found in nine female ACI rats given 0.01% Trp-P-2 in the diet for 95 to 124 weeks.[21]

IQ (0.4 mmol/kg body weight) was given by gavage to female Sprague-Dawley rats from 6 weeks after birth; it was administered three times a week in experimental weeks 1

FIGURE 3.   Macroscopic appearance of mice with hemangioendothelial sarcomas in the interscapular region induced by Glu-P-1.

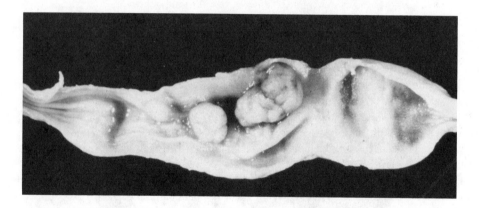

FIGURE 4.   Macroscopic appearance of a colon adenocarcinoma in a male rat given Glu-P-1.

to 4, twice a week in weeks 5 to 8, and once a week in weeks 9 to 31. All rats were killed 52 weeks after the first dose. In spite of the limited administration of IQ, high incidences of mammary tumors, liver tumors, and ear duct tumors were observed.[22]

B. OTHER ROUTES OF ADMINISTRATION
1. Subcutaneous

Trp-P-1 and Trp-P-2 were given subcutaneously to Syrian golden hamsters and F344 rats (1.5 mg per animal, once a week for 20 weeks), and the experiment was terminated in month 10. Subcutaneous sarcomas were induced at the site of injection of Trp-P-1 in 3 of 8 Syrian golden hamsters. Tumors were also induced in 5 females among 20 rats (10 males and 10 females) treated with Trp-P-1. No tumors were observed in animals treated with Trp-P-2.[23]

Newborn ICR mice were treated subcutaneously with either Glu-P-1 or Glu-P-2 (25 mg or 12.5 mg/kg body weight) within 24 hr after birth and observed for 1 year. The incidences of lung tumors in mice given Glu-P-1 or Glu-P-2 and liver tumors in male mice given Glu-P-2 were significantly higher than those in controls.[24]

TABLE 3
Potencies of Carcinogenicities of
Heterocyclic Amines

	TD$_{50}$ (mg/kg/day)		Squire
	CDF$_1$ mice	F344 rats	Score
Trp-P-1	8.8	0.1	71
Trp-P-2	2.7		71
Glu-P-1	2.7	0.8	86
Glu-P-2	4.9	5.7	86
MeAαC	5.8		71
AαC	15.8		71
IQ	14.7	0.7	81
MeIQ	8.4	0.2	81
MeIQx	11.0	0.7	81

2. Intraperitoneal

Neonatal B6C3F$_1$ mice were treated intraperitoneally with 2-amino-5-phenylpyridine, a mutagenic product of phenylalanine, at two dose levels [maximum tolerated dose (MTD) and MTD/2] on days 1, 8, 15, and 22 after birth. No treatment-related neoplastic lesions were observed in these mice.[25]

III. INITIATING ACTIVITY OF HETEROCYCLIC AMINES

For examination of the initiation activity of heterocyclic amines, these compounds were applied to the skin of CD-1 mice twice a week for 5 weeks at total doses of 5 to 100 mg, and then 12-o-tetradecanoylphorbol-13-acetate (TPA) was applied to the skin for 47 weeks. No induction of skin tumors was observed in groups given heterocyclic amines alone or TPA alone, but skin tumors developed in groups treated with both heterocyclic amines and TPA. Trp-P-1, Trp-P-2, MeAαC, and Phe-P-1 showed significant initiating activities in mouse skin.[26,27] Dietary administration of 0.1, 0.05, or 0.025% IQ for 2 weeks combined with partial hepatectomy followed by treatment with phenobarbital or 3′-methyl-4-dimethyl-aminoazobenzene resulted in dose-dependent development of neoplastic and preneoplastic lesions in the liver and thyroid of F344 rats.[28] These results showed that heterocyclic amines have initiating activity.

IV. POTENCY OF CARCINOGENICITIES OF HETEROCYCLIC AMINES

For comparison of the potencies of the carcinogenicities of heterocyclic amines with those of other carcinogens, their TD$_{50}$ values (doses for development of tumors in 50% of the animals) were calculated from the results of the experiments in mice and rats mentioned above. As shown in Table 3, the TD$_{50}$ values of heterocyclic amines were 2.7 to 15.8 mg/kg/day in CDF$_1$ mice and 0.1 to 5.7 mg/kg/day in F344 rats. These values were in a similar range to those of other carcinogens such as N,N-dimethylnitrosamine (TD$_{50}$ = 0.3 mg/kg/day) and dibenz[a,h]anthracene (TD$_{50}$ = 5.0 mg/kg/day), which show medium carcinogenic potencies.[29,30]

Squire[31] proposed a scoring system for ranking animal carcinogens according to relevant toxicological evidence derived from animal and genotoxicity studies. The factors that determine the scores include the number of different species affected, the number of histogenetically different types of neoplasms in one or more species, the spontaneous incidence of neoplasms in the control group, the cumulative oral dose of carcinogen, the malignancy of induced neoplasms, and the genotoxicity. As shown in Table 3, the scores calculated for heterocyclic amines were 71 to 86, which were similar to those of vinyl chloride (score: 90) and 2-naphthylamine (score: 81). Squire[31] reported that chemicals with scores of more than 71 should have highest priority for regulation. All the heterocyclic amines had scores of above 71.

V. DISCUSSION

All the heterocyclic amines found in cooked foods and cigarette smoke condensate have been shown to be carcinogenic to mice and rats. These heterocyclic amines have various target organs, including the liver, forestomach, lungs, blood vessels, small and large intestines, oral cavity, Zymbal glands, and skin, and all these target organs except the liver were different in mice and rats. This difference might be due to differences in metabolic activation, inactivation, and distribution of heterocyclic amines and also to differences in genetic factors in mice and rats. Recently, Adamson et al.[32] observed development of hepatocellular carcinomas with metastases to the lung and lymph nodes in cynomolgus monkeys given IQ. Their results in monkeys provide suggestive information on the target organs of heterocyclic amines in humans.

The carcinogenic risk of heterocyclic amines to humans must be evaluated. However, the discrepancy between the doses used in animal experiments and the levels to which humans are exposed makes this evaluation difficult. If a person eats 200 g of fried beef and smokes 20 cigarettes a day, the estimated daily intake of heterocyclic amines is about 3.5 μg/human/day. On the other hand, in the experiments described above, the carcinogenic dose was several milligrams per mouse or rat per day.

Recently, we studied the dose-response relationship of the carcinogenicity of MeIQx in mice.[33] When groups of 36 female CDF_1 mice were given 0.06, 0.02, 0.006, or 0.002% MeIQx in the diet for 83 weeks, their incidences of liver tumors were 82, 3, 0, and 0%, respectively. These results suggest that low doses of heterocyclic amines do not induce tumors. However, it is possible that significantly higher incidences of tumors would have been detected if larger numbers of animals had been used in the experiment. DNA adducts induced by MeIQx were also detected by the ^{32}P-postlabeling method in the liver of mice given a diet containing 0.006% MeIQx and the level of DNA adducts was about $1/_{10}$ of that in the group given 0.06% MeIQx in their diet.[33] Formation of DNA adducts in the liver of F344 rats given 0.04, 0.004, 0.0004, and 0.00004% MeIQx in their diet for 12 weeks has also been examined by ^{32}P-postlabeling analysis, and results indicated linear, dose-dependent formation of MeIQx-DNA adducts, without a threshold.[34] These results suggest that heterocyclic amines can produce DNA adducts at much lower doses than those usually used in carcinogenicity experiments. Accumulation of DNA damage and increase in the probability of occurrence of critical mutations by exposure to many environmental carcinogens might be important in the development of human cancer. Heterocyclic amines are a typical class of environmental carcinogens and should be good models for use in further studies on the mechanisms of development of human cancers.

REFERENCES

1. **Wynder, E. L. and Gori, G. B.,** Contribution of the environment to cancer incidence, *J. Natl. Cancer Inst.,* 58, 825, 1977.
2. **Doll, R.,** Strategy for detection of cancer hazards to man, *Nature (London),* 265, 589, 1977.
3. **Haenszel, W., Kurihara, M., Segi, M., and Lee, R. K. C.,** Stomach cancer among Japanese in Hawaii, *J. Natl. Cancer Inst.,* 49, 969, 1972.

4. **Sugimura, T., Nagao, M., Kawachi, T., Honda, M., Yahagi, T., Seino, Y., Sato, S., and Matsukura, N.,** Mutagen-carcinogens in foods, with special reference to highly mutagenic pyrolytic products in broiled foods, in *Origins of Human Cancer,* Hiatt, H. H., Watson, J. D., and Winsten, J. A., Eds., Cold Spring Harbor Laboratory, Cold Spring Harbor, NY, 1977, 1561.

5. **Nagao, M., Honda, M., Seino, Y., Yahagi, T., and Sugimura, T.,** Mutagenicities of smoke condensates and the charred surface of fish and meat, *Cancer Lett.,* 2, 221, 1977.

6. **Sugimura, T., Kawachi, T., Nagao, M., Yahagi, T., Seino, Y., Okamoto, T., Shudo, K., Kosuge, T., Tsuji, K., Wakabayashi, K., Iitaka, Y., and Itai, A.,** Mutagenic principle(s) in tryptophan and phenylalanine pyrolysis products, *Proc. Jpn. Acad.,* 54, 58, 1977.

7. **Yamamoto, T., Tsuji, K., Kosuge, T., Okamoto, T., Shudo, K., Takeda, K., Iitaka, Y., Yamaguchi, K., Seino, Y., Yahagi, T., Nagao, M., and Sugimura, T.,** Isolation and structure determination of mutagenic substances in L-glutamic acid pyrolysate, *Proc. Jpn. Acad. Ser. B,* 54, 248, 1978.

8. **Yoshida, D., Matsumoto, T., Yoshimura, R., and Matsuzaki, T.,** Mutagenicity of amino-α-carbolines in pyrolysis product of soybean globulin, *Biochem. Biophys. Res. Commun.,* 83, 915, 1978.

9. **Sugimura, T.,** Mutagens, carcinogens, and tumor promoters in our daily food, *Cancer,* 49, 1970, 1982.

10. **Matsukura, N., Kawachi, T., Morino, K., Ohgaki, H., Sugimura, T., and Takayama, S.,** Carcinogenicity in mice of a mutagenic compounds from a tryptophan pyrolyzate, *Science,* 213, 346, 1981.

11. **Ohgaki, H., Hasegawa, H., Suenaga, M., Kato, T., Sato, S., Takayama, S., and Sugimura, T.,** Induction of hepatocellular carcinoma and highly metastatic squamous cell carcinomas in the forestomach of mice feeding 2-amino-3,4-dimethylimidazo-[4,5-*f*]quinoline, *Carcinogenesis,* 7, 1889, 1986.

12. **Ohgaki, H., Hasegawa, H., Suenaga, M., Sato, S., Takayama, S., and Sugimura, T.,** Carcinogenicity in mice of a mutagenic compound, 2-amino-3,8-dimethylimidazo[4,5-*f*]quinoxaline (MeIQx) from cooked foods, *Carcinogenesis,* 8, 665, 1987.

13. **Ohgaki, H., Kusama, K., Matsukura, N., Morino, K., Hasegawa, H., Sato, S., Takayama, S., and Sugimura, T.,** Carcinogenicity in mice of a mutagenic compound, 2-amino-3-methylimidazo-[4,5-*F*]quinoline, from broiled sardine, cooked beef and beef extract, *Carcinogenesis,* 5, 921, 1984.

14. **Ohgaki, H., Matsukura, N., Morino, K., Kawachi, T., Sugimura, T., and Takayama, S.,** Carcinogenicity in mice of mutagenic compounds from glutamic acid and soybean globulin pyrolysates, *Carcinogenesis,* 5, 815, 1984.

15. **Degawa, M., Hashimoto, T., Yoshida, H., and Hashimoto, Y.,** Species, sex and organ differences in induction of a cytochrome P-450 isozyme responsible for carcinogen activation: effects of dietary hepatocarcinogenic tryptophan pyrolysate components in mice and rats, *Carcinogenesis,* 8, 1913, 1987.

16. **Kato, T., Migita, H., Ohgaki, H., Sato, S., Takayama, S., and Sugimura, T.,** Induction of tumors in the Zymbal gland, oral cavity, colon, skin and mammary gland of F344 rats by a mutagenic compound, 2-amino-3,4-dimethylimidazo[4,5-*f*]quinoxaline, *Carcinogenesis,* 10, 601, 1989.

17. **Kato, T., Ohgaki, H., Hasegawa, H., Sato, S., Takayama, S., and Sugimura, T.,** Carcinogenicity in rats of a mutagenic compound, 2-amino-3,8-dimethylimidazo[4,5-*f*]quinoxaline, *Carcinogenesis,* 9, 71, 1988.

18. **Takayama, S., Nakatsuru, Y., Ohgaki, H., Sato, S., and Sugimura, T.,** Carcinogenicity in rats of a mutagenic compound, 3-amino-1,4-dimethyl-5*H*-pyrido[4,3-*b*]indole, from tryptophan pyrolysate, *Jpn. J. Cancer Res.(Gann),* 76, 815, 1985.

19. **Takayama, S., Masuda, M., Mogami, M., Ohgaki, H., Sato, S., and Sugimura, T.,** Induction of cancers in the intestine, liver and various other organs of rats by feeding mutagens from glutamic acid pyrolysate, *Gann,* 75, 207, 1984.

20. **Takayama, S., Nakatsuru, Y., Masuda, M., Ohgaki, H., Sato, S., and Sugimura, T.,** Demonstration of carcinogenicity in F344 rats of 2-amino-3-methylimidazo[4,5-*f*]quinoline from broiled sardine, fried beef and beef extract, *Gann,* 75, 467, 1984.

21. **Hosaka, S., Matsushima, T., Hirono, I., and Sugimura, T.,** Carcinogenic activity of 3-amino-1-dimethyl-5*H*-pyrido[4,3-*b*]indole (Trp-P-2), a pyrolysis product of tryptophan, *Cancer Lett.,* 13, 23, 1981.

22. **Tanaka, T., Barnes, W. S., Williams, G. M., and Weisburger, J. H.,** Multipotential carcinogenicity of the fried food mutagen 2-amino-3-methylimidazo[4,5-*f*]quinoline in rats, *Jpn. J. Cancer Res. (Gann),* 76, 570, 1985.

23. **Ishikawa, T., Takayama, S., Kitagawa, T., Kawachi, T., Kinebuchi, M., Matsukura, N., Uchida, E., and Sugimura, T.,** in *Naturally Occurring Carcinogens-Mutagens and Modulators of Carcinogenesis,* Miller, E. C., Miller, J. A., Hirono, I., Sugimura, T., and Takayama, S., Eds., University Park Press, Baltimore, 1979, 159.

24. **Fujii, K., Sakai, A., Nomoto, K.-I., and Nakamura, K.,** Tumor induction in mice administered neonatally with 2-amino-6-methyldipyrido[1,2-*a*:3′,2′-*d*]imidazole or 2-amino-dipyrido-[1,2-*a*:3′,2′-*d*]imidazole, *Cancer Lett.,* 41, 75, 1988.

25. **Dooley, K. L., Stavenuiter, J. F. C., Westra, J. G., and Kadlubar, F. F.,** Comparative carcinogenicity of the food pyrolysis product, 2-amino-5-phenylpyridine, and the known human carcinogen, 4-aminobiphenyl, in the neonatal B6C3F$_1$ mouse, *Cancer Lett.,* 41, 99-103, 1988.

26. **Sato, H., Takahashi, M., Furukawa, F., Miyakawa, Y., Hasegawa, R., Toyoda, M., and Hayashi, Y.,** Initiating activity in a two-stage mouse skin model of nine mutagenic pyrolysates of amino acids, soybean globulin and proteinaceous food, *Carcinogenesis,* 8, 1231, 1987.

27. **Takahashi, M., Furukawa, F., Miyakawa, Y., Sato, H., Hasegawa, R., and Hayashi, Y.,** 3-Amino-1-methyl-5*H*-pyrido[4,3-*b*]indole initiates two-stage carcinogenesis in mouse skin but is not a complete carcinogen, *Jpn. J. Cancer Res. (Gann),* 77, 509, 1986.

28. **Tsuda, H., Asamoto, M. Ogiso, T., Inoue, T., Ito, N., and Nagao, M.,** Dose-dependent induction of liver and thyroid neoplastic lesions by short-term administration of 2-amino-3-methylimidazo[4,5-*f*]quinoline combined with partial hepatectomy followed by phenobarbital or low dose 3'-methyl-4-dimethylaminoazobenzene promotion, *Jpn. J. Cancer Res. (Gann),* 79, 691, 1988.

29. **Peto, R., Pike, M. C., Bernstein, L., Gold, L. S., and Ames, B. N.,** The TD_{50}: a proposed general convention for the numerical description of the carcinogenic potency of chemicals in chronic-exposure animal experiments, *Environ. Health Perspect.,* 58, 1, 1984.

30. **Gold, L. S., Sawyer, C. B., Magaw, R., Backman, G. M., Veciana, M., Levinson, R., Hooper, N. K., Havender, W. R., Bernstein, L., Peto, R., Pike, M. C., and Ames, B. N.,** A carcinogenic potency database of the standardized results of animal bioassays, *Environ. Health Perspect.,* 58, 9, 1984.

31. **Squire, R. A.,** Ranking animal carcinogens: a proposed regulatory approach, *Science,* 214, 877, 1981.

32. **Adamson, et al.,** personal communication.

33. **Ohgaki, H., Szentirmay, Z., Kato, S., Yamashita, K., Takayama, S., and Sugimura, T.,** Discrepancy between dose-response of carcinogenicity and DNA adduct formation in mice fed MeIQx, in preparation.

34. **Yamashita, K., Adachi, M., Nakagama, H., Sato, S., Nagao, M., and Sugimura, T.,** Exposure of rats to MeIQx: DNA modification and expression of multidrug resistant gene, in Proc. 47th Annu. Meet. Japan. Cancer Assoc., Tokyo, No. 259, 1988, 97.

Chapter 11.2

QUANTITATIVE CANCER RISK ASSESSMENT OF HETEROCYCLIC AMINES IN COOKED FOODS

David W. Gaylor and Fred F. Kadlubar

TABLE OF CONTENTS

I. INTRODUCTION

Heterocyclic amines, which can be formed during the cooking of food, represent a subclass of chemical carcinogens that are structurally and biologically related to the bicyclic and polycyclic aromatic amines. The latter, which include 2-naphthylamine and 4-amino-biphenyl, are generally recognized as strong carcinogens in both humans and dogs and, to a lesser extent, in rodent bioassays.[1] Over the last decade, several heterocyclic amines have tested positive for carcinogenicity in rats and mice in chronic feeding studies and after single subcutaneous doses in neonatal mice (see Chapter by Sugimura). These have included 3-amino-1,4-dimethyl-5*H*-pyrido-[4,3-*b*]indole (Trp-P-1), 3-amino-1-methyl-5*H*-pyrido[4,3-*b*]indole (Trp-P-2), 2-amino-6-methyldipyrido[1,2-*a*:3′,2′-*d*]imidazole (Glu-P-1), 2-aminodipyrido-[1,2-*a*:3′,2′-*d*]imidazole (Glu-P-2), 2-amino-α-carboline (AαC), 2-amino-3-methyl-α-carboline (MeAαC), 2-amino-3-methylimidazo[4,5-*f*]quinoline (IQ), 2-amino-3,4-dimethyl-3*H*-imidazo[4,5-*f*]quinoline (MeIQ), 2-amino-3,8-dimehtylimidazo[4,5-*f*]quinoxaline (MeIQx), and 2-amino-1-methyl-6-phenylimidazo[4,5-*b*]-pyridine (PhIP). The carcinogenicity of 2-amino-3,4,8-trimethylimidazo[4,5-*f*]-quinoxaline (DiMeIQx) has not yet been reported.

In the last few years, the metabolism of heterocyclic amines has been studied in some detail (see Chapter 7). Like other aromatic amines, their metabolic activation to genotoxic derivatives appears to involve *N*-hydroxylation primarily by cytochrome P-450IA2 and subsequent *O*-esterification by acetyltransferase, sulfotransferases, or aminoacyltransferases. Recently, the ability of human liver microsomal cytochrome P-450IA2 to catalyze the *N*-hydroxylation of IQ, Glu-P-1, and Trp-P-2 was examined and compared to that obtained with 4-aminobiphenyl and 2-naphthylamine (Table 1).[2] Glu-P-1 and IQ were found to be *N*-hydroxylated at rates about one half that of 4-aminobiphenyl but nearly the same as that observed with 2-naphthylamine. In contrast, Trp-P-2 was a poor substrate and was *N*-hydroxylated at only about one tenth the rate found with Glu-P-1 and IQ.

During this same period, several investigators reported the development of sensitive methods for the determination of heterocyclic amines in cooked foods, especially meats (see Chapters 5.1 and 6). As a result, a limited database has now emerged that indicates the presence of these carcinogens at parts per billion levels in the diet.

II. DOSE-RESPONSE MODELS

Thus, since certain heterocyclic amines are known to be carcinogenic in animal bioassays, to be metabolically activated by human tissue, and to be present in cooked meats, it was of interest to estimate the potential risk of cancer for humans consuming these foods. A search of the published literature provided tumor incidence data from rats or mice for ten heterocyclic amines fed in the diet. A generalized multistage model is commonly used to describe tumor dose-response data,[3]

$$P(d) = 1 - \exp[-(q_0 + q_1 d + q_2 d^2)] \tag{1}$$

where P(d) is the proportion of animals with a particular type of tumor exposed to a dose d (expressed as parts per billion in the total daily diet) of a heterocyclic amine for a significant portion of the rodent's lifetime. The q's are estimated from the bioassay data. A dose-squared term is included for MeIQ where bioassays were conducted at two dose levels plus control animals. For all of the other heterocyclic amines, only one dose level plus control animals were used in the diet. For these compounds, only a linear dose term (one-hit model) is use to fit the data. The duration of the various bioassays were long-term and differed somewhat from each other.

TABLE 1
N-Hydroxylation of Carcinogenic Heterocyclic Amines and Aromatic Amines by Human Liver Microsomes

Amine substrate	Rates of N-hydroxylation (nmol/min/mg protein)
4-Aminobiphenyl	5.00 ± 0.03
2-Naphthylamine	2.49 ± 0.32
Glu-P-1	2.42 ± 0.14
IQ	2.30 ± 0.21
Trp-P-2	0.28 ± 0.14

Taken from Butler, M. A., Iwasaki, M., Guengerich, F. P., and Kadlubar, F. F., *Proc. Natl. Acad. Sci. U.S.A.*, 86, 7696, 1989. With permission.

The generalized multistage model is dominated by the linear term at low doses. Hence, the procedure of Howe and Crump[4] provides an estimate of the upper limit on risk at low doses:

$$P^*(d) = q_1^* d \tag{2}$$

The value of q_1^* is the largest value of the low-dose slope which is compatible with the bioassay data with 95% confidence for a multistage model. This procedure provides a conservative (overestimate) of risk as long as the true dose-response is convex (curving upward) in the low-dose region. Due to a lack of information to the contrary, an implicit assumption is being made that the carcinogenic potency of heterocyclic amines is the same for both humans and rodents. If certain heterocyclic amines are not carcinogenic in humans due to their lack of metabolic activation (e.g., Trp-P-2, *vide supra*), or if there are detoxification reactions such that there are threshold doses below which a heterocyclic amine does not exhibit carcinogenic activity, then there may be no cancer risk for humans.

The interspecies, dose-scaling factor also affects estimates of risk. Dose is used here as concentration (ppb) in the total diet because this provides estimates of risk between those based on dose expressed as milligrams per kilogram body weight per day and milligrams per surface area per day, where surface area is estimated to be proportional to body weight to the two-thirds power.

The values of carcinogenic potency for the various heterocyclic amines in the diet, estimated by q_1^*, are given in Table 2. The q_1^* are estimated from the bioassay data reported by Sugimura[5] for IQ, Glu-P-1, Glu-P-2, Trp-P-1, Trp-P-2, AαC, and MeAαC; Ohgaki et al.[6] for MeIQ in mice; Kato et al.[7] for MeIQ in rats; Ohgaki et al.[8] for MeIQx in mice; Kato et al.[9] for MeIQx in rats; and Esumi et al.[10] for PhIP in mice. For each heterocyclic amine, only the tumor type producing the largest q_1^* for each sex in mice or rats is listed. In some cases, tumors are produced at four different tissue sites. In over half of the cases, the liver is the most sensitive tumor site.

III. HUMAN EXPOSURE LEVELS FROM COOKED MEATS

Levels of various heterocyclic amines reported in some cooked meats are given in Table 3. Since the conditions of cooking may affect the levels of heterocyclic amines, these data may not necessarily be typical and can only be regarded as examples. Also, not all heterocyclic amines are measured in each sample. Only examples are included for which car-

TABLE 2
Upper Limits on Estimated Carcinogenic Potency: q_1*
(Risk per ppb in Total Diet)

Chemical	Species	Sex	Tumor site/type	q_1*
IQ	Mice	M	Lung adenocarcinoma	1.8×10^{-6}
	Mice	F	Hepatocellular carcinoma	4.4×10^{-6}
	Rats	M	Zymbal gland	1.1×10^{-5}
	Rats	F	Zymbal gland	5.2×10^{-6}
MeIQ	Mice	M	Liver tumors	8.3×10^{-7}
	Mice	F	Forestomach	8.3×10^{-6}
	Rats	M	Zymbal gland	1.7×10^{-5}
	Rats	F	Zymbal gland	9.9×10^{-6}
MeIQx	Mice	M	Liver tumors	9.8×10^{-7}
	Mice	F	Liver tumors	2.0×10^{-6}
	Rats	M	Liver tumors	1.2×10^{-5}
	Rats	F	Clitoral gland	4.0×10^{-6}
Glu-P-1	Mice	M	Blood vessel sarcoma	4.4×10^{-6}
	Mice	F	Blood vessel sarcoma	3.7×10^{-6}
	Rats	M	Liver tumors	4.6×10^{-6}
	Rats	F	Liver tumors	2.4×10^{-6}
Glu-P-2	Mice	M	Blood vessel sarcoma	3.1×10^{-6}
	Mice	F	Liver carcinoma	5.0×10^{-6}
	Rats	M	Liver tumors	8.6×10^{-7}
	Rats	F	Clitoral gland	9.6×10^{-7}
Trp-P-1	Mice	M	Liver carcinoma	1.8×10^{-6}
	Mice	F	Liver carcinoma	6.7×10^{-6}
	Rats	M	Liver tumors	1.2×10^{-5}
	Rats	F	Liver tumors	1.3×10^{-5}
Trp-P-2	Mice	M	Liver carcinoma	1.5×10^{-6}
	Mice	F	Liver carcinoma	1.4×10^{-5}
	Rats	M	Liver tumors	7.9×10^{-7}
	Rats	F	Liver tumors	1.3×10^{-6}
AαC	Mice	M	Blood vessel sarcoma	1.1×10^{-6}
	Mice	F	Liver carcinoma	3.8×10^{-6}
MeAαC	Mice	M	Blood vessel sarcoma	5.4×10^{-6}
	Mice	F	Blood vessel sarcoma	3.4×10^{-6}
PhIP	Mice	M	Lymphoma	1.3×10^{-6}
	Mice	F	Lymphoma	3.3×10^{-6}

cinogenicity bioassay data exist for one or more of the heterocyclic amines measured in the sample.

IV. RISK ESTIMATION FOR DIETARY EXPOSURES

In keeping with current conventions,[3] the maximum q_1* of all tumor sites in either sex of mice or rats is used to estimate risk for each heterocyclic amines. Use of the results from the most potent tumor site in either sex of either rates or mice provides the most conservative estimate based on the available animal data and in some sense is assumed to account for the most sensitive tumor site (albeit a different tissue) in a sensitive subpopulation of humans.

The cancer risks estimated from the animal bioassays are based on the concentration of heterocyclic amines in the total diet. Obviously, cooked meats do not constitute 100% of the human diet. For ease of presentation, it is assumed that a cooked meat accounts for 10% of the total human diet. Thus, the exposure levels used in the risk estimates for the concentrations of heterocyclic amines in the total diet are 10% of those listed in Table 3. If an individual's diet consisted of 30% of a cooked meat, for example, then the estimates of risk

TABLE 3
Concentrations of Heterocyclic Amines Reported in Cooked Meats and Upper Limits of Lifetime Tumor Risks Assuming Cooked Meat is 10% of the Total Diet

Source		Heterocyclic amine (ppb)	Upper estimate of tumor risk
Fried ground beef[11]	PhIP	15	5.0×10^{-6}
	MeIQx	1	1.2×10^{-6}
	DiMeIQx	0.5	na[a]
	IQ	0.02	2.2×10^{-8}
Fried Norwegian meat[12]	MeIQx	83	1.0×10^{-4}
	PhIP	62	2.0×10^{-5}
	TMIP	24	na
	IQ	17	1.9×10^{-5}
	DiMeIQx	15	na
Heated fish (bonito)[b,13]	DiMeIQx	5.4	na
	MeIQx	5.2	6.2×10^{-6}
	IQ	?	?
	MeIQ	?	?
	Glu-P-2	?	?
Fried fish[14]	PhIP	69	2.3×10^{-5}
	MeIQx	6.4	7.7×10^{-6}
	IQ	0.16	1.8×10^{-7}
	DiMeIQx	0.10	na
	MeIQ	0.03	5.1×10^{-8}
Fried meat patty[c,15]	MeIQx	1.8	2.2×10^{-6}
	DiMeIQx	1.0	na
Fried beef patty[d,16]	MeIQx	12	1.4×10^{-5}
	DiMeIQx	3.9	na
	IQ	1.9	2.1×10^{-6}

[a] Not available due to lack of bioassay data.
[b] Levels were lower in other types of fish.
[c] Average of two patties.
[d] Cooked 15 min, lower levels in patties cooked 5 to 10 min.

would be three times higher. That is, the upper limits on risk are estimated to be directly proportional to the concentration of the heterocyclic amine in the diet.

The levels of various heterocyclic amines measured by Felton et al.[11] in fried ground beef are listed in Table 3. If fried ground beef containing these levels of heterocyclic amines constitutes 10% of the total diet for an individual, the levels in the total daily diet are 1.5 ppb PhIP, 0.1 ppb MeIQx, 0.05 ppb DiMeIQx, and 0.002 ppb IQ. No estimate of carcinogenic potency is available for DiMeIQx. The maximum q_1^* for MeIQx is obtained from liver tumors in male rats giving an upper limit estimate of risk of 1.2×10^{-5} per ppb in the total daily diet (Table 2). Assuming equal potency in humans, the upper limit estimate of risk due to average daily consumption of 0.1 ppb of MeIQx is

$$P = 1.2 \times 10^{-5} \times 0.1 = 1.2 \times 10^{-6} \qquad (3)$$

The potential risk is estimated to be less than 1.2 in a million over a long-term exposure and may be zero. For IQ, the risk is estimated to be less than:

$$P = 1.1 \times 10^{-5} \times 0.002 = 2.2 \times 10^{-8} \qquad (4)$$

TABLE 4

Upper Limits on Estimated Carcinogenic Potency from a Single Dose in Neonatal Mice: q_1* (Risk per mg/kg Subcutaneous)

Chemical	Sex	Tumor type	q_1*	Dose at maximum risk of 10^{-5} (mg/kg)
Trp-P-1	M	Liver	1.8×10^{-2}	5.5×10^{-4}
	F	Lymphoma	8.2×10^{-3}	1.2×10^{-3}
Trp-P-2	M	Liver	1.6×10^{-2}	6.3×10^{-4}
Glu-P-1	M	Lung	2.3×10^{-2}	4.3×10^{-4}
Glu-P-2	M	Lung	2.1×10^{-2}	4.9×10^{-4}

If the total diet consisted of 20% fried ground beef instead of 10%, then the risk levels would be doubled, etc.

Upper limits for the estimates of tumor risk from dietary exposure from other samples, assuming cooked meats comprise 10% of the total diet, are given in Table 3. Estimates of upper limits of lifetime tumor risks for the various heterocyclic amines measured in these samples ranged from 1.0×10^{-4} to 2.2×10^{-8}.

V. NEONATAL CARCINOGENICITY

Fujii et al.[17] gave a single injection of Trp-P-1 or Trp-P-2 subcutaneously (12.5 to 25 mg/kg) to mice 24 hr after birth. The mice were observed for 1 year. In a similar study, Fujii et al.[18] gave a single injection of Glu-P-1 or Glu-P-2 (25 mg/kg). A significant increase in tumor incidence (10 to 45%) was observed in treated animals in both studies. By comparison, 4-aminobiphenyl[19,20] and 2-naphthylamine[21] have also been tested in neonatal mice by multiple intraperitoneal or subcutaneous injections, using total doses comparable to that used for the heterocyclic amines. While 2-naphthylamine did not induce a significant incidence of tumors, 4-aminobiphenyl was strongly carcinogenic and induced a high incidence of hepatocellular carcinomas.

Upper limits on the estimated carcinogenic potency from a single dose of the heterocyclic amines tested in neonatal mice, as measured by q_1*, are given in Table 4. The proportion of animals, P, developing a tumor is estimated to be less than:

$$P = q_1 * d \qquad (5)$$

where d is the milligrams per kilogram body weight dose of a single neonatal injection. The dose corresponding to a maximum risk of 10^{-5} is $d = 10^{-5}/q_1$* (Table 4). An actual estimate of the risk to human neonates from heterocyclic amines cannot be made because the human neonatal exposure level is unknown.

An estimate of tumor risk for neonates can be made based upon an *assumed* exposure level. For purposes of illustration only, assume that the daily consumption of food by a neonate that contains heterocyclic amines is 10% of the body weight (bw), expressed in kilograms. Then, the daily intake of food containing heterocyclic amines is $0.1 \times$ bw $\times$ 10^6 mg/kg = $(10^5 \times$ bw) mg/day. For food containing 1 ppb of heterocyclic amines the daily intake of heterocyclic amines is $(10^{-9} \times 10^5 \times$ bw) mg/day or $(10^{-4}$ mg/kg bw) per day. From Table 4, the estimated upper limit on the carcinogenic risk from a single sub-cutaneous injection of Trp-P-1, Trp-P-2, Glu-P-1 or Glu-P-2 in neonatal mice is around 2 $\times$ 10^{-2} per mg/kg bw. Assuming equal carcinogenic sensitivity of mice and human neonates and equal absorption by subcutaneous and oral exposure, the upper limit of risk for a 1-day exposure of 1 ppb of Trp-P-1, Trp-P-2, Glu-P-1, or Glu-P-2 is estimated to be 2 $\times$ 10^{-2} $\times$ 10^{-4} = 10^{-6}. Presumably, the risk would be higher for repeated exposures. Again, we caution that this estimate of risk is based upon an *assumed* level of 1 ppb of Trp-P-1, Trp-P-2, Glu-P-1, or Glu-P-2 in baby food. Whether or not heterocyclic amines are present in baby food is unknown.

VI. DISCUSSION

It is generally assumed for genotoxic carcinogens that there is no threshold dose below which carcinogenicity does not occur. The linear term dominates at low doses for a multistage carcinogenic process. If the dose-response curve is convex (curving upward) at low doses, linear extrapolation from a point on the dose-response curve to zero excess tumors at zero dose provides a conservative (over) estimate of risk at low doses. The question is over what dose range is linearity a good approximation of the true dose-response curve. Unfortunately for this purpose, most of the bioassays for heterocyclic amines were conducted at one dose level, often producing tumors in more than one half of the animals. The upper limits on the low-dose slope (q_1*) are based on these relatively high dose levels. It is possible that more data at lower doses would be closer to the linear dose-response range and might result in lower estimates of risk at human exposure levels.

The risk estimates are based on assuming equal carcinogenic potency in humans and animals when the levels of the heterocyclic amines are expressed in terms of concentration (ppb) in the total diet. If the proper interspecies dose scaling factor is obtained by expressing dose in terms of body weight (mg/kg body weight), then the estimates of risk for humans based on rat data would be lower by about a factor of two. If the proper interspecies dose scaling factor should be based on surface area (mg/m²), where m² is estimated to be pro-portional to body weight to the $^2/_3$ power, then the estimates of risk for humans based on rat data would be about a factor of three higher.

Obviously, no estimates of risk could be made for those heterocyclic amines for which no bioassay data are available to estimate carcinogenic potency. Thus, it is not known whether these other heterocyclic amines pose higher or lower risks.

Cancer potency for lifetime exposure is often estimated from 2-year bioassays in rodents. Animals in most of these studies were exposed to the heterocyclic amines for less than 2 years. Most of these studies were conducted long enough to obtain relatively high proportions of animals with tumors. No doubt longer exposures would have produced more tumors. Thus, the estimates of risk may be somewhat lower than lifetime (2 year) exposures.

Of the cooked meats examined, the levels of heterocyclic amines were highest in fried Norwegian meat. Hence, this sample gave the highest estimate of the potential upper limit of lifetime risk on the order of 10^{-4}. In general, the estimated upper limits on the lifetime tumor rates for the various heterocyclic amines are on the order of 10^{-5} per ppb in the average total daily diet. This assumes equal carcinogenic potency in humans and the most sensitive tumor site in either sex of mice or rats. If several heterocyclic amines are present in the diet at 1 ppb or if one amine is present at 10 ppb of the average total diet, the potential upper limit estimate of tumor risk is on the order of 10^{-4}.

Since neonatal mice appear to be particularly sensitive to some heterocyclic amines, it might be advisable to determine the levels of exposure to human neonates to heterocyclic amines by maternal milk, if any, or by prepared cooked foods.

REFERENCES

1. **Garner, R. C., Martin, C. N., and Clayson, D. B.,** Carcinogenic aromatic amines and related compounds, in *Chemical Carcinogens,* 2nd ed., Searle, C. E., Ed., American Chemical Society, Washington, D.C., 1984, 175.

2. **Butler, M. A., Iwasaki, M., Guengerich, F. P., and Kadlubar, F. F.,** Human cytochrome P-450$_{PA}$(P-450IA2), the phenacetin O-deethylase, is primarily responsible for the hepatic 3-demethylation of caffeine and the N-oxidation of carcinogenic arylamines, *Proc. Natl. Acad. Sci. U.S.A.,* 86, 7696, 1989.

3. **Anderson, E. L.** and the Carcinogen Assessment Group of the U.S. Environmental Protection Agency, Quantitative approaches in use to assess risk, *Risk Anal.,* 3, 277, 1983.

4. **Howe, R. B. and Crump, K. S.,** *GLOBAL82: A Computer Program to Extrapolate Quantal Animal Toxicity Data to Low Doses,* K. S. Crump and Co., Ruston, LA, 1982.

5. **Sugimura, T.,** Carcinogenicity of mutagenic heterocyclic amines formed during the cooking process, *Mutat. Res.,* 150, 33, 1985.

6. **Ohgaki, H., Hasegawa, H., Suenaga, M., Kato, T., Sato, S., Takayama, S., and Sugimura, T.,** Induction of hepatocellular carcinoma and highly metastatic squamous cell carcinomas in the forestomach of mice by feeding 2-amino-3,4-dimethylimidazo[4,5-*f*]quinoline, *Carcinogenesis,* 7, 1889, 1986.

7. **Kato, T., Migita, H., Ohgaki, H., Sato, S., Takayama, S., and Sugimura, T.,** Induction of tumors in the Zymbal gland, oral cavity, colon, skin and mammary gland of F344 rats by a mutagenic compound, 2-amino-3,4-dimethylimidazo[4,5-*f*]quinoline, *Carcinogenesis,* 10, 601, 1989.

8. **Ohgaki, H., Hasegawa, H., Suenega, M., Sato, S., Takayama, S., and Sugimura, T.,** Carcinogenicity in mice of a mutagenic compound, 2-amino-3,8-dimethylimidazo[4,5-*f*]quinoxaline (MeIQx) from cooked foods, *Carcinogenesis,* 8, 665, 1987.

9. **Kato, T., Ohgaki, H., Hasegawa, H., Sato, S., Takayama, S., and Sugimura, T.,** Carcinogenicity in rats of a mutagenic compound, 2-amino-3,8-dimethylimidazo[4,5-*f*]-quinoxaline, *Carcinogenesis,* 9, 71, 1988.

10. **Esumi, H., Ohgaki, H., Kohzen, E., Takayama, S., and Sugimura, T.,** Formation of lymphoma in CDF$_1$ mice by the food mutagen, 2-amino-1-methyl-6-phenylimidazo[4,5-*b*]pyridine, *Jpn. J. Cancer Res.,* 80, 1176, 1989.

11. **Felton, J. S., Knize, M. G., Shen, N. H., Andresen, B. D., Bjeldanes, L. F., and Hatch, F. T.,** Identification of the mutagens in cooked beef. *Environ. Health Perspect.,* 67, 17, 1986.

12. **Becher, G., Knize, M. G., Nes, I. F., and Felton, J. S.,** Isolation and identification of mutagens from a fried Norwegian meat product, *Carcinogenesis,* 9, 247, 1988.

13. **Kikugawa, K. and Kato, T.,** Formation of mutagens, 2-amino-3,8-dimethylimidazo[4,5-*f*]quinoxaline (MeIQx) and 2-amino-3,4,8-trimethylimidazo-[4,5-*f*]quinoxaline (4,8-DiMeIQx), in heated fish meats, *Mutat. Res.,* 179, 5, 1987.

14. **Zhang, X.-M., Wakabayashi, K., Liu, Z.-C., Sugimura, T., and Nagao, M.,** Mutagenic and carcinogenic heterocyclic amines in Chinese cooked foods, *Mutat. Res.,* 201, 181, 1988.

15. **Murray, S., Gooderham, N. J., Boobis, A. R., and Davies, D. S.,** An assay for 2-amino-3,8-dimethylimidazo[4,5-*f*]quinoxaline and 2-amino-3,4,8-trimethylimidazo[4,5-*f*]quinoxaline in fried beef using capillary column gas chromatography electron capture negative ion chemical ionization mass spectrometry, *Biomed. Environ. Mass Spec.,* 16, 221, 1988.

16. **Turesky, R. J., Bur, H., Huynh-Ba, T., Aeschbacher, H. U., and Milon, H.,** Analysis of mutagenic heterocyclic amines in cooked beef products by high-performance liquid chromatography in combination with mass spectrometry, *Food Chem. Toxicol.,* 26, 501, 1988.

17. **Fujii, K., Nomoto, K., and Nakamura, K.,** Tumor induction in mice administered neonatally with 3-amino-1,4-dimethyl-5H-pyrido[4,3-*b*]indole or 3-amino-1-methyl-5H-pyrido[4,3-*b*]indole, *Carcinogenesis,* 8, 1721, 1987.

18. **Fujii, K., Sakai, A., Nomoto, K., and Nakamura, K.,** Tumor induction in mice administered neonatally with 2-amino-6-methyldipyrido[1,2-*a*:3′,2′-*d*]-imidazole or 2-aminodipyrido[1,2-*a*:3′,2′-*d*]imidazole, *Cancer Lett.,* 41, 75, 1988.

19. **Dooley, K. L., Stavenuiter, J. F. C., Westra, J. G., and Kadlubar, F. F.,** Comparative carcinogenicity of the food pyrolysis product, 2-amino-5-phenylpyridine, and the known human carcinogen, 4-aminobiphenyl, in the neonatal B6C3F$_1$ mouse, *Cancer Lett.,* 41, 99, 1988.

20. **Gorrod, J. W., Carter, R. L., and Roe, F. J. C.,** Induction of hepatomas by 4-aminobiphenyl and three of its hydroxylated derivatives administered to newborn mice, *J. Natl. Cancer Inst.,* 41, 403, 1968.

21. **Walters, M. A., Roe, F. J. C., Mitchley, B. C. V., and Walsh, A.,** Further tests for carcinogenesis using newborn mice: 2-naphthylamine, 2-naphthylhydroxylamine, 2-acetylaminofluorene and ethyl methane sulphonate, *Br. J. Cancer,* 21, 367, 1967.

Chapter 11.3

EPIDEMIOLOGIC STUDIES OF FECAL MUTAGENICITY, COOKED MEAT INGESTION, AND RISK OF COLORECTAL CANCER

Mark H. Schiffman, Roger Van Tassell, and A. W. Andrews

TABLE OF CONTENTS

I. INTRODUCTION

Laboratory studies have demonstrated the carcinogenic potential of heterocyclic amines found in cooked meats.[1] Epidemiologic studies are now needed to define the risk to humans of low-level, chronic ingestion of these compounds.

Based on epidemiologic studies to date, the leading candidate for a human tumor that could be related to cooked meat consumption is colorectal cancer. The evidence supporting this association is still indirect, however, deriving from investigations that were not designed to distinguish the effects of cooked meat from those of correlated dietary variables including animal fat, total fat, and total energy consumption. For example, an international comparison observed strong correlations (r = 0.8) between national per capita meat consumption and colorectal cancer incidence and mortality, but this study is usually interpreted as evidence for the carcinogenicity of a diet high in animal fat.[2] Similarly, most case-control studies of diet and colorectal cancer have focused on fat intake, although the results are also consistent with an effect of cooked meats. Most recently, an 8-year follow-up of 89,000 U.S. nurses showed a twofold prospective risk of colon cancer in women with high intakes of red meat, which again may partially reflect consumption of carcinogenic cooking products including heterocyclic amines.[20]

To discriminate in epidemiologic studies between the effects of heterocyclic amines and other meat components will require a better understanding of heterocyclic amine production and the resultant concentrations of these compounds in commonly eaten foods. Based on current understanding of heterocyclic amine production, two possibly useful discriminatory variables appear to be cooking method and meat "doneness". The heterocyclic amine content of meat rises dramatically with doneness, while the fat and caloric content do not. Boiling and broiling may result in similar caloric densities, but very different heterocyclic amine concentrations.[3]

Such distinctions might eventually permit the assessment of heterocyclic amine effects independent of other correlates of meat intake but, to date, only a few epidemiologic studies of colorectal cancer have focused on heterocyclic amine-specific variables. The results so far have failed to implicate heterocyclic amine intake as a major risk factor. A recent case-control investigation in Utah observed no significant association of colorectal cancer risk with increasing consumption of broiled and fried foods, a food group which primarily included broiled or fried beef.[4] Another case-control investigation suggested that risk of colorectal cancer increased markedly when red meats were habitually eaten "well-done",[5] but this finding could not be replicated in the large prospective study of colon cancer among nurses mentioned above.[20] It may be that the assessment by dietary questionnaire of usual cooking method or meat doneness is too crude to classify subjects correctly with regard to usual heterocyclic amine intake. If so, a methodologic step necessary to prepare for further epidemiologic work will be the creation of a more exact database of heterocyclic amine content for commonly eaten foods.

It may also be necessary to pay more attention to host susceptibility when considering the risk of heterocyclic amine ingestion. Variability in host susceptibility to colorectal cancer might derive in part from differences in the metabolic activation of ingested heterocyclic amines. Activation to genotoxic metabolites appears to be a function of hepatic P-450 enzymes[6] and bowel floral metabolism,[7] both of which may show interindividual variability with a familial component as the result of genetics and early diet. Familial influences on susceptibility to colorectal cancer occurrence have been documented for common sporadic tumors as well as for the rare familial polyposis syndromes in which nearly all affected members develop colorectal cancer. For sporadic tumors, a family history of colorectal cancer in a first-degree relative is associated with about a threefold increase in risk, and a dominant mode of inheritance has been suggested for susceptibility to colorectal neoplasia (including adenomatous polyps as well as cancer) based on data from Utah kindreds.[8] If practical methods of classifying individuals with regard to their hepatic and bowel floral metabolism of heterocyclic amines were validated, epidemiologic studies not limited entirely to exposure assessment could be broadened to investigate possible mechanisms of susceptibility as well.

Considering the current problems in measuring both heterocyclic amine exposure and host susceptibility, a very useful advance would be the validation of biomarkers of tissue-specific heterocyclic amine exposure. In other words, it would be very helpful to be able to measure the biologically effective doses of genotoxic heterocyclic amine metabolites that actually reach the colorectal mucosa, following host and bowel floral metabolism of the ingested parent compounds. Such a biomarker could be related to risk of colorectal cancer, on the one hand, and to specific dietary patterns on the other, and could serve as an intermediate marker of colorectal cancer risk in a variety of epidemiologic study designs. In this light, we have been examining fecal mutagenicity as a possible biomarker of "effective dose" of dietary genotoxins, and our recent findings from a case-control study of colorectal cancer are presented here.

II. METHODS

Fecal mutagenicity was measured in 68 patients with colorectal cancer and 114 controls, using *Salmonella* tester strains TA98 and TA100 with and without S9 activation. Samples were also tested for fecapentaenes by high-performance liquid chromatography, to permit the separation of fecapentaene and nonfecapentaene mutagenicity. Only a few details of study methodology are mentioned here; the interested reader is referred to other reports.[9,10]

We recruited patients newly diagnosed with adenocarcinoma of the colon or rectum, seen during the study period (April 1985 to June 1987) by clinical collaborators at three Washington, D.C. area hospitals. For controls, we recruited patients awaiting elective surgery for nononcologic, nongastrointestinal conditions at the three study hospitals. Controls at each hospital were frequency matched to cases on age and sex.

Cases and controls were asked to collect four 2-day stool samples using dry ice kits at home. The first sample was collected before hospitalization and treatment, which usually involved surgery. Three follow-up collections were scheduled at 1, 3, and 6 months following surgery. For this analysis, the results from the four collections were combined, as indicated below.

After collection, stools were transported to a lyophilization laboratory and freeze-dried in the individual containers without ever thawing. The lyophilate was pooled and mixed for the entire 2-day collection, then stored at $-40°C$ or colder in sealed, air-tight containers. Aliquots of the samples were sent in screw-top vials on dry ice to the two testing laboratories.

Fecapentaenes were measured at the Anaerobic Microbiology Laboratory of VPI (Blacksburg, VA). One-gram samples of the freeze-dried materials were extracted with acetone and analyzed by high-performance liquid chromatography. The reading corresponded to "total fecapentaenes", combining fecapentaene-12 and fecapentaene-14.

Salmonella/mammalian microsome mutagenicity assays were conducted at the Microbial Mutagenesis Screening Laboratory of the NCI Frederick Cancer Research Facility (Frederick, MD). Acetone extracts were tested such that each plate contained 250 mg equivalent of the original lyophilized stool sample. Each strain was tested with and without the addition of S9 mix. All experiments were conducted in duplicate, with appropriate negative controls (cells only and acetone plus cells) and a positive control (2-nitrofluorene). The mean number of revertants for duplicate test plates was compared to the mean number of revertants for the negative controls.

When a test sample yielded at least 1.5 times the mean number of spontaneous revertants (the background) a dose response was constructed using four doses of 50-, 100-, 200-, and 400-mg equivalents of lyophilate. A sample was judged mutagenic if two consecutive doses

TABLE 1
Estimated Relative Risk and 95% Confidence
Intervals Associating Colorectal Cancer and Fecal
Mutagenicity, in 68 Cases and 114 Controls[a]

	All study samples	Samples with ≥1000 ng fecapentaenes per gram dry stool deleted
TA98	2.7 (0.8—8.6)	4.4 (1.0—21.1)
TA98 + S9	1.0 (0.1—8.7)	1.1 (0.1—9.2)
TA100	0.9 (0.3—2.7)	0.7 (0.1—4.8)
TA100 + S9	1.3 (0.3—5.8)	1.5 (0.0—57.1)

[a]   Relative risks adjusted for number of samples provided.

yielded at least twice the background. Using this definition, few samples were judged mutagenic. To generate more meaningful results than the multiple comparisons and small proportions of positivity would permit, we pooled the mutagenicity data, recording for each individual the most mutagenic sample of those collected (1 = 1 to 4) during the entire study period. It was necessary to adjust, therefore, for the number of samples collected, to take into account the possibility that subjects with more samples would be more likely to have at least one mutagenic sample. For these multivariate analyses, we calculated the estimated relative risk, the ratio of colorectal cancer risk in an exposed group to that in the corresponding unexposed. A relative risk of 2.0 for fecal mutagenicity, for example, would indicate that colorectal cancer occurs twice as often among individuals with mutagenic stools than among those without; conversely, a relative risk of 0.5 would indicate that the occurrence is half as great. If the 95% confidence interval computed for the relative risk estimate excludes 1.0, this corresponds roughly to statistical significance at the level of $p = 0.05$.

To control for fecapentaene effects, samples with fecapentaene concentrations of 1000 + ng/g dry stool were termed "high-fecapentaene samples" and were excluded. The choice of 1000 ng/g as the exclusion cut-off was based on earlier methodological work, which indicated that samples containing over 1000 ng fecapentaene per gram were almost three times as likely to be mutagenic as samples with lower fecapentaene concentrations, presumably reflecting fecapentaene mutagenicity.

III. RESULTS

The cases and controls were similar in demographic features and parameters of stool sampling. We compared cases and controls with regard to the strongest mutagenic response detected in TA98 and TA100 with or without S9 activation (Table 1). When results were tabulated without consideration of fecapentaene concentration, no significant differences were found. When high-fecepentaene samples were excluded, residual TA100 mutagenicity was rare, resulting in very broad confidence intervals for the relevant estimates in Table 1. However, nonfecapentaene TA98 mutagenicity was associated with a fourfold excess in risk that achieved marginal statistical significance (lower limit of the confidence interval was 1.0). This elevation in risk persisted when the relative risk estimate was adjusted sequentially for age, sex, hospital of admission, and race.

The remainder of the analysis focused on TA98 mutagenicity, in an attempt to clarify the association with risk of colorectal cancer. Considering the original group of samples before the deletion of high-fecapentaene specimens, 10 of 68 cases (14%) and 11 of 114 controls (10%) had at least one sample definitely mutagenic in TA98 without S9. Deletion of high-fecapentaene samples eliminated two cases and seven controls from this group, including one control with two TA98 mutagenic samples, leaving the eight cases and four controls noted in Table 1. All of the excluded subjects exhibited TA100 mutagenicity as well as TA98, whereas only 2 of the remaining 12 subjects, with TA98 mutagenicity not easily explainable by high fecapentanenes, had any TA100 positive samples. Most subjects had only 1 mutagenic stool out of several collected; only 2 cases and 2 controls in the group of 12 had more than 1 mutagenic stool and none of the 12 had consistently mutagenic samples.

IV. DISCUSSION

In an investigation limited by small percentages of mutagenic samples, we have observed an association between colorectal cancer risk and nonfecapentaene mutagenicity, detected in acetone-based fecal extracts by *Salmonella* tester strain TA98 without S9 activation. At this point, we can only speculate regarding the origins of the TA98 mutagenicity. Case-control differences in fecal mutagenicity might theoretically have been produced by the cancer itself or the clinical procedures used to diagnose it, but in separate methodologic investigations, we could not demonstrate such disease effects. Moreover, we were immediately able to exclude smoking as a possible source of mutagenicity since only 1 of the 12 subjects demonstrating nonfecapentaene TA98 mutagenicity smoked in the days preceding the collection of a mutagenic stool sample. In contrast, the 12 subjects all reported eating broiled, fried, or baked meat immediately prior to the start of the collection of their mutagenic 2-day stool samples, as well as prior to most of their nonmutagenic stools.

It should be noted that virtually all other study subjects also reported recent intake of cooked meats; thus we conclude that cooked meat ingestion does not invariably lead to detectable TA98 fecal mutagenicity, even in individuals who sometimes exhibit it. Still, there is reason to believe that fecal mutagenicity can sometimes result from ingestion of cooked meats. Hayatsu et al. reported TA98 fecal mutagenicity arising in three volunteers following the ingestion of fried ground beef.[11] The mutagens (which required S9 activation, unlike our results) resembled closely but were not identical to ingested heterocyclic amines. It is known that some heterocyclic amines can be metabolized by bacterial flora of the human colon, leading to direct-acting mutagens whose activity is reduced by the addition of S9.[7] Correspondingly, the TA98 mutagenicity associated with risk of colorectal cancer in our data was detected in the absence of S9 activation, perhaps representing a particular metabolic handling of some ingested heterocyclic amines.

The results of three previous correlational studies of colorectal cancer support our study findings.[12-14] Each of these studies compared the prevalence of fecal mutagenicity in a population at high risk of colorectal cancer with the corresponding prevalence in a low-risk population. All three studies found higher prevalence of fecal mutagenicity using TA98 without S9 in the high-risk groups. On the other hand, a small case-control investigation of colorectal cancer observed virtually no increase in TA98 fecal mutagenicity among cases, but they studied a total of only 16 patients.[15] Similarly, no increase in TA98 fecal mutagenicity was observed among six volunteers fed fried meat for 3 weeks in a dietary trial.[16]

As our results demonstrate, a major difficulty in studying fecal mutagenicity is the insensitivity of standard fecal extraction and assay methods in detecting mutagens in stool.[17] Methodologic insensitivity, combined with very small sample sizes and the confounding effects of fecapentaenes, may explain some of the negative results found in previous studies attempting to correlate diet, fecal mutagenicity, and colorectal cancer. Fortunately, the monoclonal antibody and mass spectrometric techniques that have recently been developed should permit more practical and sensitive epidemiologic studies of heterocyclic amine metabolism and excretion.[18,19] In future work, we will employ these techniques to determine whether the TA98 fecal mutagenicity that we found elevated in colorectal cancer patients truly reflects heterocyclic amine metabolites derived from cooked meats.

ACKNOWLEDGMENTS

This paper was originally presented at the Fifth International Conference on Environmental Mutagens, Cleveland, Ohio, July 10 to 15, 1989.

REFERENCES

1. **Sugimura, T.,** Past, present, and future of mutagens in cooked foods, *Environ. Health Perspect.,* 67, 5, 1986.
2. **Armstrong, B. and Doll, R.,** Environmental factors and cancer incidence and mortality in different countries, with special reference to dietary practices, *Int. J. Cancer,* 15, 617, 1975.
3. **Bjeldanes, L. F., Morris, M. M., Timourian, H., and Hatch, F. T.,** Effects of meat composition and cooking conditions on mutagen formation in fried ground beef, *J. Agric. Food Chem.,* 31, 18, 1983.
4. **Lyon, J. L. and Mahoney, A. W.,** Fried foods and the risk of colon cancer, *Am. J. Epidemiol.,* 128, 1000, 1988.
5. **Schiffman, M. H. and Felton, J.,** Re: fried foods and the risk of colon cancer, *Am. J. Epidemiol.,* 131, 376—378, 1990.
6. **Snyderwine, E. G. and Battula, N.,** Selective mutagenic activation by cytochrome B_3-450 of carcinogenic arylamines found in foods. *J. Natl. Cancer Inst.,* 81, 223, 1989.
7. **Carman, R. J., Van Tassell, R. L., Kingston, D. G. I., Bashir, M., and Wilkins, T. D.,** Conversion of IQ, a dietary pyrolysis carcinogen to a direct acting mutagen by normal intestinal bacteria of humans, *Mutat. Res.,* 206, 335, 1988.
8. **Cannon-Albright, L. A., Skolnick, M. H., Bishop, D. T., Lee, R. G., and Burt, R. W.,** Common inheritance of susceptibility to colonic adenomatous polyps and associated colorectal cancers, *New Engl. J. Med.,* 319, 533, 1988.
9. **Schiffman, M. H., Van Tassell, R. L., Andrews, A. W., Wacholder, S., Daniel, J., Robinson, A., Smith, L., Nair, P. P., and Wilkins, T. D.,** Fecapentaene concentration and mutagenicity in 718 North American stool samples, *Mutat. Res.,* 222, 351, 1989.
10. **Schiffman, M. H., Andrews, A. W., Van Tassell, R. L., Smith, L., Daniel, J., Robinson, A., Hoover, R. N., Rosenthal, J., Weil, R., Nair, P. P., Schwartz, S., Pettigrew, H., Batist, G., Shaw, R., and Wilkins, T. D.,** Case-control study of colorectal cancer and fecal mutagenicity, *Cancer Res.,* 49, 3420, 1989.
11. **Hayatsu, H., Hayatsu, T., Wataya, Y., and Mower, H. F.,** Fecal mutagenicity arising from ingestion of fried ground beef in the human, *Mutat. Res.,* 143, 207, 1985.
12. **Ehrich, M., Aswell, J. L., Van Tassell, R. L., and Wilkins, T. D.,** Mutagens in the feces of 3 South African populations at different levels of risk for colon cancer, *Mutat. Res.,* 64, 231, 1979.
13. **Reddy, B. S., Sharma, C., Darby, L., Laakso, K., and Wynder, E. L.,** Metabolic epidemiology of large bowel cancer. Fecal mutagens in high and low-risk populations for colon cancer, *Mutat. Res.,* 72, 511, 1980.
14. **Mower, H. F., Ichinotsubo, D., Wang, L. W., Mandel, M., Stemmermann, G., Nomura, A., Heilbrun, L., Kamiyama, S., and Shimada, A.,** Fecal mutations in two Japanese populations with different colon cancer risks, *Cancer Res.,* 42, 1164, 1982.
15. **Askew, A. R., Ward, M., Green, M. K., and Reiber, O.,** Faecal mutagenesis and colonic cancer, *Aust. N.Z. J. Surg.,* 52, 27, 1982.
16. **de Vet, H. L., Sharma, C., and Reddy, B. S.,** Effect of dietary fried meat on fecal mutagenic and co-mutagenic activity in humans, *Nutr. Rep. Int.,* 23, 653, 1981.
17. **Schiffman, M. H.,** Epidemiology of fecal mutagenicity, *Epidemiol. Rev.,* 8, 92, 1986.
18. **Vanderlaan, M., Watkins, B. E., Hwang, M., Knize, M. G., and Felton, J. S.,** Monoclonal antibodies for the immunoassay of mutagenic compounds produced by cooking beef, *Carcinogenesis,* 9, 153, 1988.
19. **Murray, S., Gooderham, N. J., Boobis, A. R., and Davies, D. S.,** Measurement of MeIQx and DiMeIQx in fried beef by capillary column gas chromatography electron capture negative ion chemical ionisation mass spectrometry, *Carcinogenesis,* 9, 321, 1988.
20. **Willett, W.,** personal communication.

Chapter 11.4

CANCER RISK ASSESSMENT OF FOOD ADDITIVES AND FOOD CONTAMINANTS

Yuzo Hayashi

TABLE OF CONTENTS

I. INTRODUCTION

Involvement of foods and dietary habits as risk factors for human cancers has been established and emphasized by two types of evidence. Epidemiological studies in human populations have disclosed association between occurrence of various forms of cancers and ingestions of certain food components or food consumption patterns, such as increased incidence of liver cancer and ingestion of aflatoxin B_1,[1] increased occurrence of gastric cancer and intake of foods with high concentration of sodium chloride,[2] and increased incidence of breast cancer of colon cancer and high fat consumption.[3,4] Laboratory studies, using long-term animal tests according to the maximum tolerable dose (MTD) principle, have demonstrated the existence of various carcinogens in food components which include food contaminants, food additives, and chemical substances produced during the cooking processes of foods.[5,6] The convergence of these data suggests that diet is a major factor in the etiology of specific types of human cancer.[7] Thus, it is regarded as a step toward attainment of primary cancer prevention to properly assess, on the basis of available scientific information, individual carcinogenic substances found in foods for potential cancer risk in man. The aim of this chapter is to describe the current status of risk assessment conducted on food additives and food contaminants which have been shown to produce cancers in rodents.

II. DIVERSITY OF CARCINOGENS

With the objective of detecting causative principles for human cancers, a large number of chemicals have been tested for potential carcinogenicity in animals using the standard test protocol and result evaluation criteria proposed by the World Health Organization.[8] This proposal indicates that a test compound is judged positive for carcinogenicity when the test results meet one of the following three criteria:

1. A significant increase in the incidence of the same types of neoplasm found in control animals
2. The occurrence of types of neoplasm not observed in control animals
3. A decreased latent period for the production of neoplasm in comparison with that in control animals

These criteria are still useful for the evaluation of carcinogenicity tests. However, it must be realized that carcinogenicity tests according to the MTD principle have been designed primarily to detect weak carcinogenicity or noncarcinogenicity of test compounds, so that, theoretically, they can provide data to show whether or not the test compounds have the potential to induce tumors in experimental animals.[9]

According to the database of the National Institute of Environmental Health Sciences, approximately 2500 tests on 975 compounds had been completed throughout the world by 1986.[10] By analysis of the accumulated test data it has become increasingly apparent that carcinogens designated on the basis of animal tests are quite diverse in nature, in terms of both mechanism of action and potency. For example, there appears to be no definite correlation between mutagenicity test results and animal carcinogenicity test results; some chemicals with significant mutagenic activity such as quercetin are incapable of producing tumors in animals,[11,12] while some other chemicals with no mutagenic activity can.[13] Regarding the potency of carcinogens, it should be noted that there is a 10 million-fold difference in the 50% tumor-inducing dose (TD_{50}) value between the strongest carcinogen (aflatoxin B_1) and the weakest carcinogen (sodium saccharin) (Figure 1).[14] Considering this diversity of carcinogenic activity, it seems to be neither scientific nor practical to treat all carcinogens detected in long-term animal tests as being similarly hazardous to man.

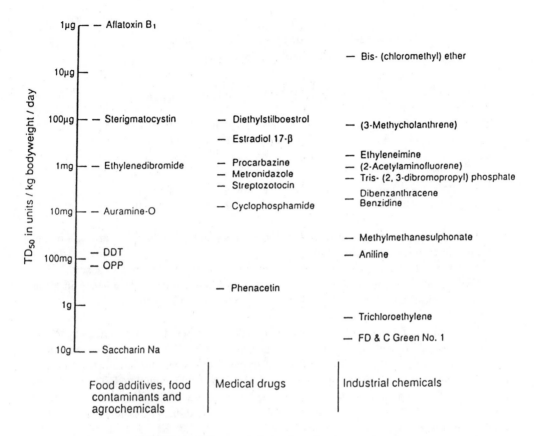

FIGURE 1. 50% tumor-inducing dose of various carcinogens.

III. PROCEDURES OF RISK ASSESSMENT

Assessment of human cancer risk associated with any particular specified chemical exposure requires a complicated scientific procedure, starting with careful review of all pertinent information on the chemical, derived from experimental, epidemiological and/or clinical studies.[15]

It is generally agreed within the scientific community that there are four steps or components which are typically involved in carcinogenic risk assessment. The first step, which is referred to as hazard identification, entails a qualitative evaluation of data concerning the potential of the chemical to produce a carcinogenic effect in man. At this step, it is important to consider which type of carcinogen the test compound belongs to, namely, primary carcinogen (genotoxic carcinogen) or secondary carcinogen (epigenetic carcinogen).[14] The second step, exposure assessment, is the process of measuring or estimating real or hypothetical human exposure to the chemical of interest. The third step, dose-response assessment, is the evaluation of both hazard and exposure[15] information to estimate the mathematical probability that the carcinogenic potential associated with the agent will be realized in the human population under defined conditions of exposure. In the final step, referred to as risk characterization, all relevant information from the first three steps is integrated to characterize the carcinogenic risk associated with expected human exposure to the chemical of interest.

The process of carcinogenic risk assessment, therefore, relies upon the availability and quality of information of the chemical, which includes:[14]

1. Long-term animal test results or epidemiological evidence concerning carcinogenic potential in man

TABLE 1
TD_{50} and VSD at 10^{-6} Risk Level of
Representative Carcinogens

Compounds	TD_{50} (mg/kg/day)	VSD (mg/kg/day)	TD_{50}/VSD
Urethane	2.8×10^0	2.3×10^{-5}	1.2×10^5
DMNA	6.0×10^{-1}	3.5×10^{-5}	1.5×10^4
ENU	2.0×10^{-1}	3.5×10^{-6}	6.0×10^4
Aflatoxin B_1	8.0×10^{-4}	1.6×10^{-7}	5.0×10^3
$KBrO_3$	6.0×10^0	3.8×10^{-3}	1.6×10^3
Saccharin-Na	9.0×10^3	2.1×10^1	4.0×10^2
OPP-Na	8.0×10^3	7.3×10^1	1.1×10^2

2. Scientific data on the mechanism of action, comparative metabolism, pharmacokinetics and structure-activity relationships
3. Biochemical or epidemiological data to estimate the level of human exposure.
4. Experimental or epidemiological data concerning the carcinogenic potency in terms of a specific risk estimate or an upper limit to the underlying risk

The execution of any given risk assessment may be hampered if uncertainties arise due to deficiences or critical gaps in the necessary information.[15] On such occasions, it is necessary to make assumptions on the basis of the available scientific information, to take account of these uncertainties so that risk assessment can still be completed. Therefore, the plausibility of these assumptions is also regarded as a critical factor influencing the result of risk assessment. Thus, inappropriate assumptions may yield inappropriate assessment due to either overestimation or underestimation of risk.

IV. METHODS FOR QUANTITATIVE ESTIMATION OF CARCINOGENICITY

Two types of methods have been used or proposed for quantitative estimation of carcinogenicity of chemicals: safety factor approach to allocation of acceptable daily intake (ADI) and estimation of relative potency index such as TD_{50} or virtually safe dose (VSD) based on dose-response data in animals. The safety factor approach has been very rarely used for evaluation of carcinogenicity because of various statistical and biological problems. For example, the observation of no treatment-related effects at given dose level (no observed effect level, NOEL) may depend on the number of animals exposed at that particular level, and, also, this approach assumes the existence of a true population threshold below which no adverse effects can occur.

$$ADI = \frac{NOEL}{Safety\ factor}$$

TD_{50} is a useful parameter for comparing carcinogenic potencies among various compounds but may offer little insight into the relative risk at the low dose level because this ignores the shape or steepness of the dose-response curve, a critical determinant for the low dose risk. VSD is defined as a value corresponding to the dose level or range which can induce tumors at extremely low rates such as 10^{-6} or 10^{-8}. This value can be obtained by downward extrapolation (low-dose extrapolation) of animal dose-response data by use of proper mathematical models.

Table 1 shows the TD_{50} and VSD of five carcinogens or tumor-inducing substances.

The values of these two indices appear to be paralleled among the compounds. However it should be noted that the ratio of TD_{50} and VSD is varied for each chemical ranging from 400 for sodium saccharin to 120,000 for urethane. These variation may be mainly attributable to the differences in the steepness or shape of dose-response curve for each chemical. Therefore, it is said that TD_{50} is a useful index for the relative potency of chemicals, but this index, when used alone, cannot properly assess the risk of chemicals at low-dose exposure.[16]

The VSD is often used as a parameter for the assessment of human cancer risk associated with exposure to a potent carcinogen, such as aflatoxin B_1, benzo(a)pyrene, or dimethyl-nitrosamine. This kind of procedure is based on the assumption that man is the animal species most sensitive to the carcinogen, and therefore it can also be applied to the case of other strong genotoxic carcinogens, either synthetic or natural. This assumption, however, is not necessarily plausible for all carcinogens designated on the basis of long-term animal tests (carcinogens according to the criteria of WHO), particularly for epigenetic or secondary carcinogens. At present, our knowledge is still insufficient to clearly distinguish between the genotoxic carcinogen and the epigenetic carcinogen. Perhaps, by a mechanism different from that of genotoxic carcinogens, epigenetic carcinogens may induce tumors in animals without any direct effects on DNA in target cells, but through an unknown secondary mechanism, such as sustained effects on hormonal, metabolic, or immunological function or repetitive necrosis and regeneration in the target tissues. Therefore, carcinogenic risk for these compounds can be assessed on the basis of proper mechanistic and dose-response studies of the effects of each compound leading to secondary tumor formation.[16]

V. PROBLEMS IN VSD ESTIMATION

Although VSD is regarded as an appropriate parameter for low-dose risk, many practical problems still exist in the process of estimation.[16] The first is the selection of dose levels in long-term animal bioassays. For example, often in dose-response studies with strong carcinogens, tumors appear at high rates in all treated groups. In such cases, the dose-response curves exhibit a hyperlinearity which results in too conservative estimates of VSD. In contrast, when tumors appear at very low rates in all treated groups, it may be difficult to select proper mathematical models on the basis of goodness-of-fit criteria.

Mathematical models proposed for low-dose extrapolation include tolerance distribution models, mechanistic models, pharmacokinetic models, and time-to-tumor models.[15] At present, however, no single model is recognized as the most appropriate or no definite criteria are established for selection of mathematical models. Biologically, the pharmacokinetic models appear to be the most realistic of all, but they are rarely used because of insufficiency of available information concerning the metabolic fate of the chemicals.

Estimation of VSD requires proper dose-response data from large-scale, long-term animal tests consisting of four or more treated groups, which constitute a major hazard to conduct studies on low-dose extrapolation. To overcome this problem, two types of methods may be proposed.[16] The first is a sequential design where VSD is estimated by combining the dose-response data obtained from two or more series of sequentially conducted experiments. The second is a cross-validation whereby VSD is estimated based on the data of the high-dose groups while the estimated value is validated with the data of low-dose groups. These methods may permit estimation of VSD by combined use of dose-response data from routine carcinogenicity bioassays and subsequently conducted additional studies on the same compounds.

VI. EXAMPLES OF RISK ASSESSMENT AND REGULATORY DECISION MAKING

Regulatory decisions on the safety of chemicals rely upon the information obtained from proper risk assessment for the compounds. However, regulatory decision making is also a complex matter with implications far beyond simple scientific judgment. It must be a decision to protect all interests. In each particular case, therefore, the final judgment should be based on integration of the information obtained from risk assessment with societal factors relating the use of the compounds so that a feasible policy for risk reduction can be achieved. An important element in the final judgment is also consideration of whether there are any feasible measures which could be undertaken to eliminate or minimize exposure to the compound, including possible use of alternatives. In other words, a multidisciplinary approach is required for regulatory decision making. For that purpose, various expert committees have been established in each country as well as in the international organization.[17] Some examples of risk assessment or regulatory decision for food additives conducted by the Japanese Food Sanitation Council or Joint FAO/WHO Expert Committee on Food Additives (JECFA) will be shown in this section.

A. α-2-FURYL-5-NITRO-2-FURANACRYLAMIDE (AF-2)

AF-2 was approved for use as a food preservative in Japan in 1965. Subsequently, in 1973 clastogenicity and mutagenicity of this compound was reported and carcinogenicity in mice was demonstrated in 1974.[17] In accordance, AF-2 was prohibited from use as a food additive in 1974. This decision was based on the prevailing paradigms at the time regarding carcinogenic risk, namely no existence of threshold for carcinogens, high sensitivity of man to any carcinogen, and additivity of multifactorial effects in carcinogenesis.

B. POTASSIUM BROMATE

Potassium bromate ($KBrO_3$) has been used mainly as an agent for the treatment of flour utilizing its oxidizing properties. Previous results from a battery of short-term tests, including the Ames test, chromosome aberration test, and micronucleus test, revealed this compound to be genotoxic. Subsequently Kurokawa et al. have conducted the carcinogenicity test for this compound in male and female F344 rats using oral administration at doses of 500 and 250 ppm in the drinking water for 110 weeks.[18] As a result, significantly higher incidence of renal cell tumors in both sexes given 500 and 250 ppm was observed. Also, weak carcinogenicity of potassium bromate in male Syrian golden hamsters, as evidenced by induction of renal cell tumors, was noted after oral administration for 89 weeks at concentrations in the range of 125 and 2000 ppm.[19] At present, the mode of action of potassium bromate in renal carcinogenesis is unknown, but the oxidizing properties of this compound are thought to be responsible, at least in part, for its carcinogenic effects. Consistent with this assumption, Kasai et al. reported that following an oral administration of potassium bromate to rats at a dose of 400 mg/kg, a significant increase of 8-hydroxyguanosine (8-OH-dG) in the kidney DNA was observed.[20] In the liver, a nontarget tissue, the increase of 8-OH-dG was not significant. The noncarcinogenic oxidants NaClO or $NaClO_2$ had no effect on 8-OH-dG formation in kidney DNA. These results suggest that formation of 8-OH-dG in tissue DNA is closely related to potassium bromate carcinogenesis. Another possibility can be raised for the mechanism of kidney tumor formation by potassium bromate. Administration of this compound to rats can induce hemolysis and cause hemoglobinuric nephrosis which results in overloading of iron in the tubular epithelial cells. Sustained high concentration of iron may cause renal cell tumors either by nonspecific cellular injuries or increased formation of active oxygen species in loco.

Considering its widespread use and unique biological activity, it seemed important and

TABLE 2
Incidence of Renal Cell Tumors in Rats Treated with Various Concentrations of Potassium Bromate in Drinking Water for 104 Weeks

Group	Effective no. of rats	No. of rats (%) bearing		
		Adenocarcinoma	Adenoma	Renal cell tumors
500 ppm	20	3 (15)	6 (30)[a]	9 (45)[b]
250 ppm	20	0	5 (25)[a]	5 (25)[b]
125 ppm	24	0	5 (21)[a]	5 (21)[b]
60 ppm	24	0	1 (4)	1 (4)
30 ppm	20	0	0	0
15 ppm	19	0	0	0
0 ppm	19	0	0	0

[a] $p < 0.05$ (by Fisher's exact probability test). [b] $p < 0.001$ (by Fisher's exact probability test).

TABLE 3
VSD of Potassium Bromate for Renal Cell Tumors at a Risk Level of 10^{-6}

	Probit[a]	Logit[a]	Weibull[a]	Gamma-multihit[a]
Chi-square value	1.627	2.155	2.472	2.693
p-value	0.898	0.827	0.781	0.747
VSD[b]	0.950	0.160×10^{-1}	0.481×10^{-2}	0.182×10^{-2}

[a] All models are with independent background.
[b] Concentration in ppm in the drinking water.

informative to conduct the dose-response carcinogenicity studies at low concentration for further risk assessment of potassium bromate. Seven groups of male F344 rats were given potassium bromate in drinking water at concentrations of 0, 15, 30, 60, 125, 250, and 500 ppm for 104 weeks.[21] From the data of water consumption recorded twice weekly throughout the experimental period, the daily intakes of potassium bromate (mg/kg/day) were calculated to be as follows: group 1 (500 ppm) — 43.4, group 2 (250 ppm) — 16.0, group 3 (125 ppm) — 7.3, group 4 (60 ppm) — 3.3, group 5 (30 ppm) — 1.7, and group 6 (15 ppm) — 0.9. The mean survival time of the animals given 500 ppm (82.8 ± 11.7 weeks) was significantly shorter than that of controls (103.1 ± 3.3 weeks) while the survival rates of the lower dose groups were comparable to that of the control group.

Table 2 shows the incidences of renal cell tumors in each group. Renal adenocarcinomas developed only in 3 of 20 rats of the 500-ppm group. The incidence of renal adenomas was significantly increased in the 125-, 250-, and 500-ppm groups with a definite dose-response relationship. On the basis of combined incidences of renal adenocarcinomas and adenomas, the VSD at a risk level of 10^{-6} was estimated by application of four mathematical models: probit model, logit model, Weibull model, and gamma-multihit model (Table 3). The p-value indicates that probit model best fits the experimental data, and, therefore, it seems reasonable to adopt the value estimated by this model (0.95 ppm) as the VSD of potassium bromate.

Potassium bromate is known to be degraded both *in vivo* and *in vitro* to potassium bromide (KBr), which is much less toxic than $KBrO_3$. Pharmacokinetic studies by Fujii et al. showed that BrO_3^- could be detected in the urine of rats after a single oral administration of $KBrO_3$ at a dose of 5 mg/kg or more whereas in rats given 2.5 mg/kg or less, no BrO_3^-

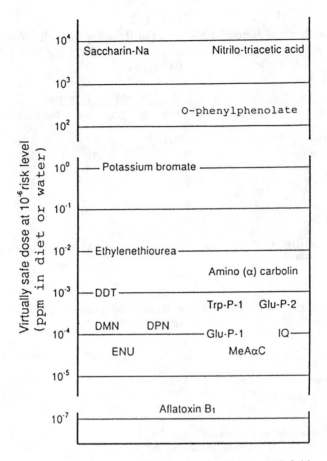

FIGURE 2.   Virtually safe dose of various carcinogens at 10^{-6} risk level.

was detectable.[22] In our study,[2] the daily intake of $KBrO_3$ was larger than 5 mg/kg when the $KBrO_3$ concentration in the water was higher than 125 ppm. In these cases, the incidences of renal cell tumors were significantly higher than the control values. This fact suggests that the contact of BrO_3^- with the renal tubular epithelium is essential for induction of renal cell tumors by potassium bromate.

Meanwhile, it is known that almost all $KBrO_3$ added to flour is converted to KBr during the normal British baking process. When potassium bromate was added to flour at 150 ppm, the residual level was reported to be lower than 5 ppm in the final bread.[23] In fact, no evidence of carcinogenicity and toxicity was reported in mice and rats given a bread-based diet made from flour treated with 50 and 75 ppm potassium bromate.[24] Recent studies on the fate of bromate in bread flour indicated that at levels of up to 62.5 mg/kg flour, no residues were detected in the resulting bread and the principal breakdown product was identified as bromide, while at levels of treatment of 75 mg/kg flour or higher, detectable residues of bromate were found in the bread.[23] As a result of this scientific evidence, the JECFA decided at the 1988 meeting to allocate the acceptable level of treatment of flour for bread making to 0 to 60 mg of potassium bromate per kilogram of flour.[23] The Ministry of Health and Welfare of Japan also permitted the use of potassium bromate at a maximum concentration of 30 mg/kg flour for baking, provided that no residue was detectable in the final product.[17] In particular, this decision considered the scientific evidence that the detectable limit of potassium bromate by chemical analysis is far below the VSD value.

C. BUTYLATED HYDROXYANISOLE

The antioxidant butylated hydroxyanisole (BHA) has been widely used as a food additive over the world. BHA is not mutagenic in several assay systems and, until recently, was considered, from the toxicological viewpoint, to be the most desirable of a range of phenolic antioxidants for food additive use. However, studies by Ito, et al. (1983) demonstrated that the long-term feeding of a diet supplemented with BHA at a concentration of 2.0% resulted in the induction of squamous cell carcinomas and papillomas of the forestomach in F344 rats.[25] They have further shown that the feeding of a 2% BHA diet to hamsters can produce hyperplasia or papillomas of the forestomach within a period of 6 months.[26]

These findings indicate that BHA has a carcinogenic capability preferentially affecting the forestomach epithelium. However, as proposed by the JECFA (1983), further information is necessary to evaluate whether or not BHA administration can exert toxicological effects on the alimentary tract epithelium in animal species not possessing a forestomach,[27] since the target tissue specificity of carcinogens is known to vary according to species, strain, or route of administration.

In response to the JECFA's proposal, three studies using either beagle dogs or cyno-mologus monkeys were reported. Two studies of beagle dogs showed that feeding of BHA-supplemented diets at palatable concentrations (1.0 or 1.2%) for 6 months had no pathological effects on the stomach, esophagus, duodenum, or liver.[28] Iverson et al. reported that oral administration of BHA in daily doses of 125 or 500 mg/kg body weight 5 days/week for 85 days was not associated with any histopathological lesions in the stomach or esophagus in cynomolgus monkeys, other than a slight elevation of the mitotic index of the esophageal epithelial cells in the highest dose group.[29] The discrepancy between the dog studies and the monkey study concerning the effects on esophageal epithelial cells may be attributable to differences in the method of BHA administration and in the dose used. In dog studies, BHA was given to animals in diets at palatable concentrations, while in Iverson's study on monkeys corn oil solutions of BHA at concentrations up to 50% were given to animals by gavage directly into the distal esophagus.

Additional information on the proliferative changes observed in the forestomach of rats fed BHA have been reported, which indicates that continuous exposure of the rat forestomach to BHA at 20 g/kg in the diet for 6 to 12 months is necessary to produce squamous cell carcinoma. The data also show that the induction of hyperplasia can occur at BHA levels of 1.25 g/kg in the diet, but not at 1.0 g/kg.[23] Based on these lines of evidence, JECFA decided at the meeting in 1988 to allocate the acceptable daily intake of BHA at 0 to 0.5 mg/kg of body weight. Ito et al. have mentioned in a review article on carcinogenicity of antioxidants that the significance of BHA in the development of human cancer is questionable since: (1) man does not have a forestomach, (2) carcinogenicity by BHA is limited to the forestomach, and (3) man ingests both synthetic and naturally occurring antioxidants from foodstuffs, but the estimated intake of synthetic phenolic antioxidants by humans is very low (less than 20 mg per person per day) compared to the doses used in most carcinogenicity or promotion studies.[30] Ito et al. have also reported that two naturally occurring phenolic antioxidants, caffeic acid and sesamol, can induce forestomach cancers in F344 rats after long-term feeding at a concentration of 2% in the diet.[30]

At present, the mechanism of BHA in causing forestomach cancer is still obscure. Since the chemical reactivity of BHA is not high enough to damage biomolecules, its carcinogenic effects may arise from its metabolites rather than the parent compound. Quinonoid metab-olites of BHA, namely *tert*-butyl-*p*-benzoquinone (BQ) and 3-*tert*-butyl-5-methoxy-1,2-benzoquinone (3-BHA-o-Q), have high electrophilic activity toward SH-compounds to form Michael type adducts (BHQSR).[31] Actually, it has been reported that BHA binds to fores-tomach protein but not DNA after oral administration to rats.[32] Recently, Miyata et al. have reported that BHQSRs are readily oxidized to the corresponding quinones accompanied by

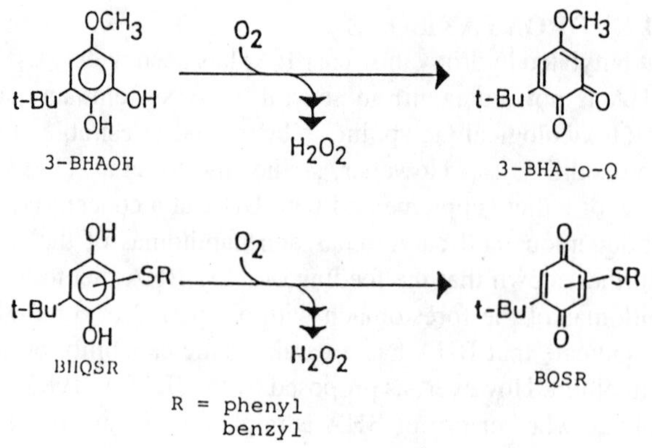

FIGURE 3.   Generation of hydrogen peroxide from 3-BHAOH and BHQSR.

hydrogen peroxide generation (Figure 3).[31] They suggest that generation of active oxygen species is responsible for the carcinogenic effects of BHA.

D. *o*-PHENYLPHENOL

o-Phenylphenol (OPP) and sodium *o*-phenylphenolate (OPP-Na) have been used as fungicides for citrus fruits. These two compounds have no genotoxicity in *in vitro* short-term tests such as assays for bacterial and mammalian mutagenesis, unscheduled DNA synthesis, or chromosomal tests, although there are a few reports that OPP has equivocal mutagenic effects in CHO-KI cells and in a human cell strain, RSa. Hiraga and Fujii reported that dietary administration of 2.0% OPP-Na or 1.25 % OPP was associated with a high incidence of urinary bladder tumors in male F344 rats.[33] It was also shown that the incidence of urinary bladder tumors after OPP-Na was significantly higher in males than in females. Fujii et al. reported further that OPP-Na was more carcinogenic than OPP, due to the higher alkalinity of OPP-Na.[34]

A schematic outline of the metabolic pathways of OPP-Na is presented in Figure 4. It was shown in a metabolic study using rats[35] that approximately 85 to 90% of the administered OPP or OPP-Na was recovered in the urine. Nakao et al.[36] and Reitz et al.[37] reported that for both OPP and OPP-Na, more than 98% of urinary metabolites were identified as conjugate forms, either glucuronide conjugates or sulfate ester conjugates, and less than 2% as free forms. It was further shown that OPP, 2-phenyl-1,4-benzoquinone (PBQ), and phenyl-hydroquinone (PHQ) could be detected in the urine of rats after administration of various levels of OPP-Na. Morimoto et al. studied DNA damage of the urinary bladder epithelium of rats by alkaline elution assay after intravesical injections of OPP, PBQ, or PHQ.[38] PBQ revealed a weak DNA-damaging activity in both sexes at 0.05 to 0.1%. OPP and PHQ had no effects at the same level (Table 4). Histopathologically, a single intravesical injection of 0.1% PBQ induced epithelial hyperplasia of the bladder epithelium on day 5, but PHQ or OPP did not induce it. Feeding studies with OPP-Na were also performed to examine the correlation between urinary PBQ levels and DNA damage in bladder epithelium. Slight DNA damage was observed in males given 1.0 and 2.0% OPP-Na in the diet for 3 to 5 months. Dose-response curves of OPP-Na dietary levels to DNA damage coincided with the incidence of tumors of the bladder reported by Hiraga et al (Figure 5). The amounts of OPP, PHQ, and PBQ in urine were greater in male rats than female rats given tumor induction by OPP-Na (Table 5). These findings indicate that the metabolite PBQ is the reactive species for the initiation steps of bladder tumors by OPP and OPP-Na.

In summary, long-term feeding of high concentrations of OPP-Na or OPP can induce tumors of the bladder in rats, and no proliferative changes of the bladder epithelium appear

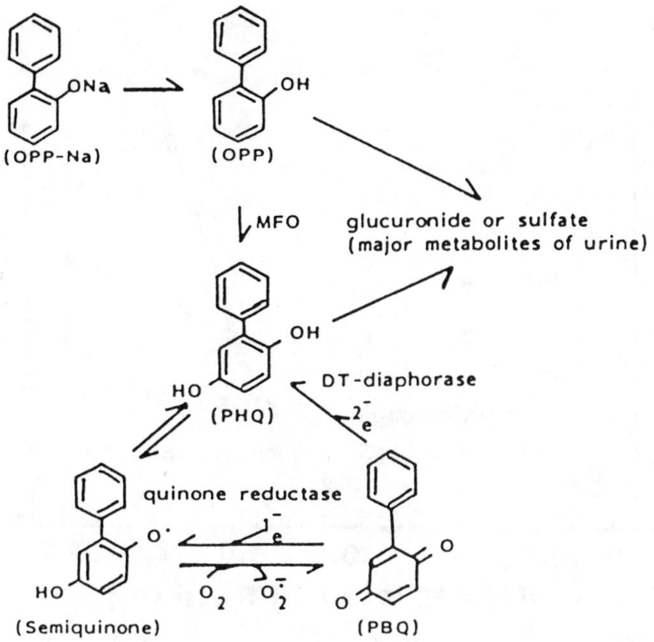

FIGURE 4. Metabolic pathways of OPP-Na in rats.

TABLE 4
Alkaline Elution Rates of DNA in Bladder Epithelium of F344 Rats Treated with OPP or Its Metabolites

Compound and concentration[a]		Number of experiments[b]	Elution rate constant[c] K (ml^{-1}) × 10^2
Male			
Saline		6 (12)	1.24 ± 0.12
PBQ	0.0005%	2 (4)	0.93
PBQ	0.005%	2 (4)	1.40
PBQ	0.05%	9 (18)	4.80 ± 2.86[d]
PBQ	0.1%	3 (6)	9.11 ± 2.57[e]
OPP	0.05%	3 (6)	1.50 ± 0.99
PHQ	0.05%	3 (6)	0.79 ± 0.44
NaHCO$_3$	0.4%	3 (6)	1.20 ± 0.38
Female			
PBQ	0.05%	4 (8)	10.20 ± 2.77[f]
PBQ	0.1%	4 (8)	11.26 ± 7.23

[a] OPP and its metabolites were injected into the bladder intravesically through the bladder wall and exposed for 10 min. Rats were killed immediately after the exposure.

[b] Number of animals in parentheses.

[c] Data show mean values ± SD.

[d] $p < 0.01$ from the groups of saline and 0.05% PHQ.

[e] $p < 0.01$ from the groups of saline, 0.05% OPP and PHQ in males, $p < 0.05$ from the group of 0.05% PBQ in males.

[f] $p < 0.01$ from the groups of saline, 0.05% PBQ, OPP, and PHQ in males.

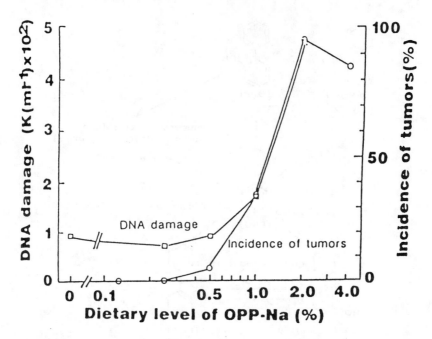

FIGURE 5. Dose-response relationship of OPP-Na dietary levels to DNA damage and carcinogenicity.

TABLE 5
Levels of PBQ and PHQ in the Urine of Rats Fed OPP-Na in the Diet in Month 5

Sex	Diet level (%)	nmol/ml[a] PBQ	nmol/ml[a] PHQ
M	0.5	12.7 ± 2.0	171.6 ± 110.1
M	1.0	13.1 ± 1.4	574.5 ± 338.4[c]
M	2.0	17.8 ± 2.6[b]	1506.6 ± 1034.8[d]
F	2.0	9.6 ± 2.2	62.3 ± 80.7

[a] Data show mean values ± SD from five rats [0.5% and 1.0% male (M)] or 10 rats [2% M and female (F)].
[b] $p < 0.001$ from the group of 2.0% F, $p < 0.01$ from the group of 0.5% M.
[c] $p < 0.05$ from the groups of 0.5% M and 2.0% F.
[d] $p < 0.001$ from the group of 0.5% M, $p < 0.01$ from the group of 2.0% F.

at levels less than 2000 ppm in the diet (corresponding to a daily dose level of 100 mg/kg body weight). Based on these data, a temporary ADI of OPP-Na was allocated at 0 to 0.02 mg/kg body weight by the JMPR in 1985.[39] Recently, a dose-response study of OPP-Na in male F344 rats for induction of bladder cancer was performed in our institute using six levels in the diet ranging from 0.25 to 2.0%. On the basis of the results, VSD at risk level of 10^{-6} was estimated as 147 ppm in the diet (7 mg/kg body weight) according to the Weibull model.

E. RETINYL ACETATE AND SORBITOL

One of the problems encountered in the assessment of carcinogenic risk of chemicals on the basis of long-term animal tests concerns understanding the meaning of enhancement of tumors which occur spontaneously in untreated rats or mice. Occurrence of adrenal medullary lesions, such as hyperplasia and pheochromocytoma, in rats provides a good example.[40]

Long-term administration of retinyl acetate or sorbitol at high concentrations to rats is associated with a significantly high incidence of adrenal pheochromocytoma in the treated groups compared to the concurrent control groups. Do these data indicate the carcinogenic risk of these compounds to man in normal use?

Large doses of retinyl acetate or sorbitol are known to elevate serum calcium concentrations in rats, either by release from the bone (retinyl acetate) or increased absorption from the intestine (sorbitol). Increase of serum calcium level may promote or interfere with catecholamine metabolism in the adrenal medulla. Consideration of the similarity to tumorigenesis of the other endocrine organs in rodents leads one to assume the likelihood that sustained hyperfunction or disfunction produced by long-term administration of the chemical may predispose the adrenal medulla to hyperplasia, eventually resulting in a higher incidence of pheochromocytoma. This evidence and assumption lead to the possibility that retinyl acetate or sorbitol does not present a carcinogenic risk to man under the normal conditions of daily use.[41]

VII. SUMMARY

Epidemiological studies suggest that diet is a major factor in the etiology of specific types of human cancer. With the objective of detecting causative principles for human cancers, a large number of chemicals including food-related chemicals have been tested for potential carcinogenicity in animals using the standard test protocol according to the MTD principle. By analysis of the accumulated test data it has become apparent that carcinogens designated on the basis of animal tests are quite diverse in nature, in terms of both mechanism of action and potency. Thus, it is regarded as an important step toward attainment of primary cancer prevention to assess individual carcinogenic substances for potential cancer risk in man on the basis of available scientific information. At the same time, it should also be considered that the risk of cancer is closely linked to a lifetime dietary pattern and not to occasional exposure to individual foods or food constituents.

ACKNOWLEDGMENTS

The author is grateful to Dr. K. Morimoto, Dr. N. Miyata, and Miss E. Hattori, National Institute of Hygienic Sciences, for their valuable suggestions and help in the preparation of this manuscript. This work was supported in part by a Grant-in-Aid for Cancer Research from the Ministry of Health and Welfare and a Grant-in-Aid from the Ministry of Health and Welfare for Comprehensive 10-Year Strategy of Cancer Control. Figures 4 and 5 and Table 5 were cited from a paper, Correlation between the DNA damage in urinary bladder epithelium and the urinary 2-phenyl-1,4-benzoquinone levels from F344 rats fed sodium o-phenylphenate in the diet, by Morimoto et al., in *Carcinogenesis,* 10, 1823—1827, 1989 by permission of Oxford University Press.

REFERENCES

1. IARC Meeting Report, Monitoring of aflatoxins in human bodies fluids and application to field studies, *Cancer Res.,* 45, 922, 1985.
2. **Hirayama, T.,** Epidemiology of stomach cancer, Gann Monogr, *Cancer Res.,* 11, 3, 1971.

3. **Carrol, K. K., Gammal, E. B., and Plunkett, E. R.,** Dietary fat and mammary cancer, *Can. Med. Assoc. J.,* 98, 590, 1968.

4. **Zaridze, D. G.,** Environmental etiology of large bowel cancer, *J. Natl. Cancer Inst.,* 70, 389, 1983.

5. **Sugimura, T.,** Carcinogenicity of mutagenic heterocyclic amines formed during the cooking process, *Mutat. Res.,* 150, 33, 1985.

6. **Sugimura, T.,** Studies on environmental chemical carcinogenesis in Japan, *Science,* 233, 312, 1986.

7. **Doll, R. and Peto, R.,** The causes of cancer: quantitative estimates of avoidable risks of cancer in the United States today, *J. Natl. Cancer Inst.,* 66, 1193, 1981.

8. **IARC Monogr.** *The Evaluation of Carcinogenic Risk of the Chemicals to Man,* IARC Monogr. 1, International Agency for Research on Cancer, Lyon, France, 1972.

9. **Hayashi, Y.,** Initiation/promotion designs in carcinogenicity bioassays, *Toxicol. Pathol.,* 11, 143, 1983.

10. **Haseman, J. K., Huff, J. E., Zeiger, E., and McConell, E. E.,** Comparative results of 327 chemical carcinogenicity studies, *Environ. Health Perspect.,* 74, 229, 1987.

11. **Hirono, I., Ueno, I., Hosaka, S., Takahashi, H., Matsushima, T., Sugimura, T., and Natori, S.,** Carcinogenicity examination of quercetin and rutin in ACI rats, *Cancer Lett.,* 13, 15, 1981.

12. **Ito, N., Hagiwara, A., Tamano, S., Kagawa, M., Shibata, M., Kurata, Y., and Fukushima, S.,** Lack of carcinogenicity of quercetin in F344/DuCrj rats, *Jpn. J. Cancer Res.,* 80, 317, 1989.

13. **Sivak, A., Goyer, M. M., and Ricci, P. F.,** Nongenotoxic carcinogens: prologue, in *Nongenotoxic Mechanisms in Carcinogenesis,* 1st ed., Butterworth, B. E. and Slaga, T. J., Eds., Cold Spring Harbor Laboratory, Cold Spring Harbor, NY, 1987, 1.

14. **Hayashi, Y., Kurokawa, Y., and Maekawa, A.,** Evaluation of tumor-inducing substances in foods, in *Diet, Nutrition and Cancer,* 1st ed., Hayashi, Y., Nagao, M., Sugimura, T., Takayama, S., Tomatis, L., Wattenberg, L. W., and Wogan, G. N., Eds., Japan Scientific Societies Press, Tokyo, 1986, 295.

15. Chemical Carcinogenesis; A review of the Science and Its Associated Principles, Federal Register Part II, Office of Science and Technology Policy, 1985.

16. **Hayashi, Y., Maekawa, A., Kurokawa, Y., and Takahashi, M.,** Strategy of long-term animal testing for quantitative evaluation of chemical carcinogenicity, in *New Concepts and Development in Toxicology,* 1st ed., Chambers, P. L., Gehring, P., and Sakai, F., Eds., Elsevier Science, Amsterdam, 1986, 383.

17. **Hayashi, Y.,** Regulation of genotoxic substances in Japan, in Proc. Management of Risk from Genotoxic Substances in the Environment, Stockholm, October 3 to 5, 1988, 230.

18. **Kurokawa, Y., Hayashi, Y., Maekawa, A., Takahashi, M., Kokubo, T., and Odashima, S.,** Carcinogenicity of potassium bromate administered orally to F344 rats, *J. Natl. Cancer Inst.,* 71, 965, 1983.

19. **Takamura, N., Kurokawa, Y., and Maekawa, A.,** Long-term oral administration of potassium bromate in male Syrian golden hamsters, *Sci. Rep. Res. Inst. Tohoku Univ.,* 32, 43, 1985.

20. **Kasai, H., Nishimura, S., Kurokawa, Y., and Hayashi, Y.,** Oral administration of the renal carcinogen, potassium bromate, specifically produces 8-hydroxydeoxy-guanosine in rat target organ, *Carcinogenesis,* 8, 1959, 1987.

21. **Kurokawa, Y., Aoki, S., Matsushima, Y., Takamura, N., Imazawa, T., and Hayashi, Y.,** Dose-response studies on the carcinogenicity of potassium bromate in F344 rats after long-term oral administration, *J. Natl. Cancer Inst.,* 77, 977, 1986.

22. **Fujii, M., Oikawa, K., Saito, H., Fukuhara, C., Onodaka, S., and Tanaka, K.,** Metabolism of potassium bromate in rats. I. In vivo studies, *Chemosphere,* 13, 1207, 1984.

23. Evaluation of Certain Food Additives and Contaminants, 23rd Report of the Joint FAO/WHO Expert Committee on Food Additives, WHO, Geneva, 1989, 14.

24. **Fisher, N., Hutchinson, J. B., and Berry, R.,** Long-term toxicity and carcinogenicity studies of the bread improver potassium bromate. I. Studies in rats, *Food Cosmet. Toxicol.,* 17, 33, 1979.

25. **Ito, N., Fukushima, S., Hagiwara, A., Shibata, M., and Ogiso, T.,** Carcinogenicity of butylated hydroxyanisole in F344 rats, *J. Natl. Cancer Inst.,* 70, 343, 1983.

26. **Ito, N., Fukushima, S., Imaida, K., Sakata, T., and Masui, T.,** Induction of papilloma in the forestomach of hamsters by butylated hydroxyanisole, *Gann,* 74, 459, 1983.

27. Evaluation of Certain Food Additives and Contaminants, 27th Report of the Joint FAO/WHO Expert Committee on Food Additives, WHO, Geneva, 1983.

28. **Tobe, M., Furuya, T., Kawasaki, Y., Ochiai, T., Kanno, J., and Hayashi, Y.,** Six-month toxicity study of butylated hydroxyanisole in beagle dogs, *Food Cosmet. Toxicol.,* 24, 1223, 1986.

29. **Iverson, F., Truelove, J., Nera, E., Wong, J., Lok, E., and Clayson, D. B.,** An 85-day study of butylated hydroxyanisole in the cynomolgus monkey, *Cancer Lett.,* 26, 43, 1985.

30. **Ito, N. and Hirose, M.,** The role of antioxidants in chemical carcinogenesis, *Jpn. J. Cancer Res.,* 78, 1011, 1987.

31. **Miyata, N., Miyahara, M., and Kamiya, S.,** Chemical studies related to the mechanism of action of 3-BHA (antioxidant), in *Proc. 4th Biennial General Meet. Society for Free Radical Research,* Elsevier Science, Amsterdam, 1989, 497.

32. **Hirose, M., Asamoto, M., Hagiwara, S., Ito, N., Kaneko, H., Yoshitake, A., and Miyamoto, J.,** Metabolism of 2- and 3-tert-butyl-4-hydroxyanisole (2- and 3-BHA) in the rat. II. Metabolism in forestomach and covalent binding to tissue macromolecules, *Toxicology,* 45, 13, 1987.

33. **Hiraga, K. and Fujii, T.,** Induction of tumours of the urinary system in F344 rats by dietary administration of sodium o-phenylphenate, *Food Cosmet. Toxicol.,* 19, 303, 1981.

34. **Fujii, T., Nakamura, K., and Hiraga, K.,** Effects of pH on the carcinogenicity of o-phenylphenol and sodium o-phenylpheneate in the rat urinary bladder, *Food Chem. Toxicol.,* 25, 359, 1987.

35. **Sato, M., Tanaka, A., Tsuchiya, T., Yamada, T., Nakaura, S., and Tanaka, S.,** Excretion, distribution and metabolic fate of sodium o-phenylphenate and o-phenylphenol in the rate, *J. Food Hyg. Soc.,* 29, 7, 1988.

36. **Nakao, T., Ushiyama, J., Kabashima, F., Nagai, F., Nakagawa, A., Ohno, T., Ichikawa, H., Kobayashi, H., and Hiraga, K.,** The metabolic profile of sodium o-phenylphenate after subchronic oral administration to rats, *Food Chem. Toxicol.,* 21, 325, 1988.

37. **Reitz, R. H., Fox, T. R., Quast, J. F., Hermann, E. A., and Watanabe, P. G.,** Molecular mechanisms involved in the toxicity of orthophenylphenol and its sodium salt, *Chem. Biol. Interact.,* 43, 99, 1983.

38. **Morimoto, K., Sato, M., Fukuoka, M., Hasegawa, R., Takahashi, T., Tsuchiya, T., Tanaka, A., Takahashi, A., and Hayashi, Y.,** Correlation between the DNA damage in urinary bladderepithelium and the urinary 2-phenyl-1,4-benzoquinone levels from F344 rats fed sodium o-phenylphenate in the diet, *Carcinogenesis,* 10, 1823, 1989.

39. Pesticide Residues in Food — 1985. Part II — Toxicology, FAO Plant Production and Protection Paper 72/2, 1986, 129.

40. Environmental Health Criteria 70: Principles for the Safety Assessment of Food Additives and Contaminants in Food, WHO, Geneva, 1987, 42.

41. **Hayashi, Y.,** Carcinogenesis, in *Risk Assessment — The Common Ground,* International Safety Evaluation Symp., Life Science Research Ltd., England, 1987, 147.

Chapter 12

REGULATORY ASPECTS OF FOOD MUTAGENS INCLUDING FOOD ADDITIVES AND CONTAMINANTS

Motoi Ishidate, Jr.

TABLE OF CONTENTS

I. INTRODUCTION

Epidemiological data showing that the incidence of stomach cancers among Japanese people living in Hawaii who have discontinued the Japanese style of food intake is low, indicating that the cause of cancer may reside more in lifestyle than in the genetic background. The food is, however, a complex mixture containing substances which are intentionally as well as unintentionally added, e.g., natural or synthetic food additives, pesticide residues, fungi products, or container materials. Even in vegetables, biologically active components such as pyrrolidine alkaloids, flavonoids, or anthraquinones are detected as mutagens. In addition, mutagenic nitrosated compounds, heterocyclic amines, or polycyclic aromatic hydrocarbones are isolated from foods that have undergone cooking or manufacturing processes. Some components, on the other hand, may act as modulators, promoting or inhibiting such biological activities (Figure 1).

Human afflictions that may be caused directly or indirectly by such mutagenic substances are carcinogenesis as well as teratogenesis, impaired reproduction, and heritable genetic toxicity. Regulatory decision has been made on the principle that human intake of such hazardous substances should be avoided as much as possible before they elicit some adverse effect in man or before they become an issue in society.

During the past 30 years, international bodies such as the Joint Expert Committee on Food Additives (JECFA), the Food and Agriculture Organization of the United Nations (FAO), and the World Health Organization (WHO) have discussed the safety of chemical substances that are used in food or that contaminate food. As a result, a series of publications have appeared that review the basis for decision-making by the committee, the results of testing these chemicals, and the evaluation of the results. In these documents carcinogenic potentials have been emphasized and there is little discussion of genotoxic effects, probably because of the general consensus that any substance which may have carcinogenic potential is not appropriate as a food additive at any level. This consensus may be acceptable even at present, but can be criticized on several grounds. First, mutagenic evidence is significant by itself as such mutagenic events can relate not only with carcinogenesis but also with heritable genetic toxicity: second, a number of mutagens in foods of both synthetic and natural origin are always present at certain amounts and they cannot be removed completely from our daily foods. It must be noted that there are no "nontoxic" substances. The most urgent and important task is, therefore, first, to identify the substances in food which may have mutagenic properties, second, to estimate the mutagenic potential quantitatively rather than qualitatively, and, third, to eliminate such potential mutagens from our daily life as much as possible based on their risk assessment.

II. STRATEGY FOR THE DETECTION OF CHEMICAL MUTAGENS AND/OR CARCINOGENS IN THE ENVIRONMENT

An antimicrobial agent, 2-(2-furyl)-3-(5-nitro-2-furyl)acrylamide (AF-2) had been used in Japan as a food additive since 1965 and was finally banned in 1974. It was highly mutagenic to bacteria and induced chromosomal aberrations in cultured mammalian cells, and eventually was proven to be carcinogenic in experimental animals. Before the issue of AF-2, sodium cyclamate or potassium cyclamate, which had been approved in 1956, was prohibited from use in 1969, since cyclohexylamine, one of their metabolites, induced clastogenic effects in germ cells or somatic cells and was also carcinogenic in animals. At that time, the prevailing idea on the carcinogenicity or genetic toxicity of a chemical was that there would be no threshold for the effect and that even a small intake of these chemicals would cause adverse effects in man.

In 1973, a national collaborative study group was organized under the project of the Ministry of Health and Welfare of Japan to find reliable short-term tests for the screening

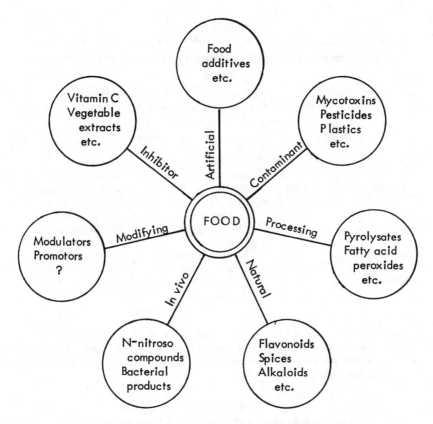

FIGURE 1. Possible mutagens derived from different sources in food.

of environmental carcinogens.[1] The group consisted of three subgroups: one for the selection of chemicals to be tested, one for screening tests using mutagenicity tests with different genetic endpoints, and one for long-term carcinogenicity tests using rats and mice on the chemicals which were detected as mutagens by the second subgroup. From this collaborative study, which lasted 10 years, several important findings were obtained, as follows:

1. More than 90% of well-known carcinogens were found to be mutagens in the reverse point mutation assay with *Salmonella typhimurium* (Ames test) and also clastogens (induction of chromosomal aberrations) in cultured mammalian cells.

2. Mutagenic or clastogenic potential varied extensively among chemicals and the difference ranged up to 10^6-fold.

3. Some kinds of carcinogens were negative in the Ames test, but positive in the chromosomal aberration test.

4. Only 30% of chemicals which were positive in at least one type of short-term test were carcinogenic in rodents, rats, or mice, as shown in Table 1.

5. The *in vitro* mutagens or clastogens are not necessarily positive in *in vivo* whole-animal tests, e.g., the sex-linked recessive lethal test in Drosophila, the bone marrow cytogenetics test, or the micronucleus test in rodents.

During the past decade it has been confirmed that the majority of well-known carcinogens are very positive in the Ames test when the test is combined with a metabolic activation system with mammalian microsome fraction (S9 mix). As shown in Table 2, about 80% of carcinogens were positive in this test system, although the values may vary greatly among investigators.

TABLE 1
Collaborative Studies on Short-Term and Long-Term Tests for Environmental Carcinogens (Ministry of Health and Welfare, Japan)

Carcinogenicity tests in rats (R) and mice (M)

Positive		Negative	
AF-2	(R, M)	Acid Red	Sodium nitrite
		BHT	Sodium erythrobate
Barbital	(M)	Caffeine	Sodium hypochlorite
		Caramel	
BHA	(R)	Erythrosine	Thiram
Hydrogen peroxide	(M)	*i*-Butyl *p*-hydroxyben-	DDVP (Dichlorvos)
		zoate	Aspirin
Phenacetin	(R, M)	Potassium sorbate	Acetaminophen
Potassium bromate	(R)	Potassium metabisulfite	Nitrofurantoin
Sulpyrin	(M)	Sodium benzoate	
		Sodium nitrate	
7 (28%)		18 (72%)	

Note: All compounds shown in the table gave positive results in at least one type of mutagenicity test. Three other compounds (diphenyl, sodium propionate and *n*-butyl *p*-hydroxybenzoate) which were not mutagenic were all negative in the carcinogenicity test.

TABLE 2
Positive Response of Carcinogens (Sensitivity) in *Salmonella*/Microsome (Ames) Test

		No. of test substances	
Investigator		Positive/carcinogens tested	
McCann et al.	(1975)	157/175	(89.7%)
Nagao et al.	(1978)	136/160	(85.0%)
Purchase et al.	(1978)	53/58	(91.4%)
Bartsch et al.	(1980)	62/82	(75.6%)
Ishidate et al.	(1988)	78/109	(71.6%)

An international collaborative study organized by the IPCS/WHO in 1982 was challenged to find possible alternatives to the Ames test, since some known rodent carcinogens were either not detected or only detected with considerable difficulty by the *Salmonella* reversion assay. In this study, eight carcinogens — *o*-toluidine, hexamethylphosphoramide (HMPA), safrole, acrylonitrile, benzene, diethylhexylphthalate (DEHP), phenobarbital, and diethylstilbestrol (DES) — were tested. It was found that the chromosomal aberration test with cultured mammalian cells (the *in vitro* chromosome test) was one of the most sensitive and reliable tests complementary to the Ames assay.[2] The strategy to combine two test systems with different genetic endpoints, gene mutation, and chromosomal damage is also recommended in the OECD guideline for genetic toxicology.[3]

From the regulatory point of view, it is important what and how many assays should be performed in a minimum set of mutagenicity tests on a specific chemical substance. The OECD Expert Committee has prepared guidelines for 15 different assays based on genetic toxicity and has classified the assays according to their utility and application, as shown in Table 3. The Committee has not specified, however, any minimum set of assays as a recommended combination.

TABLE 3
Utility and Application of Assays

I. Assays that may be used for mutagen and carcinogen screening
    Salmonella typhimurium reverse mutation assay (471)
    Escherichia coli reverse mutation assay (472)
    Gene mutation in mammalian cells in culture (476)
    Gene mutation in *Saccharomyces cerevisiae* (480)
    In vitro cytogenetics assay (473)
    Unscheduled DNA synthesis *in vitro* (482)
    In vitro sister chromatid exchange assay (479)
    Mitotic recombination in *Saccharomyces cerevisiae* (481)
    In vivo cytogenetics assay (475)
    Micronucleus test (474)
    Drosophila sex-linked recessive lethal test (477)
II. Assays that confirm *in vitro* activity
    In vivo cytogenetics assay (475)
    Micronucleus test (474)
    Mouse spot test (484)
    Drosophila sex-linked recessive lethal test (477)
III. Assays that assess effects on germ cells and that are applicable for estimating genetic risk
    Dominant lethal assay (478)
    Heritable translocation assay (485)
    Mammalian germ cell cytogenetic assay (483)

Note: Numbers in parenthesis indicate the OECD guideline number.

EEC has issued a genotoxicity guideline for chemicals, as shown in Table 4, that could be used for testing food additives, pharmaceuticals, pesticides, and other new industrial chemicals.[4] They recommend one test from each category, but the selection of tests to be performed is left to the applicant.

In Japan, the Ministry of Labour has set a policy requiring manufacturers to perform mutagenicity tests in registering new chemicals. This policy is incorporated in the Industrial Safety and Health Law, in order to prevent occupational cancers among people working in industries.[5] In this law a sequential approach is employed, requiring first a bacterial test using *S. typhimurium* strains (TA1535, TA1537, TA98, and TA100) and an *E. coli* strain (WP2 *uvr*A) and, second, the *in vitro* chromosomal aberration test if the chemical has a relatively high mutagenic potential (e.g., more than 1000 induced revertants per milligram) in the bacterial assay. If the chemical was positive in both tests, the agency will advise the manufacturer to carefully handle the chemical in the work place.

The Ministry of Health and Welfare of Japan requests that the manufacturer perform mutagenicity tests on new medical drugs and new industrial chemicals. For medical drugs, two *in vitro* tests (the Ames test and the chromosomal aberration test) and one *in vivo* test (the micronucleus test in mice) have been recommended as a minimum data set.[6] For industrial chemicals, only the two *in vitro* tests above are recommended. Those chemicals that have evaluated as potent mutagens (or clastogens) are classified as "Designated Chemical Substances".[7]

For food additives or food-related substances, the regulatory control should be more strict than for medical drugs or industrial chemicals. In Japan, a guideline for food additives was prepared in 1975 by experts in genetic toxicology. The principle of the tests in this guideline, however, is more sequential rather than battery type. It includes mutation assays in Drosophila (or silk worm), dominant lethal test in rodents, and specific locus test in mice as well. A more practical rather than theoretical guideline is now required.

More recently, in the U.K. the Committee on Mutagenicity of Chemicals in Food, Consumer Products and the Environment has published a revised document on "Guidelines

TABLE 4
EEC Recommendation for Genotoxicity Testing of Chemical Substances

Genetic endpoint	Assay system
Gene mutation	Reverse mutation assay with *Salmonella* or *E. coli*
Chromosomal aberration *in vitro*	Chinese hamster cells or human lymphocytes
Somatic mutation *in vitro*	Chinese hamster cells, mouse lymphoma cells, or human fibroblasts
In vivo test	Micronucleus test, Drosophila, bone marrow cytogenetics, heritable translocation, or dominant lethal test

for the Testing of Chemicals for Mutagenicity".[8] The recommended strategy for mutagenicity testing is based on an essentially hierarchical approach. The first stage, an *in vitro* screening, is designed to ensure a high probability of detecting mutagenic potential. This is followed by *in vivo* studies to ascertain whether any activity seen in Stage 1 can be expressed in the whole animal. The third phase consists of germ cell assays. These assays are required in order to obtain data for assessing the risk of heritable effects. In this strategy, two tests, bacterial mutation assay and clastogenicity test in mammalian cells, are required. An additional test for gene mutation in mammalian cells (for example the L5178Y TK +/− assay) test is necessary when human exposure to the chemical would be expected to be extensive and/or sustained and difficult to avoid. If the compound was positive in one or more tests in Stage 1, and where high or moderate prolonged levels of human exposure are anticipated, *in vivo* bone marrow assay for chromosome damage (metaphase analysis or micronucleus test) is recommended. If the compound is negative in the *in vivo* test and positive in any of the *in vitro* tests, test(s) to examine whether mutagenicity or evidence of DNA damage can be demonstrated in other organs (e.g., liver, gut, etc.) should be added. The guidelines also propose that if risk assessment for germ cell effects is justified (on the basis of properties including pharmacokinetics, use, and anticipated exposure), any *in vivo* tests for germ cell effects should be carried out (e.g., dominant lethal assay, cytogenetics in spermatogonia, etc.).

Testing strategy proposed by the government for mutagenic hazard assessment on chemicals may vary between individual countries. It is important, however, that a flexible approach is adopted and that each test compound be considered on a case-by-case basis with regard to the selection of tests and their protocols when the results are evaluated. Only qualitative evaluation of data, whether positive or negative, may often be imprudent when sensitivity of the test is discussed in comparison with other test systems. Quantitative rather than qualitative comparison seems to be more important in the evaluation of data.

III. QUANTITATIVE EVALUATION OF MUTAGENIC POTENTIAL OF CHEMICALS

Mutagenic activities of various chemicals found in the Ames test can be compared by number of revertants induced per milligram. Well-known carcinogens generally induce more revertants than other chemicals, as shown in Table 5.

Food additives, on the other hand, show relatively low mutagenic activities when compared with carcinogens (Table 5). For example, cinnamic aldehyde (a flavoring agent) induces 1790 revertants/mg in *S. typhimirium* TA100 in the absence of S9 mix, and calcium hypochlorite (a bleaching agent) induces 491 revertants in TA100 with S9 mix. Natural food additives such as Beet Red or cacao pigment are also positive, but their specific activities are extremely low.[9]

Similarly, clastogenic activities of chemicals in the chromosomal aberration test using a Chinese hamster cell line (CHL) can be compared by the D_{20} value which indicates the

TABLE 5
Mutagenic Activities in the Ames Test — A Comparison Between Carcinogens and Food Additives[9,19]

	Carcinogens				Food additives		
	Test strain	S9 mix	Revertants (no./mg)		Test strain	S9 mix	Revertants (no./mg)
MeIQ	TA98	+	$661,000 \times 10^3$	Cinnamic aldehyde	TA100	−	1,790
1,8-DNP	TA98	−	$488,200 \times 10^3$	Hydrogen peroxide	TA100	−	535
IQ	TA98	+	$433,000 \times 10^3$	Calcium hypochlorite	TA100	+	491
AF-2	TA100	−	$42,000 \times 10^3$	Chlorine deoxide	TA100	−	428
Trp-P-1	TA98	+	$39,000 \times 10^3$	Sodium chlorite	TA100	+	293
Aflatoxin B₁	TA100	+	$28,000 \times 10^3$	L-Cysteine monohydrochloride	TA100	+	291
4-NQO	TA100	−	$9,900 \times 10^3$	Sodium nitrite	TA100	+	47
MNNG	TA100	−	$1,300 \times 10^3$	Sodium hypochlorite	TA100	+	46
B(a)P	TA100	+	660×10^3	Potassium bromate	TA100	+	44
2-AAF	TA100	+	30×10^3	Fast green FCF	TA100	+	35
MNU	TA100	−	20×10^3	Erythorbic acid	TA100	+	4
DMN	TA100	+	230	Beet Red	TA100	+	3
				Cacaopigment	TA1537	−	3
				Caramel	TA100	−	2

Note: MeIQ: 2-amino-3,4-dimethylimidazo-(4,5-f)quinoline, 1,8-DNP: 1,8-dinitropyrene, IQ: 2-amino-3-methylimidazo-(4,5-f)quinoline, AF-2: 2-(2-furyl)-3-(5-nitro-2-furyl)acrylamide, Trp-P-1: 1,4-dimethyl-5H-pyride(4,3-b)-indole, 4-NQO: 4-nitroquinoline 1-oxide, MNNG: N-methyl-N'-nitro-N-nitrosoguanidine, B(a)P: benzo-(a)pyrene, 2-AAF: 2-acetylaminofluorene, MNU: N-methyl-N-nitrosourea, and DMN: dimethylnitrosamine

TABLE 6

Clastogenic Potential (D_{20} Values) of Carcinogens and Food Additives in Cultured CHL Cells[9,10]

Carcinogen[a]	S9 mix	D_{20}(mg/ml)	Food additive	S9 mix	D_{20}(mg/ml)
Mitomycin C	−	0.00001	Cinnamic aldehyde	−	0.01
Actinomycin D	−	0.00002	Propyl gallate	−	0.02
4-NQO	−	0.0003	Sodium chlorite	−	0.02
Trp-P-1	−	0.001	Curcumin	−	0.02
6-Mercaptopurine	−	0.001	1-Perilla aldehyde	−	0.04
MNNG	−	0.003	Calcium hypochlorite	−	0.08
AF-2	−	0.006	Potassium bromate	−	0.1
Endoxan	+	0.006	Hydrogen peroxide	−	0.1
Trp-P-2	+	0.006	Amaranth	−	0.2
Benzo(*a*)pyrene	+	0.01	Caffeine	−	0.4
Captan	−	0.01	Sodium nitrite	−	0.4
ENUR	−	0.01	Sodium hypochlorite	−	0.5
β-Propiolactone	−	0.01	Ammonium chloride	−	1.2
DMBA	+	0.02	Perilla pigment	−	1.2
MNU	−	0.03	Lacchaic acid	−	1.2
Acrylonitrile	−	0.03	Sodium benzoate	−	1.4
Propanesulfone	−	0.04	Tartrazine	−	2.1
Epichlohydrin	−	0.04	Cochineal	−	6.1
BNU	−	0.06	Saccharin sodium	−	6.6
3-MC	+	0.1	Acid red	−	7.0
Safrole	+	0.2	Eugenol	−	14.8
DMN	+	0.3	Propylene glycol	−	22.3
Quinoline	+	0.5	Acetone	−	36.9
2-AAF	−	0.8			
Phenacetin	+	0.8			
o-Toluidine	+	1.1			
Isoniazid	−	2.1			
Benzene	+	2.7			
Barbital	+	3.2			
Urethane	−	10.7			

[a]   4-NQO: 4-nitroquinoline 1-oxide, ENUR: *N*-ethyl-*N*-nitrosourethane, DMBA: 7,12-Dimethyl-benz(*a*)anthracene, MNU: *N*-methyl-*N*-nitrosourea, BNU: *N*-butyl-*N*-nitrosourea, 3-MC: 3-methylcholanthrene, DMN: dimethylnitrosamine, 2-AAF: 2-acetylaminofluorene.

dose (mg/ml) at which structural aberrations are induced in 20% of metaphases. The D_{20} values calculated among known carcinogens are shown in Table 6. Antibiotics such as mitomycin C or actinomycin D show a D_{20} value of 0.00001 and 0.00002, respectively, without S9 mix and can be classified as the most potential clastogens. Isoniazid, benzene, barbital, and urethane are positive, but their D_{20} values are relatively high and they can be classified as relatively weak clastogens.

On the other hand, the D_{20} values found in food additives are relatively high (less clastogenic) (Table 6). Cinnamic aldehyde and sodium chlorite are both positive in the chromosome test and show relatively low D_{20} values, 0.01 and 0.02, respectively. Natural food additives such as cacao pigment, cochineal, or caramel show a value of 1.7 with S9 mix and 6.1 or 25.0 without S9 mix, respectively.[10]

If a comparison is made between chemicals which are positive in both the Ames test and in the *in vitro* chromosome test, a nearly linear relationship can be found between mutagenic activity and clastogenic potential, as shown in Figure 2. It indicates that the potential in the Ames test varies from 10^{-8} to 10^0, while that in the chromosomal aberration tests varies from 10^{-4} to 10^2.

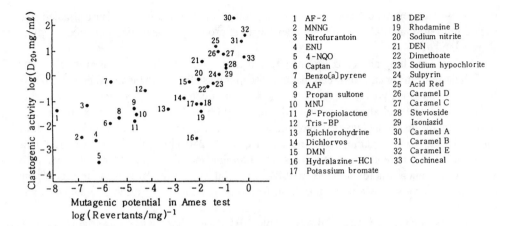

FIGURE 2. Quantitative relationship between mutagenic activity in the Ames test and clastogenic potential in the chromosomal aberration test with a Chinese hamster cell line, CHL. D_{20}: the dose at which aberrations were detected in 20% of metaphases (mg/ml).

In general, *in vitro* assays are more sensitive than *in vivo* tests, and, therefore, the chemicals which are genotoxic *in vitro* in bacteria or mammalian cells often fail to give positive results when tested *in vivo* in experimental animals. If the micronucleus test in mice is applied to such food additives positive in the *in vitro* assay, only a few substances will give a positive result. For example, potassium bromide was positive when animals were treated once intraperitoneally or by gavage, 25 or 200 mg/kg, respectively; sodium chlorite and sodium dehydroacetate were positive when treated once intraperitoneally with 25 and 125 mg/kg, respectively. One possible explanation for such a discrepancy between the *in vitro* and *in vivo* test is as follows. Detoxication or excretion processes can be expected only in *in vivo* tests; the concentration of test chemicals used *in vitro* may not be tolerated in experimental animals; or because of lack of absorption, active metabolite(s) may not reach DNA in target tissues or organs. Such *in vivo* data, however, are essential before definite conclusions can be drawn on a given substance being potentially hazardous to man. Pharmacokinetics or other toxicological data in short- or long-term animal experiments should be taken into consideration in parallel.

IV. SPECIFIC PROBLEMS OF MUTAGENS IN FOOD

As previously mentioned, AF-2 showed high mutagenic activity and finally proved to be carcinogenic in rodents. It induced 6,500 or 42,000 revertants/μg in *S. typhimurium* TA98 and TA100, respectively, and the activity is equivalent to that inducible by a well-known mutagen, aflatoxin B_1. This activity is 20- to 50-fold higher than that of benzo(*a*)pyrene. In Japan, the Food Sanitation Council has consulted on its mutagenic potential as well as carcinogenic effects under the Food Sanitation Law. The regulatory decision not to be used as a food additive is not only because of its biological nature but also because of possible replacement by other substances.

A bleaching and antimicrobial agent, hydrogen peroxide, has been widely used since 1948 in Japan. Its mutagenicity was difficult to detect with the standard Ames test, but clearly positive results were obtainable with the *in vitro* cytogenetic test for several different cell lines.[11] It induced duodenal tumors in mice by administration in drinking water at a dose of 0.4%.[12] Clastogenic activity, however, was greatly reduced by simultaneous administration of catalase. This suppression indicates that active oxygen radicals are involved in the clastogenic action. A regulatory decision was made that it can be used but it must be decomposed or removed from final products before marketing. It was also estimated that such a high dose as 0.4% in animal experiments is probably unrealistic compared to actual human intake.

A flour bleaching and oxidizing agent, potassium bromate, has been used in fish paste and wheat flour in Japan. For bread making, 50 ppm was allowed to be added before baking. It was weakly positive in the Ames test (44 revertants/mg in TA100 with S9 mix), but induced chromosomal aberrations in CHL cells at 0.06 mg/ml without S9 mix. It was also positive in the micronucleus test with mice at 25 mg/kg (i.p.).[13] It also induced kidney tumors in rats, when administered even at a dose of 250 ppm in drinking water.[14] This agent, however, may change into potassium bromide in the process of baking, which is known to be much less toxic. Thus in 1984, a regulatory decision was made to set 30 ppm of potassium bromate as a maximum limit for bread making and it cannot be added to any other food.

An antioxidant, butylated hydroxyanisole (BHA), has been widely used not only as a food additive but also for other purposes. It was negative in both the Ames test and the chromosome tests *in vitro,* but was weakly positive in a DNA repair test with *Bacillus subtilis* (Rec-assay).[15] It induced forestomach tumors in rats when the animals orally received 2.0% BHA in their diet.[16] Similar effects were also found in Syrian hamsters. This agent may be a cancer-promoting agent since it has no clear genotoxic effects *in vitro* as well as *in vivo*. In 1982, the Ministry of Health and Welfare of Japan proposed an amendment of the standard for this agent so as to restrict its use only for crude palm oil and palm kernel oil. However, it was decided to postpone the proposal in 1983. The FAO/WHO Expert Committee was also independently involved in this problem of evaluating its carcinogenic hazard to man. The use of BHA in food is now under self-control, but it can be used at maximum limits of 200 ppm in butter, dried or salted fish and shellfish, cooking fats and oils, and mashed potato and 1000 ppm in frozen fish and shellfish. A human being has no forestomach and further experiments using animals which have no forestomach have also been carried out. According to the report from the JECFA Committee on Food Additives (1987),[17] induction of hyperplasia in the forestomach of the rat is dose-dependent and can be reversed when BHA is removed from the diet. In dogs, levels of BHA that produced effects in the rat forestomach had no effects on either the stomach or the esophagus. However, in monkeys and pigs, there was some evidence that BHA produced pathological effects on the esophagus. This problem is still in question, but more detailed studies on the mechanisms involved are required before any clear-cut scientific decision is made in the future.

Each government should have responsibility for risk assessment of food additives currently used in its own country. Problems are not limited to food additives but include pesticide residues which have contaminated foods through plants and feed animals. Some other industrial chemicals can also pollute the environment, e.g., air and water. It has been known that some toxic chemicals can accumulate in fish and shellfish. For example, tributyltin (TBT) and triphenyltin (TPT), which have been widely used as antiparasitic agents for ships and fish nets, are both classified in Japan as Designated Chemical Substances. Detection of such minute contaminants in environmental waters should require painstaking experimentation. If these compounds are found in water, they must be checked for adverse effects and mutagenic and/or carcinogenic activity, as soon as possible so that possible hazard to humans can be evaluated. In my judgment, the decisions made so far by the Japanese government on environmentally hazardous chemicals have been unfailing.

V. FUTURE PROBLEMS

Regulatory decisions for hazardous substances in foods or in the environment should be based on the principle that human intake of these materials can be avoided as much as possible based on their risk assessment. Risk assessment consists of four phases: (1) hazard identification, (2) exposure assessment, (3) dose-response assessment, and (4) risk characterization.[18] All information concerning each phase of risk assessment must be subjected to discussion at both national and international levels. Usually a regulatory decision is made on a case-by-case basis by respective laws depending upon the chemical. The decision making should be a scientific process and not a political one.

When any adverse effects are suspected for a particular chemical substance, several implications should be taken into account from a regulatory point of view.

1. Mechanisms involved in the expression of adverse effects should be clarified. For its genotoxicity, for example, the effect may be due to induction of gene mutation, chromosomal aberration, or DNA damage. Some chemicals may be nongenotoxic carcinogens, e.g., promotor or stimulator of peroxisome proliferation, etc. Information on the biological activity of other substances which have a similar chemical structure may also be useful.

2. Specific activities of the substance in each genetic endpoint should be calculated. Mutagenic as well as carcinogenic potential can be expressed quantitatively rather than qualitatively and should be compared with that found in known mutagens or carcinogens.

3. Dose-response and reproducibility of data should be taken into consideration. Usually the concentration of the test substance used is relatively high, and no effects may be detected at lower levels. The minimum effective dose should be determined and compared with the dose to which humans are expected to be exposed. In general, it may be difficult to find the exact threshold dose for any chemical substance.

4. Combination of small amounts of different components may lead to additive, cumulative, multiple, or, sometimes, inhibitory effects. Recently, so-called "health foods" or "functional foods" have become available in the market, some of which contain vitamins, minerals, and for medicinal substances. Before these materials are approved, extensive studies should be done to evaluate both their positive and adverse effects.

5. Epidemiological studies on the incidence of cancers among people have been compiled during the past decades. However, there are very few studies on the heritable genetic hazard among people. Chemical substances which show relatively high mutagenic potential may cause effects genetically transmissible to the next generation through germ cells. More systematic and careful monitoring systems for these possible hazards should be developed at an international level in the future.

REFERENCES

1. **Odashima, S.,** Cooperative programme on long-term assays for carcinogenicity in Japan, in *Molecular and Cellular Aspects of Carcinogen Screening Tests,* Montesano, R., et al., Eds., IARC Sci. Publ. No. 27, International Agency for Research on Cancer, Lyon, France, 1980, 315.
2. **Ashby, J., de Serres, F. J., Draper, M., Ishidate, M., Jr., Margolin, B. H., Matter, B. E., and Shelby, M. D.,** *Evaluation of Short-Term Tests for Carcinogens,* Elsevier Science, Amsterdam, 1985.
3. **OECD,** *Introduction of the OECD Guidelines on Genetic Toxicology Testing and Guidance on the Selection and Application of Assays,* OECD, 1988.
4. **EEC,** Commission Directive of 25th April, 1984, adopting for the 6th time Council Directive 67/548/EEC on the approximations of laws, regulations and administrative provisions relating to the classification, packaging and labelling of dangerous substances. Part B. Methods for the determination of toxicity, *Off. J. Eur. Comm.,* L 251, 271, 1, 1984.
5. On the Standards of the Mutagenicity Test Using Micro-Organisms, The Labour Safety and Hygiene Law, The Labour Standard Bureau, Ministry of Labor, Japan, 1979.

6. Information on the Guidelines of Toxicity Studies Required for Applications for Approval to Manufacture (Import) Drugs (Part 1), Notification No. 118, The Pharmaceutical Affairs Bureau, Ministry of Health and Welfare, Japan, 1984 (revised in 1989).

7. Guideline for Screening Toxicity Testings of Chemicals. III. Mutagenicity Tests, Ministry of Health and Welfare, Ministry of International Trade and Industry, and Environmental Agency, Japan, 1987.

8. Guidelines for the Testing of Chemicals for Mutagenicity of Chemicals in Food, Consumer Products and the Environment, Report on Health and Social Subjects, Department of Health, Her Majesty's Stationery Office, 1989.

9. **Ishidate, M., Jr., Sofuni, T., Yoshikawa, K., Hayashi, M., Nohmi, T., Sawada, M., and Matsuoka, A.,** Primary mutagenicity screening of food additives currently used in Japan, *Food Chem. Toxicol.,* 22, 623, 1984.

10. **Ishidate, M., Jr., Ed.,** *Data Book of Chromosomal Aberration Test In Vitro,* Elsevier Science, Amsterdam, 1988.

11. **Ishidate, M., Jr., Harnois, M. C., and Sofuni, T.,** A comparative analysis of data on the clastogenicity of 951 chemical substances tested in mammalian cell cultures, *Mutat. Res.,* 195, 151, 1988.

12. **Ito, N., Watanabe, H., Naito, and Naito, Y.,** Induction of duodenal tumours in mice by oral administration of hydrogen peroxide, *Gann,* 72, 174, 1981.

13. **Hayashi, M., Kishi, M., Sofuni, T., and Ishidate, M., Jr.,** Micronucleus tests in mice on 39 food additives and eight miscellaneous chemicals, *Food Chem. Toxicol.,* 26, 487, 1988.

14. **Kurokawa, Y., Hayashi, Y., Maekawa, A., Takahashi, M., and Kokubo, T.,** Induction of renal cell tumours in F-344 rats by oral administration of potassium bromate, a food additive, *Gann,* 73, 335, 1982.

15. **Kawachi, T., Yahagi, T., Kada, T., Tazima, Y., Ishidate, M., Jr., Sasaki, M., and Sugiyama, T.,** Cooperative programme on short-term assays for carcinogenicity in Japan, in *Molecular and Cellular Aspects of Carcinogen Screening Tests,* Montesano, R., et al., Eds., IARC Sci. Publ. No. 27, International Agency for Research on Cancer, Lyon, France, 1980, 327.

16. **Ito, N., Fukushima, S., Hagiwara, A., Shibata, M., and Ogiso, T.,** Carcinogenicity of butylated hydroxyanisole in F344 rats, *J. Natl Cancer Inst.,* 70, 343, 1983.

17. **FAO/WHO,** *Evaluation of Certain Food Additives and Contaminants,* 13th Report of the Joint FAO/WHO Expert Committee on Food Additives, *Tech. Rep. Ser. 751,* 3.1.1, 12, 1987.

18. **Sato, S.,** Regulatory Perspective: Japan, presented at the Satellite Symposia of the 5th ICEM on Mutagens and Carcinogens in the Diet, Madison, WI, July 5 to 7, 1989.

19. **Sugimura, T., Nagao, M., Kawachi, T., Honda, M., Yahagi, T., et al.,** Mutagens-carcinogens in food with special reference to highly mutagenic pyrolytic products in broiled foods in *Origins of Human Cancer,* Book C., Hiatt, H. H., Watson, J. D., and Winsten, J. A., Eds., Cold Spring Harbor Laboratory, Cold Spring Harbor, NY, 1977, 1561.

Chapter 13

PERSPECTIVES IN FOOD MUTAGEN RESEARCH

Takashi Sugimura

Development of cancer in humans is closely related with diet.[1,2] This idea is supported by the facts that the main organs in which cancers develop differ in different countries and that the incidences of cancers in different organs in immigrants becomes similar to those of the general populations in the countries to which they have immigrated. Factors such as the contents of fat, total calories, fiber, vitamins, and minerals in the diet are related to cancer development.[3] Mutagens in foods have also been considered to be important in cancer development, because cancer cells are produced by genetical alteration of normal somatic cells.

Carcinogens such as aflatoxin B_1, N-nitrosodimethylamine, and benzo(a)pyrene, which may be found in food, have been shown to be mutagens.[4,5] Aflatoxin B_1 is reported to be responsible for the development of human cancer.[6] Rare carcinogenic chemicals, such as cycasin in cycad nuts[7] and ptaquiloside (aquilide A) in bracken fern,[8,9] were also shown to be mutagenic.

The most abundant mutagens in foods are probably flavonoids.[10-13] However, the carcinogenicities of flavonoids have been almost completely disproved,[14-18] although there is one report showing that quercetin is carcinogenic.[19] The fact that flavonoids are not carcinogenic indicates that the presence of mutagenicity in food components is not itself sufficient to predict the presence of carcinogenicity. Most of the mutagenicities of vegetables and fruits are due to their flavonoid contents. Flavonol glycosides are converted to mutagenic flavonols by glycosidase treatment.[12,13,20] After finding mutagenicity in a food, especially of plant origin, it is crucial to exclude the contribution of flavonoids to the mutagenicity. Then the structure of the mutagenic principle in the food must be determined, and its carcinogenicity must be demonstrated in long-term experiments in rodents. Reports of mutagenicity only in foods cause confusion in understanding the importance of these foods in development of human cancer.

Mutagenic and carcinogenic polycyclic aromatic hydrocarbons, including benzo(a)pyrene, are found in the neutral fraction of cooked foods.[21] Nitroarenes such as mononitropyrene and dinitropyrenes, which show mutagenicity, have also been isolated from the neutral fraction of cooked meat.[22] The carcinogenicities of dinitropyrenes have been demonstrated,[23-27] but that of 1-nitropyrene is still controversial.[23,25,27-30]

A series of heterocyclic amines have been found in the basic fractions of heated proteinaceous foods.[31-33] These heterocyclic amines can be produced on charring or prolonged boiling in normal cooking conditions. Heterocyclic amines in foods could be divided into two classes: those including IQ, MeIQ, MeIQx, 4,8-DiMeIQx, and PhIP and those including Trp-P-1, Trp-P-2, Glu-P-1, Glu-P-2, AαC, and MeAαC (see Chapter 2.2). The former class is produced by heating precursors such as creatinine, sugars, and amino acids present in fish and meat and the other by heating amino acids and proteins. All mutagenic heterocyclic amines tested have been found to be carcinogenic (Chapter 11).

The carcinogenicities of food mutagens can be calculated from the results of long-term feeding experiments on rodents. The intakes of individual carcinogens are not in themselves sufficient to explain the incidences of cancers in the general population. Humans are continuously exposed to a variety of mutagenic and carcinogenic agents: chemicals, ionizing radiation, and UV rays. Species of active oxygen produced in ordinary metabolic conditions must also contribute to cancer development. The contribution of each factor singly may be very minute, but the integrated effects of many factors may be sufficient to explain human carcinogenesis.

Mutagens whose carcinogenicity has been proved attract much attention from scientists, regulatory agencies, and the general public. However, in general, radical action to avoid exposure of humans to a certain chemical may not be warranted. Instead, improvement of the diet by a holistic approach may be more practical. Modest improvement of the diet and lifestyle only are recommended.

This is especially true because mutagens/carcinogens in food cause genetic alterations in somatic cells such as point mutations, rearrangements, and gene amplifications,[34,35] and these genetically altered cells are subject to proliferative pressures by tumor promoters or tumor-promoting conditions. The promotion step results in monoclonal expansion of initiated cells that have some genetic alteration.

Familial retinoblastoma, which arises from cells with an abnormality in one allele of the *RB* gene, is more frequent than sporadic retinoblastoma, which arises from cells with two intact *RB* genes. Since the *RB* gene is a tumor-suppressive gene, the involvement of two allelic *RB* genes is required to convert retinoblasts to retinoblastomas.[36] Human cancers often have several genetic alterations. For instance, in small cell lung carcinomas, loss of heterozygosity has been found in chromosomes 3p, 13q (*RB* gene), and 17p (p53 protein).[37] In colon carcinogenesis, alterations in multiple genes, such as *ras*-oncogene activation and allelic deletions in chromosomes 5q, 17p, and 18q, are often found.[38]

The human body has precursor lesions for full-blown malignant cancers, so calculation of risk from data from experiments on young, healthy rodents may not necessarily be valid. The search for further mutagens in foods is necessary, but evaluation of their risk needs careful reconsideration based on recent knowledge of multiple genetic alterations in human cancers and multiple steps in human carcinogenesis. Realistic and practical improvement of diet and lifestyle may be a more useful approach to cancer prevention.

REFERENCES

1. **Wynder, E. L. and Gori, G. B.,** Contribution of the environment to cancer incidence: an epidemiologic exercise, *J. Natl. Cancer Inst.,* 58, 825, 1977.
2. **Doll, R. and Peto, R.,** The causes of cancer: quantitative estimates of avoidable risks of cancer in the United States today, *J. Natl. Cancer Inst.,* 66, 1191, 1981.
3. *Diet, Nutrition, and Cancer,* Committee on Diet, Nutrition, and Cancer, Assembly of Life Sciences, National Research Council, National Academy Press, Washington, D.C., 1982.
4. **McCann, J., Choi, E., Yamasaki, E., and Ames, B. N.,** Detection of carcinogens as mutagens in the *Salmonella*/microsome test: assay of 300 chemicals, *Proc. Natl. Acad. Sci. U.S.A.,* 72, 5135, 1975.
5. **Yahagi, T., Nagao, M., Seino, Y., Matsushima, T., Sugimura, T., and Okada, M.,** Mutagenicities of *N*-nitrosamines on Salmonella, *Mutat. Res.,* 48, 121, 1977.
6. *IARC Monographs on the Evaluation of Carcinogenic Risks to Humans,* Overall Evaluations of Carcinogenicity: An Updating of IARC Monographs Vols. 1 to 42, Suppl. 7, International Agency for Research on Cancer, Lyon, 1987, 83.
7. **Matsushima, T., Matsumoto, H., Shirai, A., Sawamura, M., and Sugimura, T.,** Mutagenicity of the naturally occurring carcinogen cycasin and synthetic methylazoxymethanol conjugates in *Salmonella typhimurium, Cancer Res.,* 39, 3780, 1979.
8. **van der Hoeven, J. C. M., Lagerweij, W. J., Posthumus, M. A., van Veldhuizen, A., and Holterman, H. A. J.,** Aquilide A, a new mutagenic compound isolated from bracken fern (*Pteridium aquilinum* (L.) Kuhn), *Carcinogenesis,* 4, 1587, 1983.
9. **Hirono, I., Yamada, K., Niwa, H., Shizuri, Y., Ojika, M., Hosaka, S., Yamaji, T., Wakamatsu, K., Kigoshi, H., Niiyama, K., and Uosaki, Y.,** Separation of carcinogenic fraction of bracken fern, *Cancer Lett.,* 21, 239, 1984.
10. **Bjeldanes, L. S. and Chang, G. W.,** Mutagenic activity of quercetin and related compounds, *Science,* 197, 577, 1977.

11. **Sugimura, T., Nagao, M., Matsushima, T., Yahagi, T., Seino, Y., Shirai, A., Sawamura, M., Natori, S., Yoshihira, K., Fukuoka, M., and Kuroyanagi, M.,** Mutagenicity of flavone derivatives, *Proc. Jpn Acad.,* 53B, 194, 1977.

12. **Brown, J. P. and Dietrich, P. S.,** Mutagenicity of plant flavonols in the Salmonella/mammalian microsome test: activation of flavonol glycosides by mixed glycosidases from rat cecal bacteria and other sources, *Mutat. Res.,* 66, 223, 1979.

13. **Nagao, M., Morita, N., Yahagi, T., Shimizu, M., Kuroyanagi, M., Fukuoka, M., Yoshihira, K., Natori, S., Fujino, T., and Sugimura, T.,** Mutagenicities of 61 flavonoids and 11 related compounds, *Environ. Mutagen.,* 3, 401, 1981.

14. **Saito, D., Shirai, A., Matsushima, T., Sugimura, T., and Hirono, I.,** Test of carcinogenicity of quercetin, a widely distributed mutagen in food, *Teratogen. Carcinogen. Mutagen.,* 1, 213, 1980.

15. **Hirono, I., Ueno, I., Hosaka, S., Takanashi, H., Matsushima, T., Sugimura, T., and Natori, S.,** Carcinogenicity examination of quercetin and rutin in ACI rats, *Cancer Lett.,* 13, 15, 1981.

16. **Hosaka, S. and Hirono, I.,** Carcinogenicity test of quercetin by pulmonary-adenoma bioassay in strain A mice, *Gann,* 72, 327, 1981.

17. **Morino, K., Matsukura, N., Kawachi, T., Ohgaki, H., Sugimura, T., and Hirono, I.,** Carcinogenicity test of quercetin and rutin in golden hamsters by oral administration, *Carcinogenesis,* 3, 93, 1982.

18. **Takanashi, H., Aiso, S., Hirono, I., Matsushima, T., and Sugimura, T.,** Carcinogenicity test of quercetin and kaempferol in rats by oral administration, *J. Food Safety,* 5, 55, 1983.

19. **Pamukcu, A. M., Yalciner, S., Hatcher, J. F., and Bryan, G. T.,** Quercetin, a rat intestinal and bladder carcinogen present in bracken fern *(Pteridium aquilinum), Cancer Res.,* 40, 3468, 1980.

20. **Tamura, G., Gold, C., Ferro-Luzzi, A., and Ames, B. N.,** Fecalase: a model for activation of dietary glycosides to mutagens by intestinal flora, *Proc. Natl. Acad. Sci. U.S.A.,* 77, 4961, 1980.

21. **Lijinsky, W. and Shubik, P.,** Benzo(a)pyrene and other polynuclear hydrocarbons in charcoal-broiled meat, *Science,* 145, 53, 1964.

22. **Ohnishi, Y., Kinouchi, T., Tsutsui, H., Uejima, M., and Nishifuji, K.,** Mutagenic nitropyrenes in foods, in *Diet, Nutrition and Cancer,* Hayashi, Y., Nagao, M., Sugimura, T., Takayama, S., Tomatis, L., Wattenberg, L. W., and Wogan, G. N., Eds., Japan Scientific Societies Press, Tokyo, VNU Science Press, Utrecht, 1986, 107.

23. **Tokiwa, H., Otofuji, T., Horikawa, K., Kitamori, S., Otsuka, H., Manabe, Y., Kinouchi, T., and Ohnishi, Y.,** 1,6-Dinitropyrene: mutagenicity in *Salmonella* and carcinogenicity in BALB/c mice, *J. Natl. Cancer Inst.,* 73, 1359, 1984.

24. **Ohgaki, H., Negishi, C., Wakabayashi, K., Kusama, K., Sato, S., and Sugimura, T.,** Induction of sarcomas in rats by subcutaneous injection of dinitropyrenes, *Carcinogenesis,* 5, 583, 1984.

25. **Ohgaki, H., Hasegawa, H., Kato, T., Negishi, C., Sato, S., and Sugimura, T.,** Absence of carcinogenicity of 1-nitropyrene, correction of previous results, and new demonstration of carcinogenicity of 1,6-dinitropyrene in rats, *Cancer Lett.,* 25, 239, 1985.

26. **Takayama, S., Ishikawa, T., Nakajima, H., and Sato, S.,** Lung carcinoma induction in Syrian golden hamsters by intratracheal instillation of 1,6-dinitropyrene, *Jpn. J. Cancer Res. (Gann),* 76, 457, 1985.

27. **Maeda, T., Izumi, K., Otsuka, H., Manabe, Y., Kinouchi, T., and Ohnishi, Y.,** Induction of squamous cell carcinoma in the rat lung by 1,6-dinitropyrene, *J. Natl. Cancer Inst.,* 76, 693, 1986.

28. **El-Bayoumy, K., Hecht, S. S., Sackl, T., and Stoner, G. D.,** Tumorigenicity and metabolism of 1-nitropyrene in A/J mice, *Carcinogenesis,* 5, 1449, 1984.

29. **Hirose, M., Lee, M.-S., Wang, C. Y., and King, C. M.,** Induction of rat mammary gland tumors by 1-nitropyrene, a recently recognized environmental mutagen, *Cancer Res.,* 44, 1158, 1984.

30. **El-Bayoumy, K., Rivenson, A., Johnson, B., DiBello, J., Little, P., and Hecht, S. S.,** Comparative tumorigenicity of 1-nitropyrene, 1-nitrosopyrene, and 1-aminopyrene administered by gavage to Sprague-Dawley rats, *Cancer Res.,* 48, 4256, 1988.

31. **Sugimura, T.,** Mutagens, carcinogens, and tumor promoters in our daily food, *Cancer,* 49, 1970, 1982.

32. **Sugimura, T.,** Studies on environmental chemical carcinogenesis in Japan, *Science,* 233, 312, 1986.

33. **Sugimura, T.,** New environmental carcinogens in daily life, *Trends Pharmacol. Sci.,* 9, 205, 1988.

34. **Guerrero, I. and Pellicer, A.,** Mutational activation of oncogenes in animal model systems of carcinogenesis, *Mutat. Res.,* 185, 293, 1987.

35. **Suchy, B. K., Sarafoff, M., Kerler, R., and Rabes, H. M.,** Amplification, rearrangements, and enhanced expression of c-*myc* in chemically induced rat liver tumors *in vivo* and *in vitro, Cancer Res.,* 49, 6781, 1989.

36. **Knudson, A. G.,** Hereditary cancer, oncogenes, and antioncogenes, *Cancer Res.,* 45, 1437, 1985.

37. **Yokota, J., Wada, M., Shimosato, Y., Terada, M., and Sugimura, T.,** Loss of heterozygosity on chromosomes 3, 13, and 17 in small-cell carcinoma and on chromosome 3 in adenocarcinoma of the lung, *Proc. Natl. Acad. Sci. U.S.A.,* 84, 9252, 1987.

38. **Vogelstein, B., Fearon, E. R., Hamilton, S. R., Kern, S. E., Preisinger, A. C., Leppert, M., Nakamura, Y., White, R., Smits, A. M. M., and Bos, J. L.,** Genetic alterations during colorectal-tumor development, *N. Engl. J. Med.,* 319, 525, 1988.

INDEX

A